**WORLD CROPS:**

**PRODUCTION, UTILIZATION, AND DESCRIPTION**

*volume 6*

*Other volumes in this series:*

1. Stanton WR, Flach M, eds: SAGO The equatorial swamp as a natural resource. 1980. ISBN 90-247-2470-8.
2. Pollmer WG, Phipps RH, eds: Improvement of quality traits of maize for grain and silage use 1980. ISBN 90-247-2289-6.
3. Bond DA, ed: *Vicia faba*: Feeding value, processing and viruses. 1980. ISBN 90-247-2362-0.
4. Thompson R, ed: *Vicia faba*: Physiology and breeding. 1981. ISBN 90-247-2496-1.
5. Bunting ES, ed: Production and utilization of protein in oilseed crops. 1981. ISBN 90-247-2532-1.
7. Margaris N, Koedam A, Vokou D, eds: Aromatic plants: Basic and applied aspects. 1982. ISBN 90-247-2720-0.

# Faba Bean Improvement

*Proceedings of the Faba Bean Conference*
*held in Cairo, Egypt, March 7-11, 1981*

*Edited by*
## G. Hawtin and C. Webb

*The International Center for Agricultural Research*
*in the Dry Areas (ICARDA), Aleppo, Syria*

1982

MARTINUS NIJHOFF PUBLISHERS
for the ICARDA/IFAD Nile Valley Project

*Distributors:*

*for the United States and Canada*
Kluwer Boston, Inc.
190 Old Derby Street
Hingham, MA 02043
USA


*for all other countries*
Kluwer Academic Publishers Group
Distribution Center
P.O. Box 322
3300 AH Dordrecht
The Netherlands

**Library of Congress Cataloging in Publication Data**

Faba Bean Conference (1981 : Cairo, Egypt)
   Faba bean improvement.

   (World crops ; v. 6)
   Includes index.
   1. Broad bean--Congresses. 2. Broad bean--
Breeding--Congresses. I. Hawtin, Geoffrey.
II. Webb, Colin. III. Title. IV. Series: World
crops (Hague) ; v. 6.
SB205.B7F3 1981    635'.651     82-6356
                               AACR2
ISBN-13:978-94-009-7501-9     e-ISBN-13:978-94-009-7499-9
DOI: 10.1007/978-94-009-7499-9

# PREFACE

Faba beans, formerly known as broad beans, are among the oldest crops in the world. It has in fact been claimed with some justification that the Pyramids were built on faba beans ! They are today a major crop in many countries such as China, Egypt and the Sudan; and are widely grown for human food throughout the Mediterranean region, in Ethiopia and in parts of Latin America. In recent years there has been a growing interest in faba bean production as a protein source for stock feed in parts of Europe, North America and Australia.

The publication served by this preface arose from the first International Faba Bean Conference, held in Cairo, Egypt, on March 7-11, 1981 which provided a suitable forum for the review of many scientifically important aspects of the improvement of the crop. Leading faba bean specialists from four continents who participated were able not only to contribute from their personal expertise in relevant subjects, but in return to gain from their experience of Nile Valley conditions and from close contact with so many of the world's faba bean scientists.

The conference was supported in the main by the ICARDA/IFAD Nile Valley Faba Bean Project. Additional support was received from a number of other organisations and institutions whose help is gladly acknowledged. These included the Agricultural Research Council (ARC) of the Egyptian Ministry of Agriculture; G.T.Z. of Germany; IDRC of Canada; the National Research Center of Egypt; and Cairo University.

The conference brought together leading faba bean specialists from Egypt and the Sudan; together with their counterparts from other Mediterranean, West Asian, European and North American countries. From the considerable number (nearly 150) of participants present, formal presentations were made by more than 50 contributors. These provide a basis of technical strength which justifies the claim that "Faba Bean Improvement" is the best current reference book on the subject. No doubt it will be superseded as fresh knowledge becomes available. For the present, however, I have no hesitation in commending this publication for use by all who are interested in the production of this important crop.

The opportunity is gladly taken to express sincere thanks to all contributors and to acknowledge the many and varied inputs from the Nile Valley Project without which this book could not have been published.

Finally, no apology is offered for reminding the reader that "Faba Bean Improvement" is the second reference book produced by ICARDA on a food legume crop of world importance. The companion volume "Lentils" was published earlier in 1981. The Consultative Group for International Agricultural Research has entusted ICARDA with a world mandate for both crops and these books are published in partial fulfillment of this responsibility.

June 1981.

Harry S. Darling
Director-General
ICARDA

Editors' Note

We wish to acknowledge especially the help of several people who contributed to this publication.

They are Dr. Bhup Bhardwaj and his staff at ICARDA's Cairo, Egypt, office for their very effective local arrangements for the first international faba bean conference from which this book emanated; the scientists who wrote the papers; Dr. Habib Ibrahim, Acting Leader of ICARDA's Training and Communications Program; Dr. Richard Stewart for his relationships with the publishers and his assistance in the editing; Dr. Saxena for his work on the manuscripts; Mrs. Lina Khayyat and Mrs. Elizabeth Deedy who prepared the camera-ready copy; and Mrs. Hassan Khirallah who designed the figures.

December 1981.

Geoffrey Hawtin
Colin Webb

# Contents

# 1. GENETIC RESOURCES OF FABA BEANS

JOHN R. WITCOMBE

*FAO/IBPGR Plant Genetic Resources Officer (SW Asia Program), ICARDA, Aleppo, Syria*

*Vicia faba* L. is a diploid species, 2n=12, with its centre of origin in S.W. Asia. Botanically it has been divided on the basis of seed size into varieties *minor, equina* and *major*, even though there is no discontinuity in seed size between them.

The derivation of *V. faba* is unclear, and its immediate ancestor is not known. From morphological and geographical considerations, ancestors have been proposed within the species complex of *V. sativa* L., such as *V. angustifolia* L. (2n=12), and within the species complex *V. narbonensis* L. (2n=14). Of these *V. narbonensis*, Narbonne Vetch, which is found in the USSR in East Georgia and the Mediterranean region, often as a weed in cereal fields, has received the greatest attention as a putative ancestor of *V. faba*. However, in view of its different chromosome number, its very different karyotype and the total failure to produce hybrids between it and *V. faba*, it cannot be regarded as a direct ancestor of *V. faba*. It is possible that *V. faba* and *V. narbonensis* share a common ancestor.

Since *V. faba* does not produce fertile hybrids with any other species its gene pool is restricted to itself. The use of *V. narbonensis*, or other *Vicia* species, as sources of genes for *V. faba* improvement using conventional breeding techniques is considered unlikely.

With regard to variability, the most comprehensive work is that of Muratova (11). However her study was restricted to a few basic traits. No one region can be designated as a centre of diversity, and material from Europe, N. Africa, S.W. Asia, India and China are equally important. Germplasm materials from outside of this area e.g. North and South America and Australia, are likely to be of recent origin and therefore not primitive cultivars. Such advanced cultivars are usually part of breeders' working collections and, as such, are not threatened by genetic erosion.

*Existing collections*

The International Board for Plant Genetic Resources (IBPGR) is developing a network of Plant Genetic Resources Centres, each with the responsibility of maintaining the world collection of a particular crop. No institute has yet accepted this responsibility for faba

*G. Hawtin & C. Webb (Eds.): Faba Bean Improvement*
©1982 ICARDA. ISBN -13:978-94-009-7501-9

beans. However, ICARDA has one of the largest faba beans collections in existence, even though physical facilities to house this material under ideal conditions are still in the planning stage.

In 1980, IBPGR published a Directory of Germplasm Collections of Food Legumes (1). Holders of major collections of faba bean germplasm listed in this publication are summarised in Table 1. This list is doubtless incomplete and will need future revision.

Table 1.

Holders of major collections of *V. faba*

| Country | Institute | No. Accessions |
|---|---|---|
| Syria | ICARDA, Aleppo | 1931 *V. faba* |
| USSR | N.I. Vavilov IAPI, Leningrad | 2525 *V. spp.* * |
| Italy | LG, CNR, Bari | 1469 *V. faba* |
| GDR | ZGK, Gatersleben | 746 *V. faba* |
| Netherlands | SVP, Wageningen | 700 *V. faba* |
| Czechoslovakia | PBRICL, Tumenice | 500 *V. faba* |
| FRG | IPP, Braunschweig | 804 *V. faba* |

* All *Vicia* spp. *Vicia faba* numbers unknown.

The germplasm collections held at ICARDA, Braunschweig and Bari are summarised in Table 2. It is obvious from these data that certain countries although having indigenous varieties of *V. faba* are seriously under-represented. Examples of this are Iran, China and India .

The numbers of accessions from different countries are correlated between the collections. There must be a great deal of duplication between them even though, at least in the case of the Braunschweig and ICARDA collections, there has been no direct seed exchange between them.

Data describing existing collections are incomplete. In most germplasm collections, many accessions are found in which even the country of origin is not recorded. Because of this problem and the incompleteness of evaluation data, it is not possible to describe the main characteristics of the germplasm material according to region of origin. Although such information would be useful and interesting, it is largely unavailable in the case of all crop plants.

Table 2.

Origins of accessions in ICARDA, Bari, Braunschweig (BGRC) and Gatersleben (ZGK)
and area under production in various countries

| Country | No. of accessions | | | | Area* under faba beans 1977(1000 ha) |
|---|---|---|---|---|---|
| | ICARDA | Bari | BGRC | ZGK | |
| Afghanistan | 98 | 72 | 13 | 6 | n.d. + |
| Algeria | 21 | 34 | - | - | 35 |
| Argentina | 1 | - | 1 | - | 1 |
| Australia | 2 | - | 4 | 2 | n.d. |
| Austria | 1 | - | - | 1 | n.d. |
| Bangladesh | 2 | - | - | - | n.d. |
| Belgium | - | - | - | 1 | n.d. |
| Bolivia | 1 | - | 1 | - | 11 |
| Brazil | - | - | - | - | 195 |
| Bulgaria | - | 1 | - | 4 | n.d. |
| Canada | 2 | 122 | - | - | n.d. |
| China | 9 | 7 | - | 6 | 4000 |
| Colombia | 14 | - | - | - | n.d. |
| Cyprus | - | 103 | - | - | 3 |
| Czechoslovakia | - | - | 8 | 25 | 23 |
| Dominican Republic | - | 1 | - | - | 7 |
| Ecuador | 13 | - | - | - | 14 |
| Egypt | 57 | 85 | 211 | 28 | 102 |
| Ethiopia | 370 | 95 | 211 | 73 | 271 |
| Finland | 11 | - | 1 | 4 | n.d. |
| France | 9 | 10 | 2 | 15 | 19 |
| Germany D.R. | - | - | - | 11** | 6 |
| Germany FED | 257 | - | 17 | 12** | 8 |
| Greece | 25 | 55 | 6 | 43 | 11 |
| Guatemala | - | - | - | - | 18 |
| Holland | 33 | 19 | 3 | 15 | n.d. |
| Hungary | 10 | 66 | - | 2 | n.d. |
| India | 9 | 6 | 1 | 3 | n.d. |
| Iran | 13 | 9 | 4 | 2 | n.d. |
| Iraq | 57 | 52 | 5 | 1 | 29 |
| Italy | 48 | 247 | 12 | 151 | 163 |
| Japan | 5 | 4 | 4 | - | 2 |
| Jordan | 18 | 3 | 1 | - | n.d. |
| Lebanon | 30 | 30 | - | - | 1 |
| Libya | - | - | - | - | 5 |

| Country | No. of accessions | | | | Area* under faba beans 1977(1000 ha) |
| --- | --- | --- | --- | --- | --- |
| | ICARDA | Bari | BGRC | ZGK | |
| Malta | . | . | . | . | 1 |
| Mexico | 1 | - | 1 | - | 50 |
| Mongolia | - | - | - | 1 | n.d. |
| Morocco | 15 | 31 | 109 | 12 | 190 |
| Nepal | 1 | - | 2 | - | n.d. |
| Pakistan | 7 | 3 | 1 | - | n.d. |
| Palestine | 7 | 2 | 6 | 3 | n.d. |
| Paraguay | - | - | - | - | 10 |
| Peru | 2 | - | 1 | 4 | 20 |
| Poland | 12 | 1 | 4 | 71 | n.d. |
| Portugal | 5 | 5 | - | 1 | 38 |
| Romania | - | - | 16 | - | n.d. |
| Spain | 77 | 107 | 6 | 103 | 96 |
| South Africa | 1 | - | - | - | n.d. |
| Sri Lanka | 2 | - | - | - | n.d. |
| Sudan | 35 | 22 | - | 4 | 16 |
| Sweden | 10 | 4 | 1 | 1 | n.d. |
| Switzerland | 1 | - | 3 | 3 | n.d. |
| Syria | 62 | 32 | 4 | 3 | 9 |
| Tunisia | 49 | 54 | 2 | 1 | 64 |
| Turkey | 120 | 72 | 19 | 16 | 30 |
| U.K. | 88 | 56 | 11 | 32 | 40 |
| Uruguay | 1 | - | - | - | n.d. |
| USA | 2 | 2 | 1 | - | n.d. |
| USSR | 21 | 6 | 21 | 47 | n.d. |
| Yemen | 6 | 26 | 18 | - | n.d. |
| Yugoslavia | 14 | 17 | 10 | 6 | n.d. |
| N. Europe | 82 | - | - | - | n.d. |
| Unknown | 192 | - | 63 | 27** * | n.d. |
| TOTAL | 1931 | 1469 | 804 | 746 | |

\*     From FAO Production Year Book, 1977.

\*\*    + 48 with origin as 'Germany'.

\*\*\*   Includes 7 from Latin American and 8 mutants.

\+      n.d. = no data available.

*Collecting expeditions*

Recent collecting expeditions have taken place in Cyprus, Egypt and Afghanistan. Over the past few years, collections of *V. faba* have been made in the Mediterranean region by the Germplasm Laboratory, Bari, Italy, and in Ethiopia by the Ethiopian Genetic Resources Centre.

The collections made by Bari include Algeria in 1976 (47 accessions),Greece, Spain and Tunisia, 1977 (with 23, 31 and 18 accessions respectively) and Crete in 1978 (38 accessions).

In Cyprus in 1980, the Agricultural Research Institute, Cyprus, in collaboration with FAO and IBPGR, collected 98 accessions of *V. faba* (5). Ninety-five collections were made in Egypt in 1978 by the International Institute of Tropical Agriculture (2). M.M.F. Abdalla (pers. comm.) has made extensive faba bean collections in Egypt; the major one being in 1979 when 200 samples were collected from all regions of Egypt. Duplicates have been sent to Braunschweig for long term storage. In Afghanistan in 1974, fifty collections of *V. faba* were made (16).

It is certain that this is not a complete list of all recent collecting expeditions of the faba bean. I know of some collecting trips that have collected very few faba bean samples. Other, more extensive, collections of faba bean have been made, but the material has not been distributed outside of the country of collection. There must also be collecting expeditions of which I am unaware.

ICARDA is attempting to stimulate, with assistance from the IBPGR, the collecting of faba bean germplasm in India and various other regions where collections are under-represented.

*Multiplication and rejuvenation of V. faba germplasm*

The partially cross-pollinated habit of *V. faba* has profound genetic consequences and affects the method of maintaining its germplasm. Before examining two simple genetic models of the consequences of out-crossing, we must first define some terms.

Out-crossing is the rate at which individual plants cross-pollinate, or open-pollinate, to other plants (instead of selfing). *V. faba* shows a high degree of out-crossing. Figures for open-pollination in faba beans vary widely but maxima of 70 % (10) and 61 % (14) have been reported. However, in most circumstances, out-crossing is likely to be in the range of 5 to 50%.

Inter-crossing is the rate of pollination between different accessions (entries) of germplasm material. The result of inter-crossing is that, after a number of generations, the identity of the individual accessions is lost.

Hence the rate at which the identity of accessions is lost, is directly related to the rate of inter-crossing. In turn, this depends on the rate of out-crossing and the degree of isolation between entries. An extreme example of lack of isolation would be when each plant of an accession was totally surrounded by foreign material, so that the rate of inter-crossing would equal that of out-crossing. In practice, since the plants of an accession are always planted together in rows or plots, inter-crossing will always be less than out-crossing. It is worth noting that the worst planting pattern from the view point of inter-crossing is single, long, one-plant-width rows of each accession with a small distance between rows.

6

Normally the greatest rate of inter-crossing between any two accessions will be about half the rate of out-crossing, since at least 50% of a plant's neighbours will be members of its own accession. We then have a maximum rate of inter-crossing of 50% (half of the maximum out-crossing rate of 100%).

We can consider a simple model in which two accessions differ completely for a single locus so that the initial difference in gene frequency will be 1. Even in the extreme case of 50% inter-crossing, it will take six generations for the two accessions to become identical (Fig. 1a). Nevertheless, after three generations of such inter-crossing, the difference in gene-frequency as this locus is reduced from 1 to 0.1.

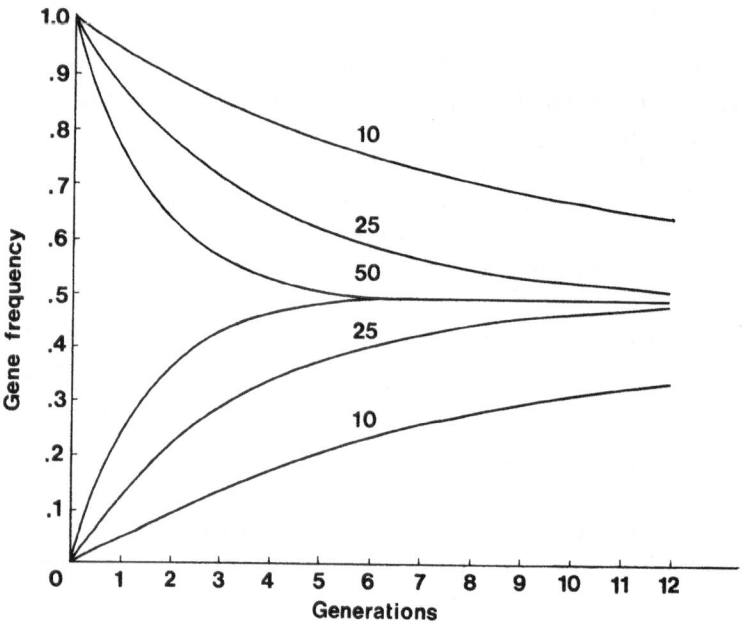

Fig. 1a. The rate at which two populations, initially differing for a gene, become identical at 50, 25 and 10 % inter-crossing

In the case of faba beans, gene-flow will always be much less than 50%, since rates of out-crossing are lower than 100%, and there will always be greater separation between the accessions than 50% of a plants' neighbours being foreign. Progress towards genetic identity at 10% and 25% inter-crossing is shown in Fig. 1a. It can be seen that when inter-crossing is reduced, it takes many generations for the accessions to become identical.

Another situation is an accession which differs from all others by uniquely having allele A at a particular locus. We assume that this accession '10' has the unique allele 'A' at a frequency of 1 and inter-crosses with the other accessions at a rate of 10%. After one generation of inter-crossing, the allele 'A' would be found in other accessions and could be present in the foreign pollen that inter-crosses with accession '10'. This possibility is disregarded since it is present at such a low frequency. Under the above conditions, the allele is lost slowly from the population (Fig. 1b). It is still present at a frequency of 0.1 after 21 generations, but is in great danger of being lost due to random drift. Moreover, if it is recessive, it is becoming increasingly hard to recover as after 20 generations double recessives only comprise 1% of the genotypes in accession '10'.

It can be seen from the models that, although the affects of inter-crossing are profound, in the absence of random drift they occur slowly at the levels of inter-crossing to be found in *V. faba*. A certain degree of inter-crossing can be tolerated in germplasm collections without losing the identity of the individual lines.

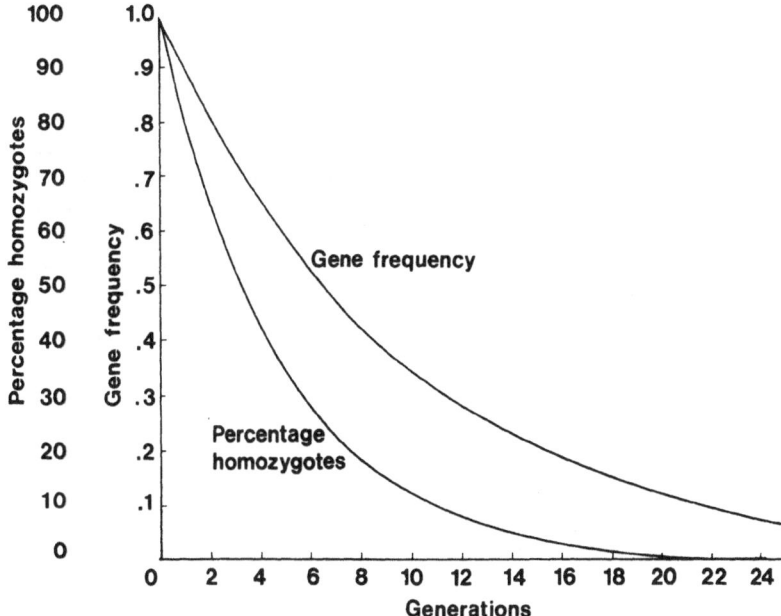

Fig. 1b.    The rate at which a population loses a gene when all other populations are different at that locus, at 10 % inter-crossing

As a consequence of the partial out-crossing of *V. faba*, there are three main ways of maintaining it in germplasm collections: as populations, inbred lines and trait-specific gene pools.

*Accessions maintained as populations*

When the germplasm collection is maintained as populations, steps are taken to reduce the rate at which the collections become identical. This is done by reducing inter-crossing and by reducing generation advance over time.

*Reducing inter-crossing between accessions*

Despite the open pollination shown by faba bean, germplasm material can be successfully maintained as collections of populations.

Of course, it is possible to eliminate inter-crossing by growing each accession in an insect proof cage or bagging individual plants. But when there are a large number of accessions, this solution is expensive. Inter-crossing can also be eliminated by very large isolation distances, but this is usually impracticable.

Growing plots isolated by distance only, and discarding the border plants, is an effective method of reducing inter-crossing. Under Syrian conditions with only 10 m between the plots, inter-crossing is around 10%.

The size of the plots was very important and an increase of 12 m$^2$ (75%) in plot area had a large effect (Table 3). This is an unfortunate fact when it is considered that often only small seed samples are available, so that plot sizes are small.

Table 3.

### Inter-Crossing in Faba Beans
### (Adapted from Hawtin and Omar, 1980a).

|  | Mean % inter-crossing between plots | | |
|---|---|---|---|
|  | border | centre | whole |
| Plot size 16 m$^2$ | 14.7 | 10.1 | 14.2 |
| Plot size 28 m$^2$ | 9.5 | 6.7 | 8.9 |
| Average inter-plot distance 11.0 m | 12.3 | 10.4 | 11.6 |
| Average inter-plot distance 14.3 m | 11.8 | 6.1 | 10.9 |

To overcome this problem, small plots of V. faba can be surrounded with a species that does not inter-cross with it but attracts the same pollinating insects. In this season at ICARDA, plots of V. faba are being surrounded by oil-seed rape (Brassica campestris) to see if the pollinating insects first visit these border plants and lose their foreign faba bean pollen to them. A physical barrier of taller border plants e.g. Triticale is also being tried.

The dispersal of pollen by visiting insects will not bear a simple relationship to distance, so that equal increases in isolation distance will not have equal effect in rates of inter-crossing. If there is a sigmoid relationship between distance and rate of inter-crossing, which is likely, then there is a critical distance above which the plots should be planted. Any further increase in this distance will not have a large effect.

The data of Porceddu et al. (13) indicate that if basal pod clusters are discarded, then inter-crossing will be reduced by a factor of a half. These authors found that rates of inter-crossing between varieties was 26% in the first to fifth pod clusters and only 16% in the clusters above. These results are supported by Hanna & Lawes (6) and Poulsen (14).

*Reducing generation advance*

It is clear that with inter-crossing, the more generations that an accession is grown, the more it loses its genetic identity. The effects of inter-crossing are reduced by minimising generation advanced using careful seed management (Fig. 2). Early generation seed (foundation seed) is maintained in medium or long term storage and used conservatively by removing sub-samples from it only for the purpose of multiplication. The multiplication seed is then used for the active collection for distribution and evaluation purposes. Only when this multiplied seed is exhausted or lost, is a further sample withdrawn from the foundation seed (Fig. 2). This scheme is a more conservative method than the simple one of using the base collection, or active collection, as a source. If the base collection is so used then generation advance may be accelerated, for the base collection will need to be regenerated when its quantity falls to a minimum level.

Ultimately, generation advance cannot be eliminated, and is dependent on the longevity of the seed in the base collection. Fortunately *V. faba* seed is very long lived.

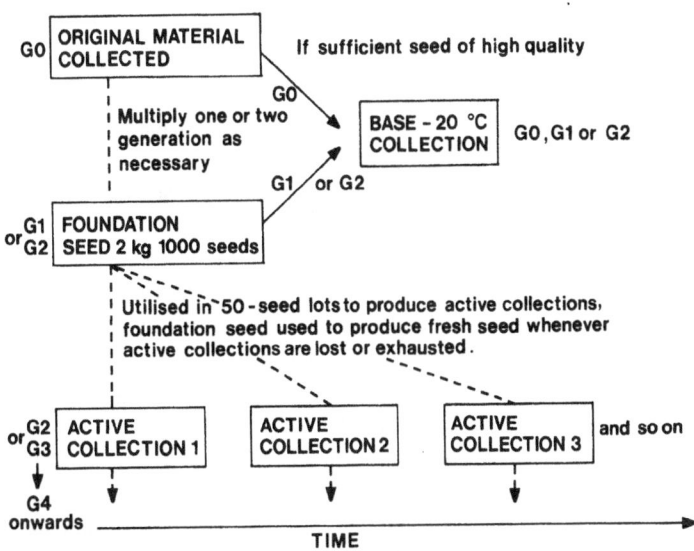

Fig. 2.    Idealised flow diagram for reducing generation advance in out-crossing germ-plasm material maintained as open-pollinated populations.

### Germplasm collection maintained as inbred lines

Burton (3) has argued that selfing (by bagging) is a good way of maintaining cross-pollinated germplasm. There are many arguments in its favour.

i.   For *V. faba*, selfing by bagging (or growing plants in a screen house in the absence of pollinating insects) requires less work than deliberate inter-crossing between plants of an accession and is less likely to produce out-crossing by error to other accessions.

ii.  After several generations, inbred lines are formed which retain their genetic identity from generation to generation. The inbred lines can be planted in rows to provide replicate plants to facilitate screening.

iii. Selfed seed will retain most of the genes from the original open-pollinated population. Due to genetic drift, because of small sample size and random segregation, some genes that occur in the open pollinated population will not be found in the inbred lines derived from it. This factor becomes less important as the sample size is increased by producing more inbred lines per accession. However, the more variable an accession is, the greater the sample size required to guard against gene loss. In any case, recessive lethals, and those genes which are closely linked to them, will be lost.

iv    Selfing uncovers genes. Material is then available for immediate screening for recessive gene characteristics. Recessive genes at low frequency in an open-pollinated population will mostly be found in heterozygotes and are therefore hidden.

Hawtin and Omar (9) describe pure line collections of *V. faba* maintained at ICARDA. From one to four single plants are selected from the original accession depending on its heterogeneity, and progeny rows from these selected plants are grown in the subsequent season in a screen house. The lines are then maintained, each generation from a single representative plant taken from the progeny row. Although some genetic variation is lost when single plants are selected from the original accession, it has the advantage, described above, that the identification of desired characters, particularly those controlled by recessives, is more efficient.

When inbred lines are produced, it is still necessary to keep base collections (and possibly foundation seed) of the open-pollinated populations from which they were derived. Further pure lines can then be produced, if desired, from the base collection or early generation seed. The inbred lines themselves are maintained in base collections for long term storage, so that they can more easily be replaced if they are lost from the active collections.

Burton (3) also describes a method of using selfed seed to simplify the distribution of open pollinated germplasm. To avoid having to measure and ship several hundred packets of germplasm, he created and registered Tift No.1 $S_1$ pearl millet germplasm. This is a mixture of equal quantities of selfed seed from 275 pearl millets and contains a great variety of the dominant and recessive genes observed in the original open-pollinated populations from which it was derived. All characters can be observed in a single space planting of a 25 g sample (c. 5000 seeds). To distribute seed, only one 25 g sample is required, instead of 275 labelled packets.

One disadvantage of Burton's method is that no pedigree data are available to the breeder who, therefore, has to base his selections on single plants which is a problem for characters of low heritability.

*Trait specific gene-pools*

Germplasm pools can be prepared by mixing seed of the accessions and growing them together in an isolated field. The plants are allowed to inter-cross naturally and are harvested as a bulk. This is not the best way to preserve germplasm. Burton (3) found that advancing five germplasm pools of pearl millet three generations narrowed the phenotypic variability of the original pool, lost genes, and obscured 'hard-to-recover characteristics'.

Loss of phenotypic variability can be overcome by dividing the germplasm material into different pools on the basis of morphological characters (traits) and region of origin. The first step is to evaluate the accessions for characters with reasonably high heritability. The results of evaluation are used to create trait specific gene-pools (TSG's) based on such characteristics as maturity (early, late), seed size, height and growth habit. The TSG's are formed from accessions with the same country or region of origin since they will tend to be similar. This also ensures that only accessions from similar latitudes with similar day length sensitivity are pooled. A high between evaluation site correlation for flowering time can then be safely predicted when the different sites in which the TSG's are subsequently grown and evaluated, have similar day length.

Multivariate analysis of evaluation data would provide the best basis of forming TSG's. Clusters of accessions in the analysis can then be pooled on the basis of a whole range of characters and the correlations between them. Multivariate analyses successfully separate accessions into geographic groups (12).

Each TSG thus consists of a number of selected accessions which are mixed in equal proportions and maintained as a bulk by growing them as a single isolated population with natural inter-crossing.

The total phenotypic variance is maintained between the germplasm pools and a collection of around 2000 accessions could be reduced to 10-200 TSG's. Within each germplasm pool, linkages are broken so that within gene pool and total phenotypic,variance may even increase. Recessive gene characters are more likely to be exposed within each trait-specific gene-pool than when a single gene-pool is created from all of the material.

For germplasm purposes, TSG's are best maintained by subjecting them to minimum selection pressures (e.g. constant seed descent where the same number of seeds are taken from each plant in the gene-pool). TSG's can then be distributed, and used as source populations for such things as improvement by single plant selection, recurrent selection and mass selection. By allowing natural selection to take place, adaptation to local conditions is improved.

It must be pointed out that trait specific gene-pools can consist of inter-crossing populations or mixtures of inbred lines or selfed seed. The former method has a lower frequency of homozygous loci, making screening for recessive characters more difficult, but has the advantages of allowing gene recombination and being easier to maintain.

Gene pooling has distinct advantages when it is realised that variability is distributed non-randomly throughout the distribution of a crop plant. Combining all early maturing varieties from Syria into a single gene-pool is not likely to lose much variability, but it can reduce the numbers of accessions enormously. A major problem with germplasm material is that it can be very repetitive, with little difference between many of the accessions, and trait specific gene-pools exploit the fact and reduce accession numbers drastically.

Gene-pools, as in the case of inbred lines, do not avoid the necessity of maintaining in base collections the original populations from which they were derived. The original populations are the source material from which the gene-pools are assembled, and contain the entire range of genes.

*Storage conditions and seed size*

To increase longevity, seeds are stored at low moisture content and low temperature. Under the long term storage conditions recommended by the IBPGR of 5-6% moisture content and -20°C, the longevity of *V. faba* seeds is very high (15). For example, under these conditions, seed samples will take decades to fall from 100% to 85% germination. 'Medium term' storage of low moisture content, and temperature of 5°C to 10°C still provide very high longevity periods. Active collections are kept at ambient temperatures, and the moisture content is allowed to equilibrate naturally. Depending on the ambient conditions, seed germination can remain high for up to five years. Problems will only be encountered in warm, humid climates.

The seed size of *V. faba* (up to nearly 2000 g per 1000 seeds) precludes the storage of samples with a large number of seeds. However, arbitrary figures of a sample size of 12,000

12

seeds per accession for heterogeneous accessions can safely be ignored. Whatever the circumstances, a considerable portion of the variability will be maintained between them. Furthermore, large sample sizes of individual accessions are only necessary if a useful gene occurs at a low frequency in a single accession and no example of this can be found in the genetic resources literature. Useful genes are likely to occur in more than one accession, and often at a high frequency.

*Variation, evaluation and utilisation*

Chapman (4) has drawn up a list of the genetic variation within *V. faba* which is available to breeders. Many of the forms described are spontaneous or induced mutations whereas some have been selected from landrace populations of cultivars. This catalogue should prove valuable to the breeder, and provide stimulus for further work.

The preponderance of mutants rather than forms from landrace populations in this list, probably reflects the work that has been done. Inbreeding to reveal recessives and produce inbred lines has revealed useful genetic variation (7). The lack of useful genetic variability for pest and disease resistance may show a need for further screening rather than a lack of variability for these characters.

There is a need to specify conditions under which plants are screened for genetic variation, and to agree on methods of evaluation. To facilate evaluation and the interchange of data, the IBPGR intends to draw up a list of descriptions for *V. faba*. This task has already been completed for many other crops.

Evaluation data can be divided into two broad categories, preliminary and full. Preliminary evaluation data, including characterisation, is produced by genetic resources centres, and consists of a short list of descriptors to provide some preliminary knowledge of the germplasm collections. Such preliminary evaluation data can be of immediate use as it is essential in the pooling of germplasm material. Breeders, when screening germplasm material, will produce evaluation data that would normally lie outside the scope of preliminary evaluation. Information on biochemical quality and race-specific disease resistance would enter into this category.

The characters looked-for in faba bean breeding and upon which a descriptor list could be based include:

*Plant morphology:* pod density, distribution and orientation; stem formation; seeds per pod and per node; seed size and seed coat thickness; seed shattering; height; determinate growth habit; seed, flower and stem colour.

*Agronomic characters:* earliness; cold resistance; drought resistance; yield; fertilizer response.

*Quality characters:* tannin levels; favism; protein levels.

*Reproductive characters:* percentage of selfing; male sterility.

*Disease pest and parasite resistance:* disease resistance e.g. to *Botrytis fabae* and *Ascochyta fabae*; resistance to viruses; resistance to pests e.g. to *Ditylenchus dipsaci* (stem nematode) and *Aphis fabae* (aphids); resistance to parasitic plants e.g. to *Orobanche crenata* (broomrape).

Faba bean breeding has often utilised germplasm material for such characters as those listed above but this extensive topic is better dealt with in a discussion on faba bean breeding and breeding methodology.

*Acknowledgements*

I would like to thank M.M.F. Abdalla of Cairo University, and Ali Abdel Aziz of the Agricultural Research Centre, Egypt, for sending me details of Egyptian faba bean germplasm. Lothar Seidewitz sent me a listing of the faba bean collection at Braunschweig, and Pietro Perrino that of Bari and Christien Lehman that of Gatersleben. Their help is gratefully acknowledged. Geoff Hawtin gave many helpful suggestions at the manuscript stage.

REFERENCES

1. Ayad, G, Anishetty, M. 1980. Directory of Germplasm Collections 1. Food Legumes. IBPGR, FAO, Rome.
2. Badra, T. 1978. Egypt. in Genetic Resources Unit Exploration 1978. pp. 123-125 International Institute of Tropical Agriculture, Ibadan, Nigeria.
3. Burton, G.W. 1979. Handling cross-pollinated germplasm efficiently. Crop Science, 19, 685-690.
4. Chapman, G.P. 1981. Genetic Variation in *Vicia faba*. FABIS . (in press).
5. Della, A. 1981. Report on the exploration and collection of broad beans (*V. faba*) in Cyprus. Plant Genetic Resources Newsletter, FAO, Rome. (in press).
6. Hanna, A.S., Lawes, D.A. 1967. Studies on pollination and fertilisation in the field bean (*V. faba* L.). Annals of Applied Biology, 59, 289-295.
7. Hanounik, S., Hawtin, G.C., 1981. Breeding for disease resistance to chocolate spot caused by *Botrytis fabae.* In 'Faba Bean Improvement' p. 243.
8. Hawtin, G.C., Omar, M. 1980a. Estimation of out-crossing between isolation plots of faba beans. FABIS, 2, 28-29.
9. Hawtin, G. C., Omar, M. 1980b. International faba bean germplasm collection at ICARDA. FABIS, 2, 20-22.
10. Holden, F.H.W., Bond, D.A. 1960. Studies on the breeding system of field beans (*V. faba* L.). Heredity, 15, 175-192.
11. Muratova, V. 1931. Common beans - *Vicia faba* L. Supplement 50th Bulletin Applied Botany, 1-298.
12. Murphy, P.J. and Witcombe, J.R. 1981. Variation in Himalayan Barley and the concept of Centres of Diversity. Paper presented at the 4th International Barley Genetics Symposium, Edinburgh, 1981.
13. Porceddu, E., Monti, L.M., Frusciante, L., Valpe, N. 1980. Analysis of cross-pollination in *Vicia faba* L. Zeitschrift Pflanzenzuchtung, 84, 313-322.
14. Poulsen, M.H. 1975. Pollination, seed-setting, cross-fertilisation and inbreeding in *V. faba* L. Zeitschrift Pflanzenzuchtung, 74, 97-118.
15. Roberts, E.H. 1975. Problems of long-term storage of seed and pollen for genetic resources conservation. in Crop Genetic Resources for Today and Tommorrow. Eds. O.H. Frankel and J. G. Hawkes. pp. 269-295. IBP, 2. Cambridge University Press, Cambridge.
16. Solh, M., Rashid, K., Hawtin, G. C. 1974. Food Legume Collection, Afghanistan, July-August 1974. ALAD, Ford Foundation, mimeo.

# 2. THE GENETIC IMPROVEMENT OF FABA BEAN
An Overview of Breeding Methods.

G. C. HAWTIN

*Food Legume Improvement Program, ICARDA, P.O.Box 5466, Aleppo, Syria*

Faba beans are widely believed to have originated in the Mediterranean-West Asia Region, probably in the late Neolithic period (16)(9). Throughout their long history as a cultivated crop, faba beans have been subjected to both natural selection and selection by farmers in the different environments in which the crop has been grown. This has resulted in the wide range of genetic variation in the species today. In spite of centuries of such selection, *Vicia faba* retains vestiges of its wild past and, in certain respects, can be regarded as an incompletely domesticated species. The indeterminate nature of the growth habit and the existence of dehiscent pods in many populations (31) can be cited as examples. The breeding system of the species, which stands between full autogamy and full allogamy, may be another.

It has been argued (e.g. 76) that autogamy can develop in response to selection pressure for a character, or characters, of immediate fitness. Although a shift from allogamy appears to have occurred in many crop species, including certain legumes, the process has been incomplete in the case of faba beans.

The primitive wild form of *Vicia faba* was probably allogamous and may have been closely related to the present day *paucijuga* types (47). The shift from allogamy towards autogamy may have occurred comparatively recently (45). The intermediate breeding system of the species may combine some of the benefits of full autogamy (e.g. good immediate fitness; independence from pollen vectors) with some of the advantages of allogamy e.g. greater flexibility to changes in environment through the maintenance of heterogeneity within populations; the possibility of generating better adapted recombinants; the exploitation of heterosis.

While a fuller description of aspects of the breeding system is given by Monti in this volume (46), it will be described briefly here to put the subsequent discussion of breeding methodologies into perspective.

## The Breeding System

*Cross-pollination:* The importance of pollinating insects has been recognised for many

*G. Hawtin & C. Webb (Eds.): Faba Bean Improvement*
©1982 ICARDA. ISBN -13:978-94-009-7501-9

years. Numerous reports from Europe have indicated that 5 –80 % cross-pollination occurs under field conditions. The majority of estimates fall between 20 and 50% e.g. (73) (28) (57) (69) (29) (19) (32) (62) (74). Although comparatively few estimates of cross-pollination have been reported in the Nile Valley region, the situation appears similar. Kambal (40) reported 35% cross-pollination near Khartoum, Sudan; El-Sherbeeny (22) found between 57 and 71 % at Giza, and Hawtin (34) reported estimates of 33% and more than 40 % at two locations in Egypt in the 1975-76 season.

In Europe, the long-tongued bumble bees (e.g. *Bombus hortorum* and *B. agrorum*) appear to be the main pollen vectors (75). Honey bees are generally considered to be less effective (12) although they can adequately pollinate the crop if present in sufficient numbers (62). In the Nile Valley, honey bees may be more important. In one study in Egypt, about 80% of the insects visiting faba bean flowers were honey bees (80).

*Autofertility:* It has been known for many years that insect pollination in faba beans can result in increased seed set. Charles Darwin reported in 1876 (18) that a total of 135 seeds was produced on 17 uncovered plants compared to only 40 seeds on 17 plants from which pollinating insects were excluded. On entering the flower, an insect may cause the release of the stamens and style from the keel petal (a process known as tripping) and thereby help the transfer of self pollen from the anthers to the stigma. Tripping may also ensure good contact between the pollen grains and stigma, through crushing the minute stigmatic papillae (69) and may result in scarification of the stigma surface (78). Although a response to tripping is common in the species, a number of reports indicate that this is not universal. The lack of a tripping requirement is referred to as autofertility (20)(21) and lines exhibiting high levels of autofertility have been widely reported e.g. (38) (56) (32) (63) (65). In recent years, several breeding programmes have concentrated on developing autofertile cultivars (43). Such cultivars should have the advantage of being able to yield well even in the absence of adequate insect pollinators, resulting in improved yield stability. The research on autofertility has led to the release of several cultivars (Dacre, Danas and Deiniol) by the Welsh Plant Breeding Station (10).

Although autofertility has been reported in lines of European origin e.g. (63), it may be more prevalent in *paucijuga* populations from India and materials originating in Africa or the Mediterranean area (43). The Sudanese line IW (40), a 'Mediterranean type' and the Egyptian cv. Habashy (32) have all been identified as highly autofertile. A study of 188 germplasm accessions originating from 19 countries, led to the conclusion that the most autofertile lines were those which originated in Ethiopia, Egypt, Iraq and Syria (25). In another study (70) a Syrian landrace, the Egyptian cvs. Giza 3 and Giza 4, the Sudanese cv. Hudeiba 72 and the Spanish-type cv. Seville Giant all had high autofertility indices compared to the northern European cvs. Throws M.S. and Maris Bead. Table 1 shows the fertility indices for a number of inbred lines and cultivars in the 1979-80 season at the ICARDA research station near Aleppo in northern Syria.

However, not all materials originating in the West Asia and North Africa region are autofertile. The Egyptian cvs. Giza 1, Giza 2 and Rebaya 40 and the Sudanese Baladi have all responded to tripping (40) (32).

Table 1.

Autofertility Indices for 20 fifth generation inbred lines and 16 open-pollinated populations grown in a screen house at Tel Hadya, Syria, 1980

| Inbred Line | Country of Origin | Seed yield autofertility index (1) | Pods/plant autofertility index (2) |
|---|---|---|---|
| 78S 48905 | U.K. | 0.48 | 0.70 |
| 78S 48907 | Egypt | 0.63 | 0.61 |
| 78S 48911 | Sudan | 1.18 | 1.56 |
| 78S 48915 | Colombia | 1.49 | 0.68 |
| 78S 48916 | Colombia | 1.08 | 1.36 |
| 78S 48930 | Sudan | 1.25 | 1.37 |
| 78S 48931 | Turkey | 0.61 | 0.93 |
| 78S 48937 | Lebanon | 0.73 | 0.85 |
| 78S 48941 | Ethiopia | 0.59 | 0.65 |
| 78S 48942 | Ethiopia | 0.84 | 1.12 |
| 78S 48944 | Egypt | 0.46 | 0.53 |
| 78S 48945 | Egypt | 0.59 | 0.80 |
| 78S 48951 | Turkey | 0.91 | 0.97 |
| 78S 48953 | Turkey | 0.89 | 1.34 |
| 78S 48954 | Turkey | 0.97 | 0.90 |
| 78S 48957 | Turkey | 1.04 | 1.01 |
| 78S 48965 | Lebanon | 1.16 | 1.09 |
| 78S 48974 | U.K. | 0.86 | 0.92 |
| 78S 48983 | Yemmen | 1.28 | 1.28 |
| 78S 48920 | Sudan | 0.86 | 0.80 |

Open Pollinated Populations

| | | | |
|---|---|---|---|
| Small seed, ILB 1811 | Syria | 1.49 | 1.09 |
| Large seed, ILB 1814 | Syria | 0.54 | 0.73 |
| Small seed, ILB 1816 | Lebanon | 0.51 | 0.83 |
| Large seed, ILB 1817 | Lebanon | 1.10 | 1.07 |
| Small seed, ILB 1818 | Jordan | 0.76 | 0.66 |
| Large seed, ILB 1821 | Turkey | 0.80 | 1.11 |
| Violetta di Policoro | Italy | 1.29 | 1.53 |
| Hudeiba 72 | Sudan | 1.26 | 1.20 |
| Giza 3 | Egypt | 1.14 | 1.06 |
| Giza 4 | Egypt | 1.13 | 1.12 |
| New Mammoth | U.K. Spanish type | 1.51 | 1.37 |
| Seville Giant | U.K. Spanish type | 0.98 | 0.91 |
| Express | U.K. Spanish type | 0.65 | 0.91 |
| Aquadulce | U.K. Spanish type | 1.08 | 1.53 |
| Throws M.S. | U.K. | 0.20 | 0.21 |
| Maris Bead. | U.K. | 0.40 | 0.47 |
| Standard Error | | ±0.28 | ±0.22 |

(1) Seed yield per plant, non tripped ÷ Seed yield per plant, tripped. Mean of 4 reps.
(2) Pods per plant, non tripped ÷ Pods per plant, tripped. Mean of 4 reps. 5 plants were tripped and 5 non-tripped in each replicate.

Drayner (21) reported that inbreeding results in a loss in autofertility in inbred lines, but that this was fully restored in the $F_1$ between them. This finding has been confirmed subsequently by several researchers although selection for autofertility within and between inbred lines has been successful (42) (64) (77). The loss of autofertility on inbreeding may be due to the stigma requiring greater scarification (77). However, Kambal *et al.* (41) reported that, in some genotypes, the lack of autofertility may be due to a blockage of the ventral passage between the stigma and the keel petal which prevents the transfer of pollen, and the dorsal passage may be blocked by a longer style and the presence of many and long stylar hairs. The greater autofertility of $F_1$'s may be associated with an increased production of pollen (20).

*Inbreeding Depression:* Inbreeding not only reduces autofertility, and hence yield in the absence of pollinators, but also reduces yield *per se* through the loss of heterosis. Higher yields associated with heterozygosity in $F_1$ plants have been widely reported e.g. (44). In diallel crosses of inbred lines, over-dominance was found to affect seed yield positively (7) and appeared important also for several fertility-related characters including pods/node and pods/plant (45). The importance of the component of heterosis due to over-dominance, implied a superiority of the heterozygote which would not be fixable in a homozygous inbred line. Whereas it may well be possible to develop inbred lines with a yield capacity equal to that of currently available cultivars (65), there is no evidence to date that it would be possible to develop autogamous inbred lines having a yield capacity equal to that of their crossbred offspring, in the best cross combinations (63).

Thus to sum up: faba beans are partially allogamous species having an intermediate level of cross-pollination. They may (non-autofertile) or may not (autofertile) require tripping of the flowers to set maximum yield. In general, autofertile lines exhibit less outcrossing (21) (40). However all faba beans show inbreeding depression when compared to their best crossbred offspring but yield levels of inbred lines, equal to those of open-pollinated populations, have been identified.

*Breeding Objectives*

Increased green or dry seed yield and improved yield stability are the primary objectives of most faba bean breeding programmes. The low heritability and consequent limited genetic advance for yield in response to selection has lead many scientists to search for characters which are associated with yield but which are more highly heritable e.g. (53) (6) (79) (66). As with most grain legumes, correlations between the components of yield and yield itself are frequently large, however compensation between components can severely limit yield gains in response to selection for one or more components. A greater chance of success in indirect selection for yield might come from selecting for various phenological or morphological attributes. Characters such as the onset of flowering, duration of flowering, onset of grain filling, duration of grain filling, number of tillers and possibly plant height might all be used in the construction of selection indices for the improvement of yield (79). Other scientists have indicated the possibility of improving the physiological efficiency of the plant as a means of increasing yields in legumes e.g. (72).

However breeding for increased yield should, perhaps, be much bolder. If faba beans are to compete economically with other crops, substantial yield advances are necessary, and soon; as Picard (58) has said, faba bean breeding 'is in a hurry because it has been neglected for too long'. The exploitation of heterosis through synthetics and ultimately hybrids, or a substantial remodelling of the plants' growth habit to transform the current indeterminate habit to a determinate one (14) are two such approaches which could pay off in improved yield potential; while increasing autofertility, reducing flower drop, and developing resistance or tolerance to major yield-reducing factors e.g. pests, diseases, and environmental stress, could all improve the stability of yield.

Insufficient attention has been paid in the past, to developing cultivars with improved disease or pest resistance. For example, chocolate spot has long been recognized as a major cause of low yields in certain environments, yet very few sources of resistance to the disease have been identified (23). However the situation looks more promising in recent large-scale screenings of inbred lines (33) and the lack of success in the past may well be due to inadequate screening methods having been applied to too small a range of genotypes. In the U.K., although winter-sown beans yield about 30% more than the spring-sown crop, the area sown to winter beans represents only about 25% of the total (10). The risk of severe chocolate spot and damage from frost are the main factors responsible for the predominance of spring sowing. Disease and frost resistant cultivars could play a major role in reversing this situation. Delaying the planting date has frequently been associated with reduced yields in Mediterranean-type environments (5) (70) (4), but early planted crops may be more heavily attacked by certain pests and diseases such as *Orobanche, Ascochyta* blight or chocolate spot. In Sudan, early planting to escape the effects of high temperatures at the end of the season can lead to an increased incidence of seedling root rot/wilt diseases (26). The development of resistance could enable the crop to be sown earlier in such environments which in turn could lead to substantial increases in yield. If such changes in the cropping system are to succeed, the combined effort of scientists working together in multidisciplinary teams is required.

Other breeding objectives such as increased protein content, improved cooking or nutritional quality, increased nitrogen fixation, resistance or tolerance to environmental stresses e.g. cold, drought, heat, and salinity, and improved characteristics for mechanization, are all considered to be important by certain programs.

*Genetic Variation*

Considerable genetic variation has been reported within the species (1) (6) (61), and germplasm resources are still largely unexploited. Picard (58) has stressed the importance of developing appropriate screening techniques for the identification of desired genotypes. These should ideally be both quick and reliable and preferably non-destructive, to enable a large number of populations, lines or single plants to be screened. In faba beans, seedling screening techniques are particularly valuable in that selected plants can subsequently be isolated from insect pollinators to ensure selfing.

At ICARDA, a set of pure lines, selected from within each accession of the base collection, is maintained for initial screening purposes (35) (81). The pure lines are considered of particular value in that (a) screening can be done on several plants of the same genotype. e.g. on a row basis, (b) if destructive techniques are used the specific genotypes selected are still available, and (c) recessive genes, which may be 'buried' in heterozygous accessions can be identified. However the maintenance of a pure-line collection requires resources beyond those of most national programmes: more than one hectare of land is under insect-proof screen mesh at the Tel Hadya site near Aleppo. Nevertheless small samples of entries in the ICARDA pure-line collection can be made available to faba bean breeders on request.

The full exploitation of the **available** genetic resources may require hybridization to allow gene recombination. This may either be done by means of insect pollinators or, for greater control, by hand crossing. Techniques for hand crossing may vary but, in general, the bud of the female parent is emasculated before the dehiscence of the anthers, and fresh pollen is transferred from the male parent to the stigma of the female either immediately or after a few hours. The time of day of crossing, the method of emasculation and the position of the female flower on the peduncle have all been found to affect the success rate in crossing at ICARDA (52). The direction of crossing may also be important (2).

Plant breeding has often been called a 'numbers game'; within limits the larger the number of crosses and subsequent populations/lines which can be adequately handled, the greater is the likelihood of success. Faba beans are difficult to cross and only relatively low success rates have been reported in terms of seeds set per cross made. Techniques which would enable more crosses to be made ( e.g. the use of gametocides) could prove valuable. At ICARDA, about 300 crosses in faba beans are made annually in comparison, for example, to more than, 2,500 for barley.

Artificially-induced mutation has also proved to be a valuable technique for creating genetic variation in faba beans e.g. (74) (49). Its use in the future may also give rise to useful genes, especially when desired traits cannot be identified within existing germplasm collections. Polyploidy and inter-specific hybridization may also play a role in future breeding programmes. Their potential fully justifies the considerable research effort which may be needed before practical techniques for their exploitation can be developed (17).

*Cultivars*

Cultivars which breeders are currently developing fall into four broad categories : (i) open-pollinated populations, (ii) synthetics (iii) hybrids and (iv) fully autogamous lines.

Although most of the commonly grown cultivars are open-pollinated populations, synthetics are becoming increasingly important, especially in northern Europe. Hybrids are not yet produced commercially and further research is necessary before they can be developed as an economic alternative. Both synthetics and hybrids exploit heterosis which may result in both greater and more stable yields than those of non-autofertile open-pollinated populations. The current status of seed production programs in many developing countries may limit the extent to which synthetics and hybrids can be produced and distributed to farmers.

However this limitation may not be fully applicable to the countries of the Nile Valley where national seed production programs already produce faba bean seed for the farmers. In these countries, it is believed that consideration should be given at least to the development and evaluation of synthetics, in addition to the on-going improvement of populations. The current status and future possibilities with regards to synthetics and hybrids have been reviewed by Bond (11) and Picard *et al.* (59) respectively.

The question arises as to how far inbreeding should be utilized in a breeding program. The situation with respect to synthetics has been considered by Bond (11), and for components of such cultivars it would appear that well-evaluated inbred lines offer the most promise. When the end product is an open-pollinated cultivar, the value of inbreeding is less clear. It has been proposed that faba beans could, or should, be developed as a fully autogamous species e.g. (43).

A conversion of the breeding system not only offers an opportunity for improved yield stability through less dependance on pollen vectors, but would allow populations and segregants to be screened and multiplied under open pollination. This, in turn, would greatly increase both the scale on which breeding operations could be conducted and the degree of selection pressure which could be applied (10). The closed-flowered character (64), coupled with cleistogamy, could prove important in the development of fully autogamous cultivars. Although autogamy would exclude the exploitation of heterosis, except possibly through the use of male sterility, the numerous advantages of such a system fully justify considerable research effort.

A more efficient exploitation of the genetic variation within the species might stem from the use of more definitive methodologies which could be applied as a result of such a change in breeding system.

*Selection Methods*

Mass selection, either within local populations or in generations following hybridization, is probably the most widely used breeding method, but it has given rise to little improvement in yield when selection has been made on the basis of yield itself (75). It is of greatest value for improving characters of high heritability, especially if these characters are strongly correlated with yield. However the main value of mass selection may lie in improving uniformity (8).

Recurrent selection has proved a useful technique for the improvement of many allogamous species, and various schemes have been proposed for use in partially allogamous grain legumes such as faba beans (67). Such methods may have particular value in the improvement of characters which are under polygenic control. This is the case with many of the characters of importance in faba bean breeding including yield (58). A system of recurrent selection, using honey bees in cages to randomly intercross flowers in the recombination phase, is currently being investigated at ICARDA. A similar system is being used in Egypt (50) (51). Recurrent selection can give rise either to improved populations for direct release as open-pollinated cultivars, or can provide a source of genetic variability for further improvement by other methods, including the development of inbred lines.

The system at ICARDA is to develop, through isolation in segregating generations, relatively homogeneous populations with desired agro-economic characteristics. The generalized scheme is shown in Fig. 1. Mass selection is in the $F_2$ generation for characters such as disease resistance, *Orobanche* resistance, growth habit, and earliness. In the $F_3$, single plants are bagged at the beginning of flowering in the field, either randomly for yield, or based on such characters as seedling vigour, seedling cold tolerance, pest and disease resistance and early flowering. Up to 25 individual plants are bagged in each population. In the $F_4$ generation, the progenies are evaluated in progeny rows, with one or two plants (normally border plants for maximum seed set) being bagged. Selection in $F_4$ is on a family basis.

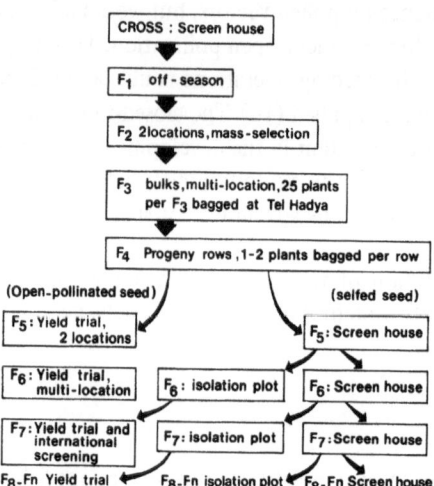

Figure 1. Flow of breeding material in the ICARDA programme.

The bulked seed of the open-pollinated plants in selected progeny rows is sown in replicated trials in $F_5$, while the progenies of the bagged plants are raised in an insect-proof screen house. Seed for multilocation testing in $F_6$ is obtained from the center of the plots in the $F_5$ yield trail, but the $F_6$ seed of the same entries, orginating in the screen house is grown both in isolation plots and in the screen house. All seeds for yield testing in $F_7$ and later generations originate from isolation plots. Seeds for sowing the isolation plots are obtained from the screen house. Thus the system involves inbreeding in the $F_3$ and $F_4$ generations, at which stage selection for autofertility is practised. Subsequent testing, with some improvement of uniformity through mass selection, is on $F_4$-derived populations. I believe that this system provides a reasonable compromise between inbreeding, which allows some precision in selection for highly heritable characters, and the identification of superior $F_4$-derived populations within which residual heterozygosity may be contributing to overall yield. The maintenance of all retained $F_4$-derived populations in the screen house during subsequent generations also enables highly inbred and autofertile lines to be selected (no hand-tripping is practised in the screen houses) and evaluated for their suitability as components of synthetic cultivars.

The basic system as has been described, can be modified in a number of ways. If facilities for bagging plants in the field are not available, remnant seed of selected plants or lines can be advanced in isolation for further evaluation/selection in the next generation. The ability to self plants in the off-season enables such a system to proceed reasonably quickly. Single plant selection can also be continued for longer. This might result in further genetic advance, but it increases the risk of loss in yield and flexibility from reduced heterozygosity. This may be less important for programmes which aim to cover a relatively narrow range of environments and when the selection environment is typical of the range. However the ICARDA programme aims at developing improved genetic stocks for a very diverse agro-

23

climatic region, from Morocco to Pakistan, and high levels of residual heterozygosity in the $F_4$-derived populations have the advantage that breeders in the national programs can improve the populations further through selection within their own environments.

Single seed descent, as proposed by Brim (13), in an insect-free environment, offers an efficient means of advancing populations without selection, while preserving genetic variance for later selection. The method is of particular interest if it is possible to raise two or more generations a year under these conditions. The method has been used successfully in other grain legume species, e.g. lentils (48).

The use of a pedigree, or bulk-pedigree system without adequate facilities for selfing is particularly dangerous in faba beans, especially for the improvement of yield. The selection of superior individuals is likely to result in the selection of many $F_1$ out-crossed plants.

*Yield Evaluation*

Selection for yield, and the multi-location testing of selected lines, is greatly complicated by genotype x environment interactions, which are frequently large.

Multi-location testing of materials, developed at ICARDA, takes place at several stages. Fig. 2 shows the generalized scheme for international evaluation of breeding material followed at ICARDA.

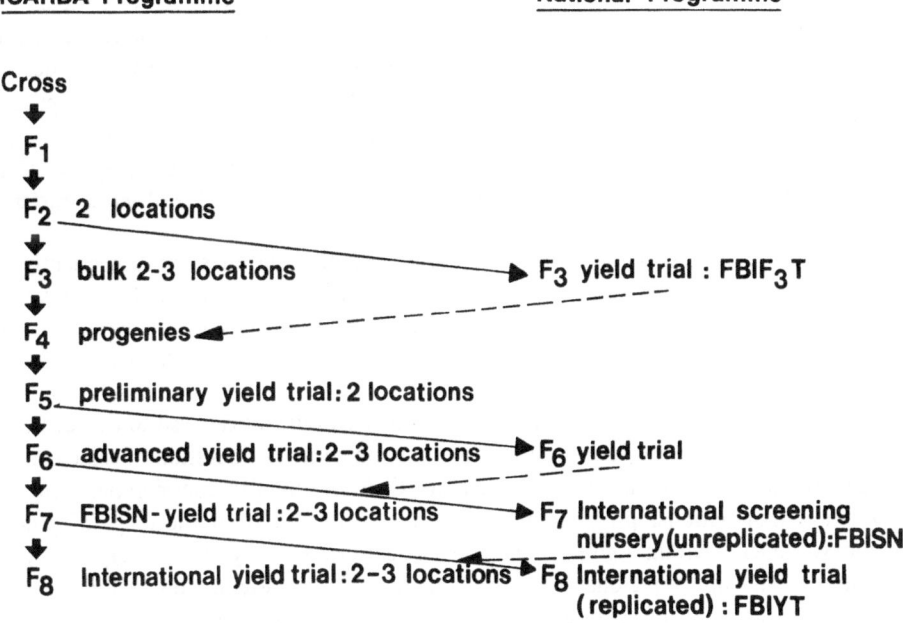

Fig. 2. Outline of flow of genetic materials to national programs and data feedback. Solid line indicates flow of genetic material; broken line indicates data feedback. National programs also contribute material to the international programs. Data from international trials are also used in selection of parents for crossing.

Where seed stocks permit, $F_2$ populations are grown in two environments: at Tel Hadya research station, primarily for an initial yield evaluation and mass selection for vigour and earliness, and at other sites for mass selection for specific characteristics. These sites include: Lattakia for resistance to *Botrytis fabae* and *Ascochtya fabae*; Kafr Antoun for resistance to *Orobanche* and Terbol, Lebanon, for screening under the higher altitude conditions of the Beqa'a Valley. After mass selection in $F_2$, the $F_3$ populations are selected for distribution to national programmes in the region, either on the basis of $F_2$ yield performance or on parental origins and performance. The use of population mean performance as a means of early generation testing has been criticised on the grounds that parental performances alone normally provide an adequate indication of subsequent cross performance. However in the ICARDA context of breeding for a wide range of environments, the multilocation evaluation of $F_3$'s provides not only an indication of mean cross performance but also a mechanism for the dissemination of promising early generation populations within which further selection can be undertaken in the environments for which the materials are ultimately intended.

Feedback of data on $F_3$ population performance provides a guide for the selection of single plants in the $F_3$ and $F_4$ generations at ICARDA; selection being biased in favour of the most promising $F_3$'s. Selections intended for environments which are substantially different from those in northern Syria are made on a random basis within the most promising populations.

$F_6$ or $F_7$ populations ($F_4$-derived lines) are subsequently distributed to national programmes for evaluation. These nurseries are normally unreplicated and use an augmented design (24). The value of non-replicated data for yield assessment, even using such designs, is currently being questioned at ICARDA. Correlations for yield between non replicated and replicated trials at the same location and year have been disappointingly low, and frequently non-significant. However, non-replicated screening nurseries may be of far greater value in the assessment of characters of high heritability, such as disease resistance, and growth habit.

Lines which show promise in the international screening nurseries, are subsequently distributed to national programmes in replicated yield trials. As indicated above, superior entries in such trials can be considered for further evaluation before direct release as open-pollinated cultivars; can be further improved by selection, or can be included in crosses or synthetics.

The extent to which selection and testing in one environment can lead to the identification of cultivars which perform well in other environments is a question of some concern to an international program such as that at ICARDA. Similar considerations are also relevant within national programs which are attempting to cover the range of production environments within a given country. Faba beans vary widely in their adaptability; cv. Aquadulce has shown reasonably wide adaptability across several sites in the region in 1978-79 and in a trial of nine cultivars (FBAT-79, Table 2) Giza 3 ranked first in Algeria, second in Egypt, Libya and Sudan, third in Aleppo, Syria and Izmir, Turkey, and fourth at Lattakia, Syria, (39). On the other hand, many other cultivars, have shown a much narrower adaptability, and the extent to which breeders should attempt to develop widely adapted cultivars is still an open question.

Table 2.

Seed Yield (kg/ha) of entries in the FBAT-79 trial at different locations during 1978-79

| Entry | | Country of Origin | ALGERIA Sidi-Bel-Abbes | EGYPT Sids | LIBYA Tajoura | SUDAN Hudeiba | SYRIA Tel Hadya | SYRIA Lattakia | TURKEY Izmir | ENTRY Mean |
|---|---|---|---|---|---|---|---|---|---|---|
| ILB | 1814 | Syria | 1519 | 2799 | 276 | 20 | 348 | 3040 | 2548 | 1507 |
| ILB | 1811 | Syria | 1560 | 5047 | 1591 | 339 | 1529 | 3258 | 1584 | 2129 |
| ILB | 1821 | Turkey | 1444 | 2221 | 432 | 9 | 436 | 3280 | 1246 | 1295 |
| ILB | 1818 | Jordan | 1765 | 4655 | 1638 | 194 | 1210 | 3845 | 1511 | 2117 |
| ILB | 1817 | Lebanon | 1857 | 5750 | 1571 | 34 | 418 | 3683 | 2402 | 2245 |
| ILB | 1816 | Lebanon | 1909 | 5004 | 1077 | 78 | 1586 | 4370 | 1794 | 2259 |
| Aquadulce | | Spain | 1880 | 5307 | 1706 | 61 | 542 | 4043 | 2130 | 2238 |
| Giza 3 | | Egypt | 2002 | 5964 | 2002 | 412 | 1104 | 4000 | 2174 | 2523 |
| Giza 4 | | Egypt | 1794 | 6954 | 2080 | 492 | 791 | 3680 | 1589 | 2483 |
| Location mean | | | 1748 | 4855 | 1375 | 182 | 885 | 3689 | 1886 | |
| S.E. entry | | | 183.1 | 445.9 | 97.9 | 57.8 | 210.1 | 401.9 | 250.9 | |

Yield correlations of a uniform set of genotypes across a range of environments have indicated the dangers of selecting in a single environment, and of inadequate multi-location testing. In a total of 44 correlations between Tel Hadya and 12 locations in 10 countries, most were very small. Table 3 shows the correlations for yield between nine entries in seven locations in the FBAT-79 trial. It appears that sites in North Africa (Sudan, Egypt, Libya, Algeria) were reasonably well correlated; but the Izmir and Lattakia sites correlated with no other location. Results, such as these, have lead ICARDA towards the sub-division of the region into separate zones for the purposes of selection and testing of breeding materials. In future, greater emphasis will be placed on developing materials within a zone rather than on centralizing breeding activities in Aleppo.

A major obstacle to the improvement of faba beans, especially major types, is the low multiplication rate; frequently not more than 10 to 15 fold. Picard (58) has asked 'could it be efficient to divide a 20 plant progeny into 4x5 ? '. Although some research on optimum plot sizes and configurations has been reported e.g. (54) (55) (68), Picard's question still remains largely unanswered. More work remains to be done on field plot designs, particularly in the development of micro-plot techniques.

*Off-Season Nurseries*

Although growing more than one generation of breeding material each year greatly speeds up the breeding process, the use of off-season nurseries has been reported only rarely in faba beans. Some breeders make use of environmentally controlled green houses for limited off-season work, but this is generally a poor substitute for an alternative location in which the crop can be grown out in the field. However, it may be possible to raise three generations each year under green house conditions; thus further increasing the rate of genetic turnover. The suggestion has been made (15) of using both hemispheres for the improvement of the spring-sown crop in Europe and the breeding programme at Manitoba has reported satisfactory results using Hawaii as a winter nursery (27). Abdallah (3) has reported the use of a location in West Germany as a summer off-season site for the breeding programme at Cairo University. The ICARDA program has an arrangement with the Ministry of Agriculture, Jordan, for the use of its high altitude station at Shawbak for raising a crop during the summer (June to October). About one hectare is sown to summer faba beans annually for advancing $F_1$'s and certain $F_3$ and $F_4$ progenies and for seed increase. It is planned to expand the off-season activities and to make the site available to other national programmes in the region.

*Isolation Plots*

Reference has already been made to the use of isolation plots for increasing or advancing faba beans under conditions which limit inter-crossing between populations. When populations are small (e.g. single plant progenies) the use of insect-proof cages, or bagging individual plants is normally the most economical way of ensuring selfing. But when lines or populations are bulked in advanced generations, such techniques may not be feasable. The literature contains several reports of the optimum distances to ensure minimum crossing between isolation plots e.g. (60) (22) (30) (35). Crossing within isolation plots is normally an advantage in that levels of heterozygosity are increased and seed harvested from such plots can be used in yield trials without inbreeding depression affecting the results.

Table 3.

Correlations between locations of the nine entries in FBAT-79 trial.

| LOCATION | EGYPT Sids | LIBYA Tajoura | SUDAN Hudeiba | SYRIA Lattakia | Tel Hadya | TURKEY Izmir |
|---|---|---|---|---|---|---|
| Algeria-Sidi-Bel-Abbes | 0.77 | 0.71 | 0.29 | 0.28 | 0.88 | 0.33 |
| Egypt - Sids | | 0.92 | 0.68 | 0.33 | 0.56 | 0.11 |
| Libya - Tajoura | | | 0.73 | 0.36 | 0.51 | -0.47 |
| Sudan - Hudeiba | | | | 0.45 | 0.10 | -0.26 |
| Syria - Lattakia | | | | | 0.44 | -0.39 |
| Syria - Tel Hadya | | | | | | 0.11 |

The literature is in general agreement that inter-plot crossing falls rapidly with increasing isolation plot distance, but that even with distances of longer than one kilometer, genetic contamination can still occur. At ICARDA, isolation plots are planted at a distance of 50 m which should ensure that genetic contamination is less than 5%. With the large number of isolation plots required by the program (about 160 plots are planted at Tel Hadya this season) it is not considered practical to increase isolation distances beyond 50 m. It appears to be possible to reduce out-crossing further by discarding the plot borders and seed produced on the lowest nodes of the plant.

*Conclusions*

It is expected that new techniques will be developed which could have a profound effect on faba bean breeding. Rapid techniques for the vegetative propagation of hybrid materials without returning to the parental lines; the culture of haploid tissues to produce homozygous lines; the transfer of genes from other species and the use of genetic engineering techniques, could all play a substantial role in the improvement of the species. However faba bean breeding still has a long way to go and can not await such developments. There is still no general agreement between breeders as to which types of cultivar offer the most promise or the most appropriate methods for developing them. In spite of this, all are agreed that if faba beans are to maintain or even increase their role in world agriculture, greater and more coordinated research efforts are needed. Remodelling the plant growth habit, increasing levels of resistance to major diseases and pests, reducing the plants' sensitivity to environmental stresses and the exploitation of heterosis, all offer promise for the future of the crop.

## REFERENCES

1. Abdalla, M.M.F., 1976. Natural variability and selection in some local and exotic populations of field beans, *Vicia faba* L. Z. Pflanzenzuchtg. 76:334-343.
2. Abdalla, M.M.F., 1977. Intraspecific unlitateral incompatibility in *Vicia faba* L. Theor. Appl. Genet. 50 227-233.
3. Abdalla, M.M.F., 1980. A cross breeding program for faba bean (*Vicia faba*). FABIS 2, 24.
4. Baldwin, B.B., 1989. Time of sowing-a major factor influencing the yields of *Vicia faba* in South Australia. FABIS 2, 39.
5. Bianco, V.V., 1979. Agronomic aspects of broad bean production in southern Italy. In: Some current research on *Vicia faba* in western Europe. Ed. D.A. Bond, G.T. Scarascia-Mugnozza and M.H. Poulsen Pub. EEC. EUR 6244 EN, Luxembourg pp. 125-143.
6. Bianco, V.V., Damato, G., Miccolis, V., Polingnano, G., Poerceddu, E., Scippa, G., 1979. Variation in a collection of *Vicia faba* L. and correlations among agronomically important characters. In: Some current research on *Vicia faba* in western Europe. Ed. D.A. Bond, G.T. Scarascia-Mugnozza and M.K. Poulsen. Pub. EEC, EUR 6244 EN Luxembourg pp. 217-250.
7. Bond, D. A., 1966. Yield and components of yield in diallel crosses between inbred lines of winter beans (*Vicia faba*). J. Agric. Sci., Camb. 67, 325-336.
8. Bond, D.A. 1971. Breeding methods in field beans (*Vicia faba* L.). Rep. Meet. EUCARPIA Fodder Crops Sec., Lusignan, 1970, 119-126.

9.  Bond, D.A., 1976. Field bean, *Vicia faba* (*Leguminosae - Papilionatae*). Chapter in: Evolution of Crop Plants. Ed. N.W. Simmonds, Pub. Longman, London.

10. Bond, D.A., 1979. Breeding work on *Vicia faba* in the U.K. FABIS 1, 5-6.

11. Bond, D.A. 1981. Development and performance of synthetic varieties of *Vicia faba* L. Paper presented at International Faba Bean Conference, Cairo.

12. Bond, D.A., Hawkins, R.P. 1967. Behaviour of bees visiting male sterile field beans (*Vicia faba*): J. Agric. Sci. Camb. 68, 243-247.

13. Brim, C.A., 1966. A modified pedigree method of selection in soybeans. Crop Sci.6,220

14. Chapman, G.P., 1977. Restructuring the Field Bean Plant, *Vicia faba* L. Scot. Hort. Res. Inst. Associ. Bull. 15, 3-9.

15. Chapman, G.P., 1979. New concepts in breeding high yield *Vicia faba* L. In: Some current research on *Vicia faba* in western Europe. Ed. D.A. Bond, G.T. Scarascia-Mugnozza and M.H. Poulsen. Pub. EEC, EUR 6244 EN Luxembourg pp. 293-302.

16. Cubero, J.I., 1974. On the evolution of *Vicia faba*. Theoret. Appl. Genet. 45, 47-51.

17. Cubero, J.I.,1981. Interspecific hybridization in *Vicia*. Paper presented at the International Faba Bean Conference, Cairo.

18. Darwin, C., 1876. The effects of cross and self-fertilization in the vegetable kingdom. Pub. Murray. London.

19. Derenne, P. 1966. Feverole. Station de recherches de l'Etat pour l'amelioration des plantes de grande culture. Gembloux, Rapport d'activite pour l'annee 1966, 67-81.

20. Drayner, J.M., 1956. Regulation of outbreeding in field beans. Nature 177, 489-490.

21. Drayner, J.M., 1959. Self-and cross-fertility in field beans (*Vicia faba* Linn.). J. Agric. Sci., Camb. 53, 387-403.

22. El-Sherbeeny, M.H., 1970. Studies on pollination, fertilization and pod-setting in the field bean and their bearing on breeding the crop. M.Sc. Thesis. Univ. of Cairo.

23. Enriquez, G.A., 1977. Chocolate spot on *Vicia faba* caused by *Botrytis fabae* Sardina and *B. cinerea* Pers.; a literature review. CATIE, Turrialba, Costa Rica, 34 pp.

24. Federer, W.T., 1956. Augmented designs, Hawaian Planters' Record 55, 191-208.

25. Filippetti, A. 1979. Breeding projects and work for the improvement of broad beans (*Vicia faba*) in Puglia. In: Some current research on *Vicia faba* in western Europe. Ed. D.A. Bond, G.T. Scarascia-Mugnozza and M.H. Poulsen, Pub. EEC, EUR 6244 EN, Luxembourg pp. 168-188.

26. Freigoun, S.O., 1980. Effect of sowing date and watering interval on the incidence of wilt and root rot diseases in faba bean. FABIS 2, 41.

27. Furgal, F.J., Evans, L.E., 1980. Faba bean breeding in Manitoba. FABIS 2, 22-23.

28. Fyfe, J.L., Bailey, N.J.T., 1951. Plant breeding studies in leguminous forage crops. I. Natural cross-breeding in winter beans. J. Agric. Sci., Camb., 41, 371-378.

29. Gottschalk, W., 1960. Untersuchungen uber die Befruchtungsverhaltnisse von *Vicia faba* mit Hilfe einer fruh erkennbaren mutante. Zuchter 30, 22-27.

30. Gottschalk, W., 1978. The breeding system of *Vicia faba*. Legume Research 1 (2), 69-76.

31. Hanelt, P., 1972. Die infraspezifische variabilitat von *Vicia faba* und ihre Gliederung. Kulturpfl. 20, 75-128.

32. Hanna, A.S., Lawes, D.A., 1967. Studies on pollination and fertilization in the field bean. (*Vicia faba* L.). Ann. Appl. Biol. 59, 289-295.

33. Hanounik S., Hawtin, G.C., 1981. Breeding for resistance to chocolate spot caused by *Botrytis fabae*. Paper presented at International Faba Bean Conference, Cairo.

34. Hawtin, G.C., 1979. Strategies for the genetic improvement of lentils, broad beans and chickpeas, with special emphasis on research at ICARDA. In: Food Legume Improvement and Development. Ed. G.C. Hawtin and G. J. Chancellor. IDRC-126e Ottawa pp. 147-154.

30

35. Hawtin, G.C., Omar, M., 1980. International faba bean germplasm collections at ICARDA. FABIS 2, 20-22.
36. Hawtin, G.C. , Omar, M., 1980. Estimation of out-crossing between isolation plots of faba beans. FABIS 2, 28-29.
37. Hayes, I. D., Hanna, A.S., 1968. Genetic studies in field beans. III. Variation in self-fertility in a diallel cross, Z. Pflanzenzucht 60, 315-326.
38. Holden, J.H., Bond, D.A., 1960. Studies on the breeding system of the field bean (*Vicia faba* Linn.). Heredity 15, 175-192.
39. ICARDA 1981. Report on the International Cooperative Program on Food Legume Improvement, 1978-79. Pub. ICARDA, Aleppo, Syria.
40. Kambal, A.E., 1969. Flower drop and fruit set in field beans, *Vicia faba* L. J. Agric. Sci., Camb. 72, 131-138.
41. Kambal, A.E., Bond, D.A., Toynbee-Clarke, G. 1976. A study on the pollination mechanism in field beans (*Vicia faba* L.). J. Agric. Sci. Camb., 87, 519-526.
42. Lawes, D.A., 1973. The development of self-fertile field beans. Welsh Pl. Breed. Stn. Rep. 1972, 163-176.
43. Lawes, D.A., 1980. Recent developments in understanding, improvement and use of *Vicia faba*. In: Advances in Legume Science. Ed. R.J. Summerfield and A.H. Bunting. Pub. H.M.S.O. pp. 625-636.
44. Lawes, D.A., Newaz, M.A., 1979. Genetical control of the distribution of seed yield in field beans In: Some current research on *Vicia faba* in western Europe. Ed. D.A. Bond, G.T. Scarascia-Mugnozza and M.H. Poulsen. Pub. EEC, EUR 6244 EN, Luxembourg. pp. 303-312.
45. Martin, A., Cubero, J.I., 1979. Inheritance of quantitative characters in *Vicia faba*. In: Some current research on *Vicia faba* in western Europe. Ed. D.A. Bond, G.T.Scarascia-Mugnozza and M.H. Poulsen. Pub. EEC, EUR 6244 EN, Luxembourg. pp. 383-385.
46. Monti, L.M., 1981. Pollination studies on faba beans. Paper presented at the International Faba Bean Conference, Cairo.
47. Moreno, M.T., 1979. A note on the systematics and evolution of broadbean (*Vicia faba*). FABIS 1, 15.
48. Muehlbauer, F.J., Slinkard, A.E., 1981. Genetics and Breeding Methodology. Chapter in Lentils, ed. C. Webb and G.C. Hawtin, pub. Commonwealth Agricultural Bureaux, U.K.
49. Nagl, K., 1979. Results of mutation and breeding work on *Vicia faba* in Austria. In: Some current research on *Vicia faba* in western Europe. Ed. D.A. Bond, G.T.Scarascia-Mugnozza and M.H. Poulsen. Pub. EEC, EUR 6244 EN, Luxembourg, pp. 355-369.
50. Nassib, A.M., Ibrahim, A.A., Khalil, S.A., 1979. Methods of population improvement in broad bean breeding in Egypt. In: Food Legume Improvement and Development. Ed. G.C. Hawtin and G.J. Chancellor, IDRC-126e Ottawa, pp. 176-178.
51. Nassib, A.M., 1981. Population improvement in faba beans. Paper presented at the International Faba Bean Conference, Cairo.
52. Omar, M., Hawtin, G.C., 1980. Hybridization techniques for crossing in faba beans. FABIS 2, 26.
53. de Pace C. 1979. Characteristics with significant correlation to seed yield in broad bean populations grown in southern Italy. In: Some current research on *Vicia faba* in western Europe. Ed. D.A. Bond, G.T. Scarascia-Mugnozza and M.H. Poulsen. Pub. EEC, EUR 6244 EN, Luxembourg, pp. 144-167.
54. Petersen, R. Omar, M., Hawtin, G.C. 1980a. Plot technique studies on faba beans 1: row length and border effects, FABIS 2, 14.
55. Petersen, R., Omar, M., Hawtin, G.C. 1980b. Plot technique studies on faba beans 11: length of row, number of rows and number of replicates. FABIS 2, 14-15.

56. Picard, J. 1960. Donnees sur l'amelioration de la Feverole de Printemps, *Vicia faba* L. Ann. Amelior. Plantes 10, 121-153.

57. Picard, J. 1963. La coloration des teguments du grain chez la feverole (*Vicia faba* L.). Etude de l' heredite des differentes colorations. Ann. Amel. Plantes 13, 97-117.

58. Picard, J. 1979. Some reflections on problems and prospects in *Vicia faba* breeding. In: Some current research on *Vicia faba* in western Europe. Ed. D.A. Bond, G.T. Scarascia-Mugnozza and M.H. Poulsen. Pub. EEC, EUR 6244 EN, Luxembourg, pp.33-34.

59. Picard, J., Berthelem P., Duc G., and Le Guen, J., 1981. Male sterility in *Vicia faba*. Future prospects for hybrid varieties. Paper presented at the International Faba Bean Conference, Cairo.

60. Pope, M., Bond, D.A., 1975. Influence of isolation distance on genetic contamination of field bean (*Vicia faba* L.) seed produced in small plots. J. Agric. Sci., Camb. 85, 509-513.

61. Porceddu, E., Bianco, V.V. Damato, G., Miccolis, V., Polignano, E. 1979. Variability of some agronomical characters in 158 Italian accessions of *Vicia faba* L. In: Some current research on *Vicia faba* in western Europe. Ed. D.A. Bond, G.T. Scarascia-Mugnozza and M.H. Poulsen. Pub. EEC, EUR 6244 EN, Luxembourg, pp. 251-265.

62. Poulsen, M. H., 1973. The frequency and foraging behaviour of honey bees and bumble bees on field beans in Denmark. J. apic. Res. 12, 75-80.

63. Poulsen, M.H., 1975. Pollination, seed setting, cross-fertilization and inbreeding in *Vicia faba* L. Z. Pflanzenzuchtg. 74, 97-118.

64. Poulsen, M.H., 1977. Obligate autogamy in *Vicia faba* L. J. Agric. Sci., 88, 253-256.

65. Poulsen, M.H., 1979. Performance of inbred populations and lines of *Vicia faba* L. ssp. minor. In: Some current research on *Vicia faba* in Western Europe. Ed. D.A. Bond, G.T. Scarascia-Mugnozza and M.H. Poulsen. Pub. EEC, EUR 6244 EN, Luxembourg, pp. 342-354.

66. Poulsen, M.H., Knudsen, J.C.N., 1980. Breeding for many small seeds per pod in *Vicia faba*. FABIS 2, 26-28.

67. Rachie, K.O., Gardner, C.O., 1975. Increasing efficiency in breeding partially outcrossing grain legumes. In: International Workshop on Grain Legumes. ICRISAT, pp. 285-297.

68. Rowland, G.G., 1980. Intergenotypic competition in field plots of faba beans. FABIS 2., 30-31.

69. Rowlands, D.G., 1958. The nature of the breeding system in the field bean (*Vicia faba* L.) and its relationship to breeding for yield. Heredity 12., 113-126.

70. Saxena, M.C., Hawtin, G.C., El-Ibrahim, H. 1980. Aspects of faba bean ideotypes for dried conditions. Paper presented in EEC Seed Legume Seminar. Wageningen.

71. Scarascia-Mugnozza, G.T., Marzi, V., 1979. Retrospective and prospective views for the *Vicia faba* crop in southern Italy. In: Some current research on *Vicia faba* crop in Western Europe. Ed. D.A. Bond, G.T. Scarascia-Mugnozza and M.H. Poulsen. Pub. EEC, EUR 6244 EN, Luxembourg pp. 7-22.

72. Sinha, S.K., 1977. Physiological aspects of yield improvement in grain legumes. In: Food Legume Crops : Improvement and production. FAO Pl. Prod. and Prot. Paper 9, 109-117.

73. Sirks, M.J., 1923. Die Verschiebung genotypischer Verhaltniszahlen innerhalb Populationen laut mathematischer Berechnung und esperimenteller Prufung. Meded. Landbouwhoogesch. Wageningen, 26.

74. Sjodin, J., 1971. Induced morphological variation in *Vicia faba* L. Hereditas 67, 155-180.

75. Sjodin, J. 1977. Methods of breeding broadbeans (*Vicia faba*). In: Food Legume Crops: Improvement and Production. FAO Pl. Prod. and Prot. Paper 9, 148-161.

76. Stebbins, G.L., 1950. Variation and evolution in plants. Colombia Univ. Press. New York.

77. Toynbee-Clarke, G., 1971. Pollination studies with highly-inbred lines of winter beans (*Vicia faba* L.) J. Agric. Sci., Camb. 77, 213-217.

78. Toynbee-Clarke, G., 1974. The response to various pollination treatments in inbred lines of horse and tick beans (*Vicia faba* L.). J. Agric. Sci., Camb. 84, 531-534.

79. de Vries, A.P., 1979. In search of characters to be used for indirect selection on grain and protein yield in *Vicia faba* L. In: Some current research on *Vicia faba* in western Europe. Ed. D.A.Bond, G.T. Scarascia-Mugnozza and M.H. Poulsen. Pub. EEC, EUR 6244 EN, Luxembourg, pp. 324-341.

80. Wafa, A.K., Ibrahim, S.H. 1957. Temperature as a factor affecting pollen-gathering activity by the honey bee in Egypt. Bull. Fac. Agric. Ain. Shams Univ., No. 162.

81. Witcombe, J., 1981. Genetic Resources of faba beans. Paper presented at the International Faba Bean Conference, Cairo.

# 3. POLLINATION STUDIES ON FABA BEANS

LUIGI M. MONTI AND LUIGI FRUSCIANTE

*Institute of Agronomy, Plant Breeding, University of Naples, Portici, Italy*

The area of faba beans (*Vicia faba minor*) in Italy is becoming progressively smaller, mainly because of low yields which make the crop no longer competitive with others (6) (10).

Current breeding programs (2) (10) (8) aim at exploiting the large potential of this pulse by developing new cultivars which have stable and satisfactory yields.

This chapter reports studies on one of the many ectotypes which are grown in Southern Italy where experiments were conducted using dark-coloured seeds from a local population of field beans.

In 1977, this material was grown both in net cages, to ensure self pollination, and in the open field. Six hundred plants were harvested from each of two replications. In 1978, the seed derived from the selfed plants was divided and tested under both self and open pollination; the same was done on the seed of the open pollinated lines.

The procedure was repeated in 1979 and 1980, resulting in a total of 16 different populations in the 1980 season. In the past three years, a split-plot design has been used with the main plot based on the system of pollination during the experiment, and the sub-plots based on the origin of the material.

For simplicity, each population will be indicated by sequence of the letters P (open pollination) and S (self pollination) to describe the pollination system in the different years, i.e. PSP indicates an open pollinated population in 1979, coming from selfed population in 1978 which, in turn, derives from an open pollinated population in 1977.

Table 1 shows that increased yields were obtained in the three years.

*G. Hawtin & C. Webb (Eds.): Faba Bean Improvement*
*©1982 ICARDA.* ISBN -13:978-94-009-7501-9

Table 1.

Seed yield (Y=g/plot) in populations and derivated sub-populations of field beans grown isolated in cages (S) or open-pollinated (P). The % are referred each year to the continuously open pollinated population.

| | 1977 Y* | % | | 1978 Y | % | | 1979 Y | % | | 1980 Y | % |
|---|---|---|---|---|---|---|---|---|---|---|---|
| P | 5980 | 100 | P | 2090 | 100 | P | 2675 | 100 | P | 3362 | 100 |
| | | | | | | | | | S | 1987 | 59 |
| | | | | | | S | 1807 | 68 | P | 2975 | 88 |
| | | | | | | | | | S | 1525 | 45 |
| | | | S | 1316 | 63 | P | 2220 | 83 | P | 3600 | 107 |
| | | | | | | | | | S | 2712 | 81 |
| | | | | | | S | 980 | 37 | P | 2700 | 80 |
| | | | | | | | | | S | 1525 | 45 |
| S | 4500 | 75 | P | 1430 | 68 | P | 2400 | 90 | P | 3262 | 97 |
| | | | | | | | | | S | 2000 | 59 |
| | | | | | | S | 1370 | 51 | P | 2950 | 88 |
| | | | | | | | | | S | 1637 | 49 |
| | | | S | 518 | 25 | P | 1995 | 75 | P | 3550 | 106 |
| | | | | | | | | | S | 2625 | 78 |
| | | | | | | S | 1015 | 38 | P | 2625 | 78 |
| | | | | | | | | | S | 1462 | 43 |
| LSD (P=0.05) | | | | 127 | | | 561 | | | 417 | |

* Yield from 600 plants

Considering only the populations which were always grown in the open, values of 2090g/plot for PP, 2675 for PPP and of 3362 for PPPP were obtained. This holds true also for populations which were always grown in isolation: figures of 518, 1015 and 1462 g were found respectively for SS,SSS and SSSS materials. Whatever their origin, the selfed populations always gave a lower yield than the open-pollinated materials.

Significant differences were found each year among the populations. Fig. 1 shows the grain yield of selfed populations as a percentage of the yields of open populations.

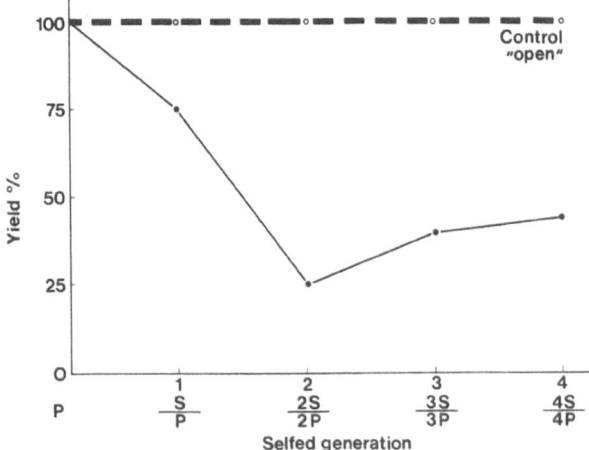

Figure 1.

The main causes of the yield reduction in field beans when the plants are grown in complete absence of pollinating insects, are (2):

i. The lack of tripping results in reduced pollination and the consequent drop of flowers.

ii. Besides this direct effect, the total absence of pollinating insects indirectly causes a yield reduction in the following generation because of the inbreeding depression due to the higher number of inbred plants which are obtained.

iii. The inbred plants have a lower ability to self-fertilize. Table 2 shows the yield reductions due to these three causes. The yield reduction due to the lack of tripping is calculated by comparing the PS and PP populations. Values of 37, 38, and 32% reduction were found in the three years respectively. The yield reduction determined by inbreeding depression, as a consequence of the complete isolation of the plants in the previous generation, is obtained by comparing the populations in which the last two generations were SP and PP. Figures of 32, 17 and 18 % were found in the respective years.

The yield reduction of the SS population in comparison with the SP population is due both to the lack of tripping and to the lower ability of the inbred plants to self-fertilize. Hence the value of this latter cause is extrapolated, taking into account the value due to the lack of tripping. An estimation of the reduced ability of the inbreds to self-fertilize can be obtained also by the difference between the SS and the PS populations, considering the cumulative effect due to the inbreeding depression of the higher proportion of inbred plants. From the means of the last two comparisons, values of 27, 19 and 12 % yield reduction can be attributed to this third cause.

Fig. 2 shows the average effect on the seed yield of one generation of selfing in an otherwise always open-pollinated population. Reduction of yield was found when selfing was applied in the last generation and also in the last but-one generation (56 % and 82 % of the open population respectively), while no effect was found when selfing was applied three or more generations previously.

Table 2.

Causes and respective percentages of the seed yield reduction in faba beans grown in cages in different generations in the different years. The pollination systems of only the last two generations are indicated. (P = open; S = self).

| | Compared populations | 1978 | | 1979 | | 1980 | |
|---|---|---|---|---|---|---|---|
| | | No. of comparisons | yield reduction (%) | No. of comparisons | yield reduction (%) | No. of comparisons | yield reduction (%) |
| Lack of tripping | PS/PP | 1 | 37 | 2 | 38 | 4 | 32 |
| Inbreeding depression | SP/PP | 1 | 32 | 2 | 17 | 4 | 18 |
| Lower self-fertilizing | SS/SP[1] | 1 | | 2 | | 4 | |
| ability of inbreds | SS/PS[2] | 1 | 27 | 2 | 19 | 4 | 12 |

1–2. Difference between the total reduction and the reduction due to lack of tripping (1) and to inbreeding depression (2).

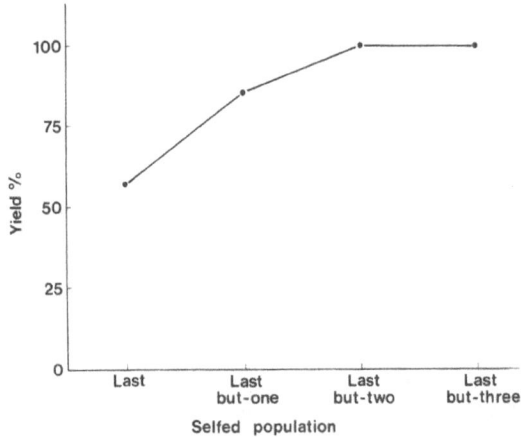

Figure 2.

In a similar way, the effect on the yield of one generation of open-pollination in an otherwise continuously selfed population was calculated (Fig.3). A clear effect was found only when the open-pollination was applied in the last generation (202 %) and in the last but-one generation (177 %), while no effect was found if open pollination was applied three or more generations previously.

The genetic structure of *Vicia faba* is complicated by the variable percentage of cross-fertilization. In experiments in the same place and with the same ecotype (8) it was found that, on average, there was 50 % selfing and 50 % out-crossing. As a consequence, with all the flowers bee-pollinated, a maximum of 50% out-crossing should be obtained in the populations. In normal field conditions, lower values than this are expected. We found an average value of 22 % in 1977.

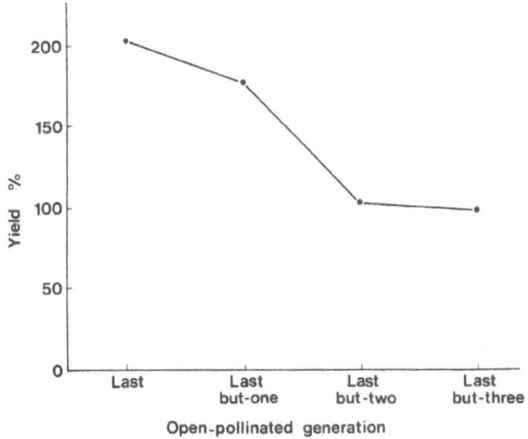

Figure 3.

When a population, which had been grown under open pollination, was grown in the next generation in isolation (PS), a mean yield reduction of 35% was found in comparison to the PP population. This reduction is a direct effect of the lack of pollinating insects. The poorer yield of plants grown in isolated cages is due to a smaller number of pods per plant (2). In fact, no difference in the number of seeds per pod was found and the seed weight was higher, though not significantly so. A lower seed weight has been reported in plants grown in open-pollination (9) and in plants in net cages with bees (2); both data being obtained in comparison with plants grown in cages without pollinating insects. Hence we can say that 35% of the yield capacity of our ecotype is bee-dependent. The reduction of this bee-dependence should be one of the objectives of the breeding work, because of the variability of bee populations in the field.

As we have shown, such a bee-dependence not only affects the present crop but also the following generation. When the seeds coming from self- and open-pollinated plants are grown in the open the yield differential provides an estimate of inbreeding depression due to the higher number of inbred plants in the SP population in comparison with the PP material. The average reduction for the three years was 22%. When comparing the SS and SP populations and the SS and PS populations over the three years, the reduced self-fertility of the inbreds is estimated to be about 20%.

Heterosis and inbreeding depression are well known phenomena in *V. faba* (5) (1) (3). Poulsen (9) found a reduction of about 11% due to inbreeding depression in a long term experiment. We found reduction of the seed weight and of the height of the plants in populations deriving from selfed plants (2). Kambal *et al.* (4) say that the lower self-fertilizing capacity of the inbreds can be explained through a lower production of pollen. In commercial seed multiplication, the amount of cross-fertilization will have an influence on the seed yield of the following crop and will be one of the causes of the yield fluctuation.

After repeated selfings, a clear effect of natural selection was found resulting in the reduction of the negative effects of both the inbreeding depression and the lower ability of inbreds to self-fertilize. A decrease in yield reduction due to the inbreeding depression of 32, 26 and 22% in the three years respectively was evident when comparing homogenous populations (SP-PP; SSP-PPP; SSSP-PPPP); in the same time, a decrease in the lower self-fertility ability of the inbreds was found from 27% (SS-SP) to 11 and 12% (SSS-SSP and SSSS-SSSP).

In many allogamous species, self fertile lines can originate as a consequence of specific selection pressure for an immediate fitness (12). In *V. faba*, the natural selection which occurs in populations which are obliged to self, increases the yield capacity by acting against the deleterious effect of inbreeding, and in favour of higher self-fertility. The possibility then exists of improving self-fertility in our open-pollinated ecotype by continuous selfing and selection, in order to get it more bee-independent. Several authors have reported lines with a good degree of self-fertility. The best inbred lines selected by Poulsen (9) reach levels of even 90% of the basic population. We have just begun a selection program to check the effect of the natural selection in our different populations.

A recurrent selection program with our open-pollinated ecotype, aims at obtaining an improved cultivar based on inbred lines with good self-fertility and with a sufficient combining ability. The pollinating system of the crop makes it necessary to set up a specific selection procedure. This research has shown that both selfing and open-pollination can influence the yield of the two following generations. Selection for self-fertility can only be done in the third generation. Good cross-pollination is ensured also by the presence of bees in net cages (2). If this method is followed for intercrossing, two generations with bee-pollinators are needed (SSP and SSPP) to allow selection for good combining ability.

# REFERENCES

1. Bond, D.A. 1966. Yield and components of yield in diallel crosses between inbred lines of winter beans (*Vicia faba*) J. agric. Sci., Camb. 67, 325-336.

2. Frusciante, L., Monti, L.M. 1980. Direct and indirect effects of insect pollination on the yield of field beans. Z. Pflanzenzuchtg 84, 323-328.

3. Hayes, J.D., Hanna, A.S, 1968, Genetic studies in field beans (*Vicia faba* L.) III. Variation in self-fertility in a diallel cross. Z. Pflanzenzuchtg, 60, 315-326.

4. Kambal, A.E., Bond, D.A., Toynbee-Clarke, Gillian. 1976. A study on the pollination mechanism in field beans (*Vicia faba* L.). J. Agric. Sci., Camb. 87, 519-526.

5. Lawes, D.A., Newaz, M.A. 1979. Genetical Control of the Distribution of Seed Yield In Field Beans. EEC Seminar, Bari (Italy) 27-29 April 1978, 303-312.

6. Monti, L.M. 1977. Le leguminose da granella per l'alimentazione animale: il miglioramento genetico della *V. faba* L. e sua importanza per l'Italia. Riv. di Agronomia 4, 229-235.

7. Porceddu, E., Scarascia-Mugnozza, G.T., Monti, L.M. 1979. Genetica e miglioramento genetico della produzione proteica delle leguminose da granella. –Atti convegno Prospettive delle Proteaginose in Italia, Perugia 1979, 161-186.

8. Porceddu, E., Monti, L.M., Frusciante, L., Volpe, N. 1980. Analysis of cross-pollination in *V. faba* L. Z. Pflanzenzuchtg 84, 313-322.

9. Poulsen, M.H.1980. Performance of Inbred Populations and Lines of *Vicia faba* L. sp. *minor*. EEC Seminar, Bari (Italy), 27-29 April 1978, 342-354.

10. Scarascia-Mugnozza. G.T., Porceddu, E., Monti, L.M. 1979a. Stato Attuale del. Miglioramento Genetico delle Leguminose da granella. Riv. di Agronomia 1, 24-40.

11. Scarascia-Mugnozza, G.T., De Pace, C. 1979b. Concepts and goals for *Vicia faba* breeding in Mediterranean environments. Israeli-Italia joint meeting on genetics and breeding of crop plants, Roma, December 11-13, 1978, 217-244.

12. Stebbins, G.L. 1950. Variation and evolution in plants. Columbia Uni. Press, New York, pp 643.

# 4. THE DEVELOPMENT AND PERFORMANCE OF SYNTHETIC VARIETIES OF VICIA FABA L.

D. A. BOND

*Plant Breeding Institute, Cambridge, U K*

Allard (1) defined a synthetic variety as ".... being sythesised from genotypes which have been tested for combining ability", but he also noted that the term had come to be used to designate a variety that is "maintained under open pollination following synthesis by hybridization in all combinations among a number of selected genotypes". The latter, more general, definition is nearer to what is appropriate to the synthetic varieties of faba beans which are currently being cultivated or developed; so for the the purpose of this chapter a synthetic variety will be defined as any population which has been constituted from a limited number of distinct and well-evaluated components.

This modified definition is necessary because (a) not all hybridizations are made when faba bean components are bee-pollinated (some selfing then being unavoidable), and (b), there may not be previous testing for combining ability, the performance of the synthetic variety itself being a measure of the combining ability of its components or else the performance of the components *per se* being assumed to be related to their general combining ability.

Two main types of synthetics are recognised: (i) those whose components or parental stocks are inbred or partially inbred lines, and (ii) those whose components are populations or themselves synthetics. The first generation after constitution is known as syn-1, and synthetic varieties based on inbreds are normally nationally tested and utilised in commerce in the syn-4 to syn-6 generations. Synthetics based on populations can be utilised in syn-1 but often syn-2 or syn-3 are commercialised to reduce the scale of multiplication of parent populations.

The number of components may vary from 3 to 10 or more but 4 or 5 are much more common, these numbers providing the best balance between the danger of too high a proportion of sibbing with few components and the inclusion of more lowly ranked components if the number is greater. This number has also been shown to be the optimum in maize (9) though the higher the parental yields the fewer the components required for greatest efficiency.

*G. Hawtin & C. Webb (Eds.): Faba Bean Improvement*
©1982 ICARDA. ISBN -13:978-94-009-7501-9

*Current use*

Throws MS, developed by RHM (Agriculture) Ltd., was the first synthetic faba bean to be grown commercially in the UK. It was first officially recommended in 1960 and has four component populations within which there is recurrent selection. Maris Beagle, a synthetic composed of five inbred lines, was released in 1974, and now most of the UK and French winter beans are synthetic varieties. One synthetic, Soravi, is an example of international cooperation in that three of its inbred components were selected by French breeders at INRA and two by UK breeders at the Plant Breeding Institute; royalties on sales are shared proportionately.

Synthetic varieties have not been employed so widely for the spring-sown crop but the variety Minden from Petkus (GFR) had six parental lines, Herra from Franck (GFR) has more lines, and two PBI synthetics, with three and six components, are currently being tested nationally in U.K. In addition to varieties already in use or under test, breeding programs with the aim of producing synthetic varieties or of investigating their potential, are in progress at Hohenheim, GFR, (10) (14), at Gotha, GDR (13), at Copenhagen, Denmark (14) and at ICARDA, Aleppo, Syria (8). These programs usually involve the selection of lines, either highly or partially inbred, and some testing of them for combining ability either by controlled crossing or a polycross under open pollination.

Synthetic varieties which are in current use have met legislative requirements for distinction, uniformity and stability. Components are maintained in relatively small quantities so that the commercial generation is often the syn-4 to syn-7.

*Performance of Synthetics*

*Yield*: $F_1$ hybrids commonly outyield sythetics composed of the same parents, and this is to be expected in view of the limited amount of crossing obtainable in a synthetic. However hybrid varieties are not yet in commercial use and their performance is described by Dr. Picard (see chapter 5).

Comparisons of synthetics with populations selected in other ways is usually not a valid evaluation of the breeding methods because the genetic material is often not the same. However some measure of the productivity of synthetics can be seen from a summary of yield data from PBI spring-sown trials between 1972 and 1979 (Table 1). The two synthetics and the population were higher yielding than the controls but a yield advantage of both the synthetics over the population was not proven, though the synthetics did produce a similar yield with a lower mean seed weight than the population.

Most winter bean varieties in the UK are synthetics, so the most useful comparisons in winter beans are between synthetics and their components. Inbred lines at PBI are normally in different trials to synthetics, so as to reduce competition between plots, but mean yields from 11 adjacent pairs of trials (Table 2) show that both the synthetics, Banner and Bourdon, outyielded all their respective components and the mean of their components, whether they were in trial with other synthetics or with the inbreds.

Table 1.

Mean yields of two synthetics and one mass-selected population in relation to controls in PBI trials 1972 to 1979.

| | | Controls | Mass Selected Population | Synthetics | |
|---|---|---|---|---|---|
| | Minor | Maris Bead | Blaze | Tiger | Stag |
| Mean Yield (% controls) | 97 | 103 | 115 | 121 | 115 |
| No. trials above control | 13 | 25 | 29 | 26 | 35 |
| No. trials below control | 25 | 13 | 4 | 2 | 3 |
| Mean 1000 seed weight (g) | 370 | 420 | 510 | 490 | 450 |

Table 2.

Mean yields (t/ha) and their variances of two synthetic varieties and of their component inbreds in 1978, 1979 and 1980.

| | Synthetic Banner | | Inbred components of Banner | | | | | |
|---|---|---|---|---|---|---|---|---|
| | (a)* | (b)* | 1 | 2 | 3 | 4 | 5 | 6 |
| No. of trials | 11 | 11 | 11 | 11 | 11 | 7 | 11 | 7 |
| Mean yield | 3.90 | 3.64 | 3.13 | 2.74 | 3.12 | 3.47 | 3.21 | 2.67 |
| SD** | 1.64 | 1.66 | 1.36 | 1.14 | 1.43 | 1.51 | 1.19 | 1.63 |
| CV | 12.7 | 13.8 | 11.3 | 12.5 | 13.8 | - | 11.2 | - |

| | Synthetic Bourdon | | Inbred components of Bourdon | | | | | |
|---|---|---|---|---|---|---|---|---|
| | (a)* | (b)* | 7 | 8 | 9 | 10 | 11 | 12 |
| No. of trials | 11 | 11 | 7 | 11 | 11 | 7 | 11 | 7 |
| Mean Yield | 4.05 | 4.02 | 3.39 | 4.01 | 3.07 | 3.45 | 3.14 | 3.12 |
| SD** | 1.63 | 1.56 | 1.68 | 1.79 | 1.59 | 1.24 | 0.87 | 0.95 |
| CV | 12.1 | 11.7 | - | 13.5 | 15.6 | - | 8.4 | - |

*(a) = with other synthetics in a trial adjacent to the trial of inbred components.

(b) = as one of the entries in the trial of inbred components.

**SD derived from trial means for each genotype, i.e. a measure of stability.

LSD (P=0.05), from pooled trial variances, = 0.15 t/ha for comparison of means from 11 trials and = 0.26 t/ha for means from 7 trials.

However, one component of Bourdon was very close to its synthetic's yield. Similarly, when synthetics were used as components of a multiple synthetic, the new synthetic often significantly outyielded the mean of the components but not the best component. For example, the syn-2 of IB16 (Table 3) significantly outyielded two of its components and the mean of its components in Trial 1 of 1973 and Trial 2 only just failed to detect a difference, but it was not significantly better than its best components. This result applied in later years though only two components were continued in the trial.

Table 3.

Mean yields (t/ha) of the synthetic variety IB16 and some of its synthetic-variety components.

|  | 1973 | | 1974 | 1975 | 1976 |
|---|---|---|---|---|---|
|  | Trial 1 | Trial 2 |  |  |  |
| Throws MS (control) | 5.01 | 3.88 | 6.13 | 2.81 | 1.92 |
| IB16    syn-1 | 5.05 |  | syn-2,7.33 | syn-3,3.51 | syn-3,2.48 |
|         syn-2 | 5.35 | 4.36 | syn-3,7.26 |  |  |
| Components of IB16 |  |  |  |  |  |
|    IB6 | 4.52 | 4.26 |  |  |  |
|    Beagle | 4.40 | 4.04 | 6.66 | 3.58 | 2.11 |
|    Bulldog | 4.89 | 4.35 | 7.06 | 4.18 | 2.42 |
|    Auto | 4.72 | 3.69 |  |  |  |
| Mean of 4 components | 4.63 | 4.08 |  |  |  |
|    LSD (P = 0.05) | 0.78 | 0.37 | 0.67 | 0.83 | 0.29 |
|    CV | 10.0 | 5.3 | 5.9 | 12.6 | 8.4 |
| SE of one variety | 0.27 | 0.13 |  |  |  |
| SE of mean of 4 varieties | 0.13 | 0.06 |  |  |  |

The 25 trials, each with 5 replications, of IB23 (Table 4) provided a sensitive test of the relative yields of components and the synthetic. Again, except in 1979, the synthetic significantly outyielded the mean of the components but not the highest-yielding component. The constitution of Throws MS winter bean is broadly analogous to that of IB16 and IB23, and a trial at PBI Cambridge in 1980, using seed kindly provided by RHM (Agriculture) Ltd., just failed to detect a yield advantage of the synthetic over the mean of its four component populations (Table 5).

Table 4.

## Mean yields (t/ha) of the synthetic variety IB23 and its synthetic-variety components.

| Year | 1977 | 1978 | 1979 | 1980 | Mean | SD*of Mean | CV |
|------|------|------|------|------|------|------|-----|
| No. of trials | 7 | 8 | 6 | 4 | 25 | | |
| IB23 syn generation | 1 | 2 | 2 | 3 | | | |
| IB23 yield | 3.95 | 5.72 | 3.54 | 3.40 | 4.33 | 1.45 | 6.70 |
| Components of IB23** | | | | | | | |
| Banner | 3.99 | 5.11 | 3.74 | 2.98 | 4.13 | 1.30 | 6.29 |
| Bourdon | 4.02 | 5.59 | 3.39 | 3.67 | 4.31 | 1.37 | 6.36 |
| Buccaneer | 3.55 | 5.36 | 3.48 | 3.03 | 4.03 | 1.29 | 6.40 |
| Bulldog | 3.29 | 4.99 | 3.66 | 3.26 | 3.92 | 1.23 | 6.28 |
| Mean of 4 components | 3.71 | 5.26 | 3.57 | 3.24 | 4.10 | | |
| LSD, P = 0.05 (from pooled variance) | 0.19 | 0.23 | 0.16 | 0.12 | 0.10 | | |

* SD derived from trial means
**All at advanced generations, syn-4 or more

Table 5.

Yields (t/ha) of Throws MS synthetic winter bean and its four components in a trial at the Plant Breeding Institute, Cambridge, in 1980.

| | Throws MS syn-1 | A | B | C | D | Mean of 4 Components |
|------|------|------|------|------|------|------|
| Yield | 40.3 | 38.9 | 39.3 | 35.4 | 37.9 | 37.8 |
| SE | 1.2 | 1.2 | 1.2 | 1.2 | 1.2 | 0.6 |

*Stability of yield over environments*

The reasons why synthetics would be expected to be more stable in yield than their more inbred components are: first, a buffering effect might be expected as a result of the mechanical mixing of different types. Secondly, a component which is the highest yielding in one trial, or year, is often outyielded by a different component in another trial or year, yet a synthetic usually equals or exceeds the yield of the highest yielding component whichever one it is. Thirdly, trials repeated with different soil types, sowing dates, plant densities and years have generally shown more highly significant interactions of genotype x environment when most of the trial entries have been inbred lines than when they have been synthetic varieties (Table 6).

Table 6.

Probability levels for significance of genotype x environment interactions in yield trials of inbred and synthetic varieties of winter beans at PBI, Cambridge.

| | Seasons Involved | No. of geno-types | No. of envir-onments | G | E | G x E (F value if small) | Regres-sion | Devia-tion |
|---|---|---|---|---|---|---|---|---|
| **Inbreds** | | | | | | | | |
| | 1976 | 16 | 3 | ** | *** | NS | * | NS |
| | 1978 | 25 | 3 | * | *** | * | * | * |
| | 1979 | 25 | 4 | ** | *** | ** | * | ** |
| | 76,77 | 13 | 4 | *** | *** | ** | ** | * |
| | 75,76,77 | 13 | 7 | *** | *** | *** | ** | * |
| **Synthetics** | | | | | | | | |
| | 1974 | 16 | 6 | ** | *** | NS | ** | NS |
| | 1975 | 16 | 7 | *** | *** | NS | NS | NS |
| | 1976 | 16 | 6 | ** | *** | NS | NS | NS |
| | 1977 | 16 | 6 | ** | *** | NS | NS | NS |
| | 1978 | 16 | 7 | *** | *** | 1.87* | ** | NS |
| | 1979 | 16 | 5 | ** | *** | 2.2* | * | * |
| | 74, 75 | 11 | 13 | *** | *** | 1.81* | * | NS |
| | 74,75,76,77 | 5 | 25 | *** | *** | 2.40** | NS | * |

Probabilities: * = 0.05, ** = 0.01, *** = 0.001

Fourthly, as $F_1$ hybrids are thought to be more stable than their inbred parents, it would be expected that synthetic varieties, having a degree of hybridity intermediate between that of inbreds and $F_1$'s, would be somewhat more stable than their inbred components.

However these expectations were not supported by the variances of the mean yields of synthetics and their components found at Cambridge. Standard deviations based on trial means were in fact lower for all the components of Banner than for Banner, for 4 out of 6 components of Bourdon than for Bourdon (Table 2); and lower for all components of IB23 than for IB23 itself (Table 4).

An explanation of this result could be that the synthetics being partially hybrid, were better able to respond to favourable environments but were almost as much depressed by poor environments as their components. The fact that the IB23 mean yield also had a greater variance than its components even though they were non-inbred suggests that mixing of synthetics had increased the degree of heterosis so that the composite was better able to exploit high-yielding environments.

Analyses by the Finlay and Wilkinson (6) method also showed that the synthetics Banner and IB23 had higher regression coefficients than all their respective components. Bourdon was intermediate between its components in regression on environmental means. However for Banner and IB23 and their components there was a positive association of means and regression coefficients or variances, suggesting that means and variances were not independent. The variation was therefore expressed in proportion to the means, i.e. as coefficients of variation (100xSE/Mean) but in the case of IB23 (Table 4) the synthetic still showed greater variation than its components. Banner did not have a higher coefficient of variation than all its components (Table 2) though only those (4 of the 6) which had been in the same number of trials could be compared for CV values. The greater variance in absolute, as opposed to relative, units remains and this may be relevant to growers budgeting against the different risks.

The high variances of Banner and Bourdon (Table 2) are unlikely to be the effect of different generations because both synthetics were in advanced generations, syn-4 or more, at the time of testing and should have reached an equilibirum. The SD of IB23 over 25 trials (Table 4) did include data for syn-1, syn-2 and syn-3 but mean yields relative to controls did not change with generation as they have been found to do with synthetics composed of inbreds(2).

## Attributes other than yield

As synthetics include differing genotypes and have a degree of hybridity with associated vigour, including for example good root growth, they are expected to offer some buffering against stress factors such as drought and disease. Data have not been recorded but it is a common observation that inbred lines become uniformly infected by a disease whereas the degree of infection on synthetics is more variable from plant to plant. In the future when major genes for disease resistance come to be evaluated, a different gene will need to be incorporated in each component of a synthetic so as to guard against the breakdown of resistance.

As autofertility is often associated with heterozygosity (5) synthetics are likely to be more autofertile than their inbred components, and possibly more so than their population components, if crossing between populations brings some extra heterozygosity. Nodulation may be increased in synthetics compared with inbreds, because of the former's vigorous root system, but specific *Rhizobium* strain/host genotype combinations as suggested by Mytton *et al.* (11) cannot be exploited easily in synthetics.

Slight defects can sometimes be tolerated in one out of a number of components, especially where other characters such as yield or fertility have overriding importance, because in advanced generations of the synthetics they have no detectable effect on performance; individual plants prone to lodging for example would, if at a low frequency in the synthetic, be supported by their stiff-strawed neighbours.

## Identifying parental populations or lines

*Populations*: Where the parents of synthetics are populations, their suitability as components can probably be tested economically only by trial and error or by systematically removing one population at a time from the synthetic. Where the component populations are undergoing recurrent selection the assumption is made that improvement in the component populations will result in a similar improvement in the synthetic constituted from them.

*Inbreds*: Inbreeding however, produces uniform lines which can be readily identified and characterised. Whether there is some selection during inbreeding or selfing is by single-seed

descent to say $S_4$ followed by evaluation of lines, is probably immaterial. What is important is that the parental population should be partly from locally adapted types and partly
from new introductions so that wide crossing and inbreeding gives the opportunity for un-
covering new recombinants.

*General and specific combining ability*: After the selection of a limited number of lines,
some breeders of synthetics have conducted tests of general (gca) and specific combining
ability (sca). Von Kittlitz (10) for example, proposed as a preliminary screening for gca
a simulated polycross, that is the use of open-pollinated seed following systematic planting,
and possibly followed by selection for both gca and sca in diallel crosses.

However, yields of crosses and parental lines reported by Bond (2) and summarised in
Table 7 showed that in six out of eight trials there was a significant regression of $F_1$ yields
on mid-parent yields and gca was of much more importance than sca. Thus if $F_1$'s can be
predicted by mid-parent values there is less reason for making the crosses. On the other
hand, four trials showed a significant residual variance of the crosses unaccountable to regression on parents, and information about this could be gained from crosses; but this has
to be balanced against the effort of making the crosses (effort which could alternatively be
employed in testing more inbreds) and against the fact that as synthetic varieties of *Vicia
faba* are only partly cross pollinated, gca and sca, in terms of $F_1$, account for only part of
the performance of the synthetic.

However, the production of $F_1$'s as a preliminary to constitution of synthetics, may be
justified where it can be combined with the crossing of the best inbreds in a recurrent selection program. Several of the inbred components of the synthetics Banner and Bourdon
had been through a second cycle of testing. Also, testing for combining ability may become
feasible when male sterility can be used to facilitate production of a series of hybrids.

Table 7.

Probability levels for significance of combining ability and of regression of crosses on
parents in terms of yield. (* = 0.05, ** = 0.01, *** = 0.001).

| Trial ref. | Number of | | GCA | SCA | Regression of $F_1$ on mid parent | Residual variance after regression removed |
|---|---|---|---|---|---|---|
| | Lines | Testers | | | | |
| 1 | 3 | 5 | * | NS | * | ** |
| 2 | 13 | 5 | * | NS | * | NS |
| 3 | 11 | 5 | *** | NS | *** | *** |
| 4 | 12 | 3 | *** | * | *** | ** |
| 5 | 14 | 1 | ** | - | ** | NS |
| 6 | 7 | 3 | *** | NS | NS | NS |
| 7 | 8 | 2 | * | * | * | *** |
| 8 | 6 | 2 | * | NS | NS | NS |

*Other factors influencing choice of parents*: If information has not come from combining ability tests, it may be possible to choose parents on the basis of their morphology, so that they complement rather than compete with each other. Also the more dissimilar they are genetically, the greater is likely to be the heterosis in their crosses.

However, selection for contrasts must not be made to the extent that characters such as maturity date are so variable that they cause harvesting difficulties. Ideally, flowering periods of parents should be synchronous so as to maximise inter-component crossing and reduce inbreeding; and the synthetic should be uniform enough to comply with requirements for national listing of varieties. In U.K. the criterion is that a candidate variety should not be more variable than those already listed in its class. It is recognised that synthetic varieties must contain a greater variation than the pure lines used as varieties in many self pollinated crops, and significant differences between means are usually accepted as evidence of distinctness.

*Autofertility*: The question as to whether components of synthetics should be autofertile or autosterile is currently being debated but few data are available. Autosterile components would be expected to increase the proportion of crossbreds in the syn-1 and to some extent in later generations. An attempt at this at Cambridge failed because of lack of synchrony of flowering periods. If autofertile components are of the type in which automatic self-pollination occurs late in the life of the flower, thus giving an opportunity for crossing if bees are present, they would be expected to provide an insurance against absence of bees. However Poulsen (14) has argued that such a synthetic would be more variable over environments in degree of cross fertilisation and therefore in yielding ability. He concluded that faba bean varieties should be high yielding, highly self pollinating pure lines or multiline varieties based on such lines. Although some high yielding inbreds have been reported (12)(4) it is not yet clear whether these are stable over environments, and whether the stage in faba been breeding has been reached where the advantages of hybridity can be circumvented.

## Methods of production

Where the components are populations which are relatively simple to multiply they may be maintained in quantities and after mixing, the syn-1 or syn-2 given to the commercial grower or, in the case of a new variety, for national testing. However, inbred components are maintained in small quantities and the synthetic is multiplied to syn-4 or syn-5 before entering commerce from syn-5 to about syn-7.

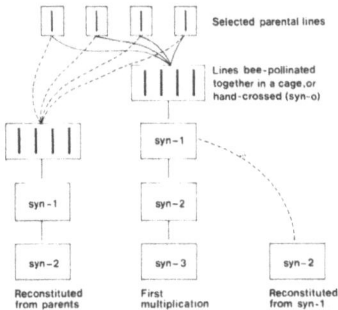

Figure 1.

Wright (15) predicted that a faba bean synthetic based on inbred parents, would on average reach an equilibrium for its inbreeding coefficient by syn-3. This had been found in practice in terms of relative yield levels (3), but Wright (15) also showed the important effect on degree of inbreeding of the level of cross pollination in the generation before commercialisation. He recommended bee hives or other measures to raise natural crossing in such seed production fields. The introduction of genetic male sterility to synthetics through one or more of the parents, so as to increase crossing has been suggested but the *ms* gene needs to be at a high frequency in the synthetic population to substantially increase crossing and it then puts the reliability of yields from advanced generations in jeopardy when bees are absent.

The method of making the parental pollinations has, as Wright (15) predicted, little effect on the expected performance of syn-3 and later generations (Table 8). Although handcrossed parents gave high yields in syn-1 and syn-2, by syn-3 there was little difference in the yield of the same synthetic constituted in the three ways. However, where a large number of synthetics are being constituted and there is a need to discard the least promising before syn-3, handcrossing of parents would allow an earlier evaluation of the maximum yield to be expected.

Table 8.

Mean yields of a synthetic variety which had been constituted from four inbred lines in three different ways: (a) inbreds crossed by hand in all combinations, then the $F_1$'s selfed but $F_2$ *et seq.* pollinated in a cage with bees  (b) as (a) but $F_1$'s pollinated in a cage with bees, and (c) inbred lines mixed and pollinated in a cage with bees.

| Method of constitution | Generation | 1973 Mean Yield | 1974 Generation | 1974 Mean Yield | 1975 Generation | 1975 Mean Yield |
|---|---|---|---|---|---|---|
| a | syn 1 = $F_2$ | 5.24 | syn 2 | 5.96 | syn 3 | 4.04 |
| b | syn 1 = $F_2$ + double crosses | 4.78 | syn 2 | 6.81 | syn 3 | 3.78 |
| c | syn 1 | 4.30 | syn 2 | 5.83 | syn 3 | 3.74 |
| LSD | | 0.78 | | 0.67 | | 0.83 |

No instability of yield over the generations syn-4 to syn-8 has been detected in PBI synthetics, but once constituted, a synthetic is a population subject to genetic drift, and therefore it must be capable of being reconstituted. Fig. 1 shows two methods of reconstituting a synthetic derived from inbred parents. In one, the inbreds are remixed to form a new syn-1, while in the other, the variety is re-formed from the syn-1 (or syn-2) of the first constitution. In the first method, it is necessary to ensure that the components enter the syn-1 in the second constitution in the same proportions (usually equal proportions), as they did in the first. Genetic markers, now including isoenzymes (7) can be used to check this. The second method has the advantage that such checking is unnecessary, and the decision to reconstitute can be deferred until performance of the syn-1 or syn-2 is known.

Successful synthetics of *V. faba*, unlike multi-line varieties, are not an end in themselves but are potential sources of further variation. Hence new lines may be selected from them, though with a low frequency of natural crossing the pressure of selection must be proportionately greater.

*Conclusion*

Synthetic varieties are higher yielding than the mean of their components even when the components are themselves synthetics but experimental evidence of the stability of yield over environments which would be expected from the buffering of one component against another is lacking at present. However, until $F_1$ hybrids and high yielding pure lines can be developed as varieties, synthetics offer a means of exploiting an intermediate level of heterosis from highly selected and characterised parents.

## REFERENCES

1. Allard, R.W. 1960. Principles of plant breeding. John Wiley & Sons Inc., New York.
2. Bond, D.A. 1967. Combining ability of winter bean (*Vicia faba* L.) inbreds. Journal of Agricultural Science, Cambridge, 68, 179-185.
3. Bond, D.A. 1970. Breeding methods in field beans (*Vicia faba* L.). Eucarpia Fodder Crops Section meeting, Lusignan, 119-126.
4. Dantuma, G. 1980. Personal communication.
5. Drayner, J.M. 1956. Regulation of outbreeding in field beans (*Vicia faba*). Nature 177, 489-490.
6. Finlay, K.W., Wilkinson, G.N. 1963. The analysis of adaptation in a plant breeding programme. Australian Journal of Agricultural Research, 14, 742-754.
7. Gates, P., Boulter, D. 1980. The use of pollen isoenzymes as an aid to the breeding of field beans (*Vicia faba* L.). New Phytologist, 84, 501-504.
8. Hawtin, G.C. 1979. Strategies for the genetic improvement of lentils, broad beans and chick-peas, with special emphasis on research at ICARDA. In Food Legume Improvement and Development (ed. by G.C. Hawtin and G.J. Chancellor) pp.147-154. ICARDA Aleppo, Syria.
9. Kinman, M.L., Sprague, G.F. 1945. Relation between number of parental lines and theoretical performance of synthetic varieties of corn. Journal of American Society of Agronomy, 27. 341-351.
10. von Kittlitz, E. 1980. Personal communication.
11. Myton, L.R., El-Sherbeeny, M.H., Lawes, D.A. 1977. Symbiotic variability in *Vicia faba*. 3. Genetic effects of host plants, Rhizobium strain and of host x strain interaction. Euphytica, 26, 785-791.
12. Poulsen, M.H. 1979. Performance of inbred populations and lines of *Vicia faba* L.ssp. minor. In some current research on *Vicia faba* in Western Europe (Ed. by D.A. Bond, G.T. Scarascia-Mugnozza, and M.H. Poulsen) EEC Seminar Bari 1978 (EUR 6244 EN) pp. 342-354.
13. Poulsen, M.H. 1980a. Survey of the breeding work on Vicia faba at VEG Saatzucht Gotha-Friedrichswerth in the GDR. In, Vicia faba: Physiology and Breeding pp. 259-265. (EEC Seminar at Wageningen) 1980. Ed. by R. Thompson, publ. by Martinus Nijhoff.
14. Poulsen, M.H. 1980b. Inbreeding. and autofertility in *Vicia faba* L. Ph.D. thesis, University of Cambridge, 1980.
15. Wright, A.J. 1977. Inbreeding in synthetic varieties of field beans (*Vicia faba* L.). Jour nal of Agricultural Science, Cambridge, 89, 495-501.

# 5. MALE STERILITY IN *VICIA FABA*

Future prospects for hybrid cultivars

J. PICARD*, P. BERTHELEM**, G. DUC*, J. LE GUEN**

\* *Station D'Amelioration Des Plantes, I.N.R.A., B.V., 21304, Dijon, Cedex, France*
\*\* *Station D'Amelioration Des Plantes, I.N.R.A., Domaine de la Motte au Vicomte, 35650, Le Rheu, France*

In a species such as *Vicia faba*, male sterility is a prerequisite to achieve hybrid varieties. Two types of male sterility have been reported (8a) (8b):

  i.    male sterility controlled by the nucleus only (genetic male sterility)
  ii.   male sterility controlled through an interaction of the cytoplasm and nucleus (geno-cytoplasmic male sterility, commonly named cytoplasmic male sterility = CMS).

Genetic male sterility was used to demonstrate, on an experimental scale, the magnitude of hybrid vigour in *Vicia faba* and the potential of hybrid varieties. Commercial production of hybrid seed through the use of this recessive gene is difficult because of the maintenance scheme in which only half of the seed is expected to be male sterile. Sophisticated systems must be devised to make it work more effectively; e.g. the identification of genetic linkage with a seedling colour gene, such as the system which has been used at the farm level in sunflower. Bond (8) has studied the linkage relationships of this male sterility gene with different characters.

In Great Britain, Bond has devoted a lot of work to CMS in order to develop hybrid cultivars. Quite a lot of work is still under way in France, however certain difficulties have made this objective very difficult to achieve. The main difficulty is the reversion of male sterility to normal fertility which makes it hard to develop male sterile lines on a commercial scale i.e. with a high enough percentage of male sterile plants to adequately exploit the potential hybrid vigour.

## CYTOPLASMIC MALE STERILITY IN VICIA FABA

Two types of CMS are known; one from Bond (8a) named 447, discovered in 1957, and the other from Berthelem (4) named 350, discovered in 1967. They were discovered independently in populations, and do not seem to originate from interspecific crosses which, in a number of cases, are the origin of CMS.

*G. Hawtin & C. Webb (Eds.): Faba Bean Improvement*
©1982 ICARDA. ISBN -13:978-94-009-7501-9

The two cytoplasms are different as they do not accept the same genotypes as restorers (among a number of restorer genotypes, only one is known to restore fertility in the two cytoplasms). Other differences exist but they have some features in common such as their instability. Most of the work has been done with 447 cytoplasm.

*Classical scheme for CMS in Vicia faba.*

Fig. 1 shows methods for transferring a genotype to a given cytoplasm, to maintain the CMS lines, and obtaining a fertility-restored single-cross hybrid. Schemes for obtaining three-way crosses, and possibly other types of hybrids, can easily be deduced.

All this is very classical. A more specific and unique situation with regard to the comparison of 447 faba bean cytoplasm with other species, is that once a male sterile plant of faba beans has been restored, either through the action of a major restorer gene or through any kind of reversion, male sterility will never reappear in the following selfed progenies. This has been indicated by Bond (9). Figure 2 shows this situation in comparison to another species i.e. maize.

At the moment, nothing is clearly known concerning the hereditary background of CMS in *Vicia faba.*

*Phenotypic aspects of CMS in Vicia faba.*

As early as 1966, Bond reported that some sterile plants produced small quantities of viable pollen, more or less all over the flowers on the plant, and that other plants produced one or more completely fertile tillers. The first type he named semi-sterile; and the second tiller-sterile.

Since then and because instability makes CMS very difficult to use at the practical level, different authors have had great interest in studying instability. They included Duc (17), Thiellement (35), Berthelem and Le Guen (4), Berthelem and Formall (5), and Benevent and Le Guen (2).

Thiellement (35) divided the level of sterility into three classes - sterile, intermediate and fertile by direct visual examination, and divided his progenies accordingly. Other workers, e.g. Duc (17) measured the level of male sterility by estimating the percentage of fertile pollen through a microscopic examination of pollen grains which had been stained by the Alexander technique.

Conclusions arising from this work include:

- Cytoplasmic sterility, at least in *Vicia faba*, cannot be reduced to a simple and clear cut situation with only two possibilities: male sterile or male fertile.
- A continuous variation seems to exist from some stainable pollen grains on a plant to tiller-sterile or completely fertile plants.
- A chimeral structure regarding male sterility is a very frequent event with not only the occurrence of tiller-sterile plants, but also in situations where all flowers from the same orthostic on a stem, some flowers within a raceme, or some stamens inside a flower being fertile or partially fertile.
- The phenomenon which gives rise to fertility seems to be unidirectional; working from sterility to fertility.

Beside the genotypic control and its influence on the frequency of the two types of expressions (a) semi-sterile plants with all intermediate situations (a slow process) and (b) tiller-sterile (a rapid process), variations are under two types of control:

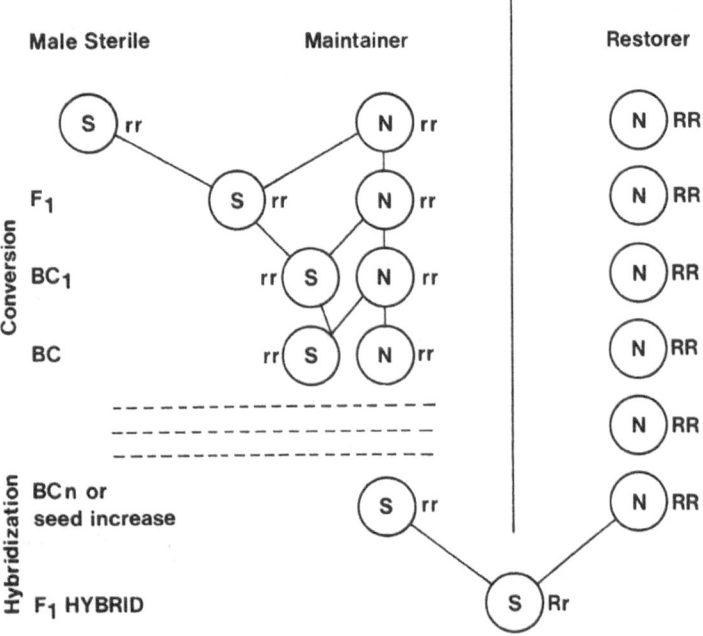

Figure 1.

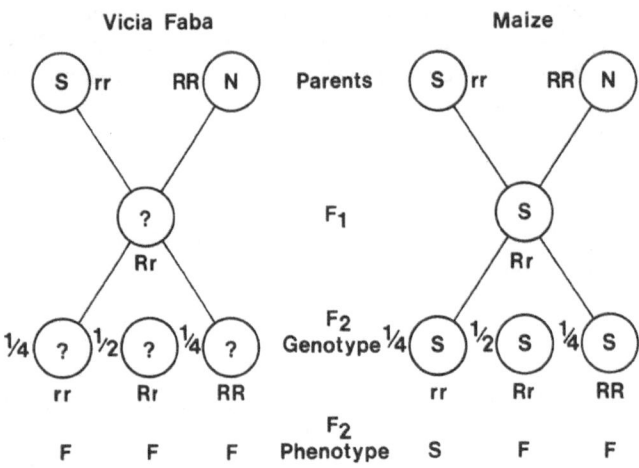

Figure 2.

56

i.  ontogenic control which determines chimeras such as tiller-sterile ones,
ii. environmental control, chiefly temperature but also light.

Similar phenotypic instabilities have been described in other species e.g. lucerne by Barnes and Gaboucheva (1), maize (S cytoplasm) by Laughnan and Gabay (26), *Pennisetum* (A$_1$ cytoplasm) by Clement (16). *Pennisetum* (A$_2$ cytoplasm) by Sequier (34), *Pennisetum* (A$_1$ · A$_2$ cytoplasm) by Burton (14), radish by Bonnet (12), sugar beet by Owen (31), sunflower by Leclercq (29), sunflower by Vear (37).

At the phenotypic level, many similar characteristics can be found in S maize cytoplasm and 447 faba bean cytoplasm. For instance reversion to fertility can occur with a high frequency in progenies (up to 10% according to Laughnan and Gabay(27));reversion can cover either the tassel as a whole or a limited sector of the tassel.

In 95% of cases, the reversion occurs with a change at the cytoplasm and no change at the genetic level. This cytoplasmic change is permanent; no sterile plant is found in the selfedprogeniesof restored plants. The frequency of reversion is very much dependent on the genotype and some stable lines are known.

*Different parameters controlling CMS instability in Vicia faba.*

A good review of the different phenotypic situations has been given by Berthelem *et al.* (7). These situations include interline variation, intraline interplant variation, and intraplant variation.

*Paternal effects:* In 1966, Bond classified genotypes as (a) non-restorers, (b) partial restorers, and (c) high level restorers.

Among the non-restorers, there appears to be a continuous variation from poor to good maintainers; leading Berthelem (3) to hypothesise the existence of recessive minor genes for restoration. However, at present, no maintainer is good enough for breeders.

Recent work by Huglo (23) studied the fertility level in a diallel between 9 CMS lines and their near-isogenic maintainer counterpart in which the number of backcross generations was five and nine. This material was known to have different levels of sterility. The chief result is that a polygenic system appears to be working, besides the known major gene pair (R/r), with dominance and epistatic interactions, and favourable genes which can be recessive in some genotypes; dominant in others.

One of the more interesting conclusions is that this polygenic system should allow recombination and, as a consequence, breeding of genotypes with better male sterility-maintaining ability. However this still has to be done.

*Maternal effects:* Different studies, using pollen from a common parent plant in controlled crosses (7) (17) or a common pollen population in open pollination (35) have shown that the fertility level of the progeny is positively correlated to the fertility level of the mother plant, the stem or a segment of three successive nodes on a stem. However in the 350 cytoplasm at the level of the flower, this correlation is very low; a selfed male fertile flower can result in sterile seeds (17). This latter fact can be understood if we assume that the phenotypic expression of male sterility is a result of some physiological process during gametogenesis, whereas the real genetic information is carried by the ovules.

*Environmental effect* : Studies under both field conditions and more or less controlled environments, have shown the effects of climatic conditions (33) (13) (17).

The most sensitive period for the phenotypic expression of male sterility appears to be around the time the flowers undergo male meiosis. Responses occur within a range of temperatures which are commonly encountered in the open. For instance Berthelem and Le Guen obtained a correlation coefficient of 0.94 between the temperature at meiosis and the percentage of viable pollen within the temperature range of 17 to 22 $^{\circ}$C.

Climatic shocks also have an effect on male sterility, and susceptibility to combinations of temperature and light intensity seem to be inherited. The relationship with cytoplasm instability has still to be verified. However Duc (17) obtained results which indicate an effect of the growing conditions during the entire cycle from sowing to anthesis, on the frequency of instabilities.

## Possible explanations of CMS instability

The situation, as briefly described, suggests the somatic segregation of cellular organelles (determinants) which control male sterility. Under selection pressure, a big range of variation can appear through the very numerous cell mitoses which occur between the zygote and the flowers which arise from this zygote.

The following factors can be suggested:
- chance repartition of a finite number of cytoplasmic determinants from the mother cell to daughter cells through mitosis can give cell lineage with too few a number of determinants or no determinant at all.
- instability of some minor genes in the polygenic system postulated. This instability could be amplified by environmental action, the end result being the elimination of the cytoplasmic determinant.
- instability at the level of the cytoplasmic determinants.

The recent results with cytoplasmic reversions of S maize cytoplasm (30) could provide a good illustration of this last point. These authors have shown that spontaneous reversion to fertility in S male sterile cytoplasm in maize is correlated with the disappearance of the mitochondrial plasmid-like DNA's $S_1$ and $S_2$, and changes in the mitochondrial chromosomal DNA. Hybridization data indicate that $S_2$ plasmid-like DNA is prominently involved in the mitochondrial DNA rearrangements.

The situation is thus: male sterility and its instability in the case of the S cytoplasm could be related to excision and integration of a piece of DNA in the mithochondrial genome.

Of course, similarities between the two CMS instabilities do not necessarily imply that the same phenomena are involved in the two situations.

## How to have a more stable CMS in Vicia faba.

Two very different approaches can be considered:
i.   to obtain a different or modified CMS
ii.  to improve existing CMS through breeding.
Three ways have been tried in relation to the first approach:
a.   the induction of mutations at the cytoplasm level.
b.   interspecific hybrids,
c.   recovery of male sterile plants in $F_1$ or $F_2$ progenies from CMS x restorer crosses.

a. *Mutations:* **Mutagens** have included chemicals such as EMS (Ethyl Methane Sulfonate) or colchicine, and also gamma rays which have succeeded in inducing CMS in barley (19), sorghum (18) and sugar beet (24). Other chemicals such as EB (ethydium bromide) are known to induce cytoplasmic mutation in microorganisms.

In France, we have treated some commercial varieties such as Ascott and Diana with EMS and a combination of EMS + EB without success. Only a genic dominant male sterility has been discovered.

b. *Interspecific hybrids: Vicia narbonensis* seems to be closer to *Vicia faba* taxonomically than other *Vicia* species. As it is *Botrytis* - resistant at the field level and male sterility can be expected from the *faba* genotype being introduced in *narbonensis* cytoplasm, some authors have tried to achieve this interspecific cross (21) (25) (15). Van Cruchten (36) has shown that *Vicia faba* pollen tubes can reach *Vicia narbonensis* ovules but nobody has succeeded in obtaining a hybrid. It may be necessary to develop an embryo culture technique or to go through somatic hybridization but, to achieve this, the problem of plant regeneration would have to be solved.

Other proximate species such as *Vicia hyaeniscyamus* or *Vicia galilaea* can also be tried.

c. *Selection of modifications of the 447 cytoplasmic determinism*. If male sterile plants are found in the selfed progeny of a single cross between a CMS line and a restorer genotype (R/R), these plants can be expected to bear some modifications and to have a more stable cytoplasm. In fact, a program under way in the Forage Breeders' Association has revealed such exceptions. Their true genetic structure has still to be studied. Another parellel program is using mutagenic treatments in the search for exceptions to restoration.

In the second approach i.e. improving existing CMS through breeding, it is necessary to take into account paternal and maternal influences.

As stated above, the role of maintainer genotypes in determining both the level of male sterility and also a given level of stability (or instability) is important because of the complex action of the polygenic system working beside the major gene pair R/r.

The search for new maintainer genotypes is a time consuming task. In the beginning, the only way was to make a number of conversion crosses and look at the $F_1$ and following back cross generations.

It seems now that a better way could be to understand more about the maintainers through diallel analyses, as reported above, and then try to combine the favourable genes from different parents. If this approach is followed, genetic male sterility could prove useful in increasing the gene recombination and would not be difficult to eliminate at the end of the work when trying to recover good maintainer lines.

Our knowledge of the existence of minor genes for maintenance indicates the importance of the choice of the tester male sterile line to be used to screen the maintainer types.

The maternal effect is probably less important; nevertheless it must not be neglected. With the same genotypic situation, a poor cytoplasm would result in a greater chance of a quick appearance of instability. Practically speaking, this means that an accurate assessment of the pollen sterility of the potential parent plants  to  be used in back crosses must be done, possibly through the Alexander technique.

The best plants and the best parts of a plant must be used for the next backcross (BC) generation. Environmental effects could possibly be used as a means of increasing selection pressure, and would allow a better chance of discarding the poorest and more environmental responsive geno-cytoplasmic combinations.

All this means very considerable work, and we must always be careful of limiting its magnitude. For instance, the information which can be obtained about the maintainer value of a given genotype is very limited at the $F_1$; limited at the $BC_1$; but becomes important in the $BC_2$ generation. In the $F_1$ and $BC_1$ a quick visual observation can be made but the $BC_2$ level demands accurate pollen observation and a strong selection to determine the genotypes whose conversion is to be continued further.

## CMS instability and hybrid seed production

Starting from a given BC generation, seed production on a farm scale needs a number of seed increase generations without any possibility of roguing fertile plants or tiller-sterile plants among the steriles. During these generations, the percentage of fertile plants in the CMS parental line of an hybrid increases exponentially as shown by Bond (11) and Berthelem (4). This is a result of:
- the unidirectional conversion of the cytoplasm increases the frequently of new fertile plants appearing at each generation.
- the progeny of restored plants is completely fertile. The yield of such plants is less dependent on pollinators and thus may build up more rapidly than the steriles, especially if the maintainer line is highly self-fertile and unresponsive to tripping.
- some mechanical mixing of maintainer seed in the male sterile lines cannot be excluded at the farm level.

It should be stressed that a situation which can be valuable for instance at the third generation is no longer so at the fourth or fifth; the number of generations we expect to be necessary at the commercial level because of the low seed increase rate of *Vicia faba*. However it is interesting to note that the CMS we have in *Vicia faba*, could be sufficient with seed increase rates such as those of rape, for instance.

It is clear that the potential yield of a supposed hybrid cultivar is related to the percentage of true hybrid plants which must, in turn, be related to the percentage of fertile plants in the CMS female parent.

Thus CMS instability clearly limits the development of hybrid cultivars.

Independently of any official regulations concerning the definition of hybrid cultivars, the percentage of fertile plants which could be acceptable in the female parent without impairing the practical results, will depend on the hybrid yield compared to that of the female parent, and also on the yield increase which commercial hybrid seed can offer in comparison with standard commercial cultivars.

## Breeding for Hybrid Cultivars.

Years before male sterile genotypes or cytoplasms were identified, some authors had tried to appraise inbreeding depression and hybrid vigor in *Vicia faba*. As early as 1951, Fyfe and Bailey (20) when studying the breeding system of *Vicia faba*, concluded that within a population, plants arising from cross pollination were more vigorous than the population mean. Lechner (28) also mentioned hybrid vigour after artificial crossing.

Using hand-made crosses and working with a limited number of individual plants, Picard (32) made two types of comparisons. When comparing $F_1$ plants to their mid parent values, he found 36 of the 38 were superior; 31 being significantly superior. When he compared $F_1$'s to the mean of parental populations from which the inbred parents were obtained, 24

60

were superior; 14 being significantly so. After this, sterility was identified and comparisons became more numerous, leading to significant experiments.

*Yielding ability in F$_1$ hybrids*

Bond (8b), working with winter beans, reported on 28 F$_1$ hybrids obtained by using genic male sterility in comparison to Garton's S.Q. in 14 experiments in different years, of 78 comparisons, 77 were in favour of the F$_1$. The maximum gain over Garton's S.Q. was 38.5%. The gain over the parental lines was, of course, much higher.

In (9b), Bond reported on comparisons with hybrids obtained through cytoplasmic male sterility. The 57 comparisons to Garton's S.Q. gave 53 positive results; 36 being significant.

The results of similar experiments from 1964-69 were presented at Gottingen (11). They are summarised in Fig. 3.

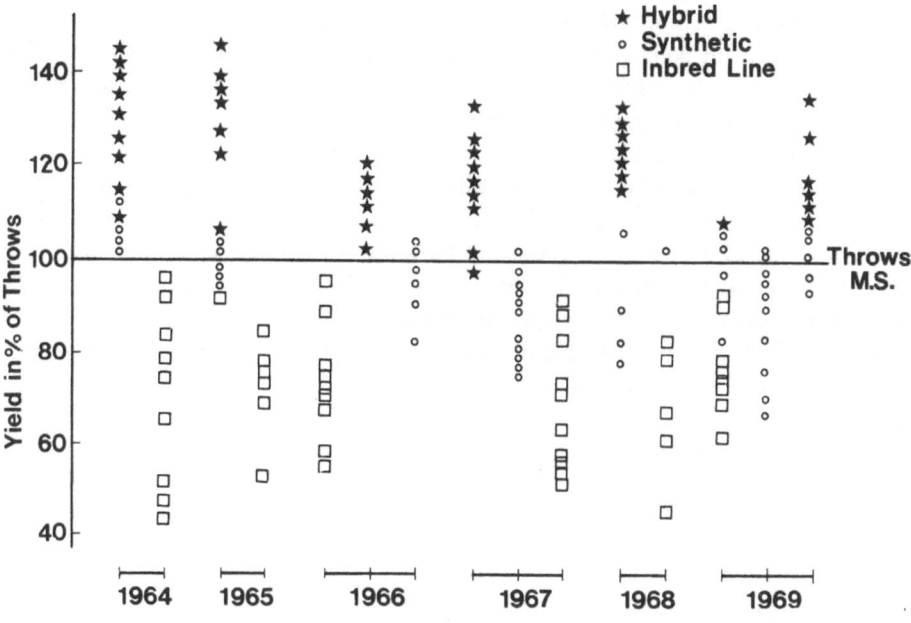

Fig. 3   Comparison of yields of hybrids, synthetics and inbred lines. After Bond (1974).

Table 1.

Yield of winter bean hybrids in q/ha and in comparison with Maris Beagle in eight locations in harvest year 1972.

| MS lines Restorer lines | 29 - D | | 29-E | | 972-D | | 972-A | | Maris Beagle | Cote d' or | Mean | LSD 5 % |
|---|---|---|---|---|---|---|---|---|---|---|---|---|
| | C.F | S.45 | C.F | S.45 | C.F | S.45 | C.F | S.45 | | | | |
| Rennes | 34.9 | 29.3 | 46.0 | 51.4 | 34.2 | 39.9 | 38.8 | 48.9 | 38.6 | 16.0 | 37.2 | 7.1 |
| Dijon | 52.2 | 57.2 | 58.0 | 64.3 | 57.6 | 68.2 | 59.1 | 58.2 | 45.6 | 39.8 | 56.4 | 6.8 |
| La Miniere | 32.8 | 38.6 | 38.1 | 44.9 | 32.2 | 39.2 | 35.8 | 48.2 | 32.1 | 19.1 | 35.7 | 4.0 |
| Boigneville | 39.2 | 42.2 | 48.4 | 51.5 | 45.0 | 51.0 | 50.3 | 54.4 | 44.0 | 28.7 | 44.8 | 3.9 |
| Fresnes | 35.8 | 38.1 | 39.4 | 44.8 | 34.2 | 42.9 | 40.8 | 45.2 | 25.2 | 20.6 | 34.4 | 3.8 |
| Chassieux | 51.2 | 49.2 | 60.2 | 59.4 | 59.1 | 59.2 | 58.2 | 57.4 | 48.4 | 39.7 | 53.7 | 5.7 |
| Marmillat | 16.0 | 17.6 | 18.2 | 19.4 | 15.0 | 15.1 | 15.0 | 19.6 | 15.4 | 14.5 | 16.6 | 2.8 |
| Greoux | 46.4 | 53.2 | 49.7 | 56.8 | 50.7 | 56.1 | 47.4 | 49.5 | 41.3 | 31.1 | 46.1 | 5.8 |
| Mean ( LSD5%= 4.1) 38.6 | | 40.7 | 44.8 | 49.1 | 41.0 | 46.5 | 43.2 | 47.7 | 36.3 | 26.2 | 41.2 | |
| Mean as % of Maris Beagle (LSD5%= 11) | 106 | 112 | 123 | 135 | 113 | 128 | 119 | 131 | 100 | 72 | | |

Experimentation to determine the value of hybrid cultivars has been conducted in France. The results on winter beans have been summarized by Berthelem (4). (Table 1).

Many experiments have also been conducted with spring beans since 1972, for example:
- In 1973, 76 comparisons gave 61 results in favour of the $F_1$ with a maximum gain over the common French cultivar Ascott of 60%.
- In 1976, 79 comparisons gave 168 results in favour of the $F_1$ (124 significant) with a maximum gain of 49% over Ascott.

Table 2 shows a comparison between $F_1$ performance and that of the parental lines.

Table 2.

Yield in q/ha of some hybrids and parent lines and increase in yield in % over the best parent, Rennes 1972.

| Parents | Hybrid Yield | Parent Yields MS | Restorer | % $F_1$ yield over best parent |
|---|---|---|---|---|
| G 78 x Po ad 74 | 53.0 | 24.1 | 23.5 | 120 |
| G 78 x G 8 | 45.4 | 24.1 | 27.6 | 64 |
| Asco x Po Major | 43.3 | 26.5 | 20.3 | 63 |
| Asco x Col. | 45.3 | 26.5 | 28.4 | 60 |
| Asco x Rinal | 41.4 | 26.5 | 14.4 | 56 |
| G 79 x Po Major | 44.6 | 28.8 | 20.3 | 55 |
| G 79 x Col | 41.2 | 28.8 | 28.4 | 43 |
| G 77 x Rinal | 39.0 | 28.8 | 14.4 | 35 |
| G 77 x Col | 46.4 | 36.6 | 28.4 | 27 |
| G 77 x G 8 | 42.6 | 36.6 | 27.6 | 16 |
| G 77 x Po ad 74 | 40.2 | 36.6 | 23.5 | 10 |

Another way of expressing hybrid vigour could be to compare the yield of hybrids with other types of cultivars with a similar, if not identical, genetic background.

Table 3 gives results from the two trials conducted in France during growing season 1978-79 with winter beans. Using three male sterile lines (48 B, 29 E, L 6) or their maintainer counterpart and one restorer line (S 45), three types of hybrids and two synthetics were developed. Unfortunately, some possible hybrids are missing (e.g. only one $F_1$ hybrid instead of the three possible and no three-way hybrids).

However it appears clearly that, with a given set of lines, a hybrid structure is much better adapted to make use of potential hybrid vigour than is a synthetic cultivar; in these experiments, third generation synthetics were tested.

Table 3.

Yield of different experimental hybrids and synthetics in comparison to Maris Beagle.

| Experimental Cultivars | | RENNES | | DIJON | |
|---|---|---|---|---|---|
| | | Q. ha$^{-1}$ | % M. Beagle | Q. ha$^{-1}$ | % M. Beagle |
| F 1 | (1) | 49.1 | 96 | 47.2 | 119 |
| Double Cross | (2) | 53.8 | 105 | 43.5 | 110 |
| Four ways | (3) | 64.1 | 125 | 47.7 | 121 |
| Syn 3 | (4) | 43.1 | 84 | 31.9 | 81 |
| Syn 3 | (5) | 40.0 | 78 | 29.7 | 75 |
| Maris Beagle | | 51.1 | 100 | 39.5 | 100 |
| L.S.D. P≤0.05 | | 5.9 | 12 | 2.9 | 7. |

Structure of the experimental cultivars:

(1)  48 B x S 45
(2)  (48 B x 29 E) x (L 6 x S 45)
(3)  ((48 B x 29 E) x L 6) x S 45
(4)  Maintainor lines 48 B + 29 E + L 6 + Restorer line S 45
(5)  Male sterile 48 B + 29 E + L 6 + Restorer line S 45

*Other aspects of hybrid cultivars*

In a species which has not been submitted to long and intensive breeding work, increasing yield is by far the most important objective. However, it would be wrong to have very high yielding types with poor performances in other characters such as winter hardiness, protein content and yield stability.

Few experiments have been done for specifically testing such characteristics and we have no complete set of diallel crosses which can give an accurate indication of combining ability values. This is because breeders who have had a very marked interest in obtaining high yielding cultivars, have more or less neglected other useful qualities. However, such information can be deduced from various experiments.

*Winter hardiness:* As an example, Table 4 gives results of counts made at Dijon during the 1972-73 growing season. It is clear that good winter hardiness can be obtained with $F_1$ hybrids.

*Protein content:* Table 5 gives the results of seed protein analyses from a trial at Dijon in 1974. Whatever the yield increase over the mid-parent (12 to 46% ), protein content was always very close to the mid-parent level. As has been shown also on other occasions, we can expect high yielding $F_1$ varieties with a high protein content.

*Yield stability:* Many authors agree that hybrids are more self fertile than inbreds. If pollination is a limiting factor for yield stability, hybrids should be more stable than synthetics or natural populations which, as a mean, would be more inbred.

*Hybrids other than $F_1$*

$F_1$ hybrids have a number of drawbacks, and thus other hybrid combinations have been studied in view of:

- the cost of the seed harvested on inbred lines
- the possibility of introducing as allomaintainers (the non restorer parent of the $F_1$ used as female), lines which are poor maintainers but have a good combining ability.
- the possibility of slowing down CMS instability through seed increase generations by modifying the genetic structure at the last step.

Table 4.

Results of plant counts made at Dijon during 1972 and 973 growing seasons.

|  | % of surviving plants | % in relation to Cote d'Or population (1) |
|---|---|---|
| 29 E x CF | 83.7 | 92 |
| 29 E x S 45 | 81.7 | 90 |
| 972 A x CF | 79.1 | 87 |
| 972 A x S 45 | 86.9 | 96 |
| 972 D x S 45 | 89.5 | 99 |
| Maris Beagle | 75.6 | 83 |
| Cote d' Or | 90.5 | 100 |

(1)     Cote d' Or is the most winter hardy faba bean in France.

Table 5.

Results of seed protein analyses from a trial in 1974 in Dijon.

|  | Protein content | | Yield in q/ha | |
|---|---|---|---|---|
|  | Hybrid | Mid parent | Hybrid | Mid parent |
| G 7 x G 8 | 31.9 | 31.8 | 56.3 | 45.5 |
| G 77 x G 8 | 32.9 | 32.1 | 51.3 | 45.3 |
| G 78 x G 8 | 34.5 | 32.9 | 49.1 | 40.1 |
| I x Col | 34.8 | 34.8 | 56.1 | 38.4 |
| I x Po Major | 34.7 | 34.4 | 46.0 | 35.7 |
| G 78 x G 107 | 33.9 | 33.4 | 50.6 | 39.1 |

Some experiments have been done to compare the yield of three-way crosses with those of $F_1$ hybrids. Berthelem has looked at the value of $F_1$ hybrids for predicting the yields of three-way crosses, using the formula:

Yield of hybrid (AxB)xC = mean yield of (AxC) + (BxC).

The results obtained in trials with winter beans at two locations in 1974, are shown in Table 6. Three-way crosses showed a small superiority over single crosses, as has been borne out in other trials.

Table 6.

Yield (q/ha) of $F_1$'s compared to predicted and actual yields of three way hybrids in two locations in 1974.

| MS lines | | Non parent $F_1$ | RENNES Three way cross Calculated | Actual | Non parent $F_1$ | DIJON Three way cross Calculated | Actual |
|---|---|---|---|---|---|---|---|
| S45 Restorer | L 6 | 43.9 | 47.2 | 52.9 | 57.3 | 56.9 | 58.0 |
| | 29 E | 45.0 | | | 56.5 | | |
| | L 6 | 43.9 | 47.5 | 47.4 | 57.3 | 56.2 | 57.3 |
| | 972 A | 51.1 | | | 55.0 | | |
| | L 6 | 43.9 | 46.3 | 48.7 | 57.3 | 55.3 | 59.1 |
| | 333 A | 48.6 | | | 53.2 | | |
| C.F. Restorer | L 6 | 46.1 | 42.9 | 39.7 | 56.3 | 55.0 | 53.9 |
| | 29 E | 39.7 | | | 53.7 | | |
| | L 6 | 46.1 | 47.8 | 50.5 | 56.3 | 55.2 | 55.1 |
| | 972 A | 49.4 | | | 54.0 | | |
| | L 6 | 46.1 | 43.0 | 43.5 | 56.3 | 54.0 | 54.0 |
| | 972 D | 39.8 | | | 51.6 | | |
| | Means | 45.6 | 45.8 | 47.1 | 54.0 | 55.4 | 56.3 |

The predictive value of $F_1$ is very good for estimating three-way cross yields: three-way crosses also appear to be efficient in making use of hybrid vigour in faba beans. The reasons for this apparent superiority are not clear. One reason may be that the seed is harvested from more vigorous mother plants.

Similar but less well structured experiments have been done with spring types. In these, three-way crosses were also found to be very promising. Some double crosses have been tried, and the small number tested have ranked quite well when compared to other hybrids.

Returning to winter beans, it appears that there is a good complementarity between material bred in France, and material from the United Kingdom. This situation should be kept in mind when developing important cultivars through reciprocal recurrent selection. A similar situation has to be sought in spring types between different geographical regions.

*Hybrid seed production*

Even if good hybrid combinations have been found, the seed still has to be produced commercially. As a first general remark, production of restorer or maintainer lines is no different from seed increases in other autogamous species. Secondly, there is no fundamental difference (except that the male parent is different) between the increase of CMS lines and hybrid seed production.

The schemes for hybrid seed production, presented in Figure 1, are similar to those in other plants such as maize. For instance, isolation must be very strict in the first seed increase generation as MS beans are very receptive to any foreign pollen.

One peculiar point arises from the fact that pollen transport needs pollinating insects. Observations everywhere have concluded that the most efficient and numerous pollinators are among the genus *Bombus*. Some of them such as *B. hortorum*, *B. agrorum*, *B. lapidarius* are well adapted to transfer pollen from fertile to sterile faba bean plants. Honey bees which collect nectar, are quite good pollinators but the best are certain solitary bees (*Eucera* sp. for instance). *B. terrestris*, one of the most common *Bombus* species, at least in France, works as a robber, making holes in the bottom of the corolla tube, inducing other species to rob nectar through these holes.

A good yield on a CMS line needs good pollination. Good pollination is also necessary for limiting the development of fertility in CMS lines. As some experiments have shown (7), poor pollination in a CMS line leads to higher fertility levels in the progeny.

Seed production at the farm level, either for increase of CMS lines or for hybrid seed production, requires that the plants be grown in alternate beds, the width of which must be in agreement with farm machinery. Experiments have been done at the farm level in winter beans in Great Britain and France, using different proportions of male and female lines. In France, the National Seed Farmers Association did many such experiments during 1972-76. Specific questions still have to be solved, but the chief problem remains CMS instability. Experiments stopped pending a better sterility.

*Conclusion*

In western European countries, faba bean is not grown on a large scale. In countries such as France, Germany and the United Kingdom, areas have been reduced by about five fold during the past century. Insufficient yields and poor yield stability were among the main reasons why farmers discarded this crop.

As previously mentioned, it is not evident that hybrids can solve the second point but among the breeding methods, the hybrid way (single crosses, three-way crosses) has appeared for at least 10 years (1965-75) to be the best and fastest way to achieve a very significant yield increase. A 30% gain by only changing the cultivar, could change the attitude of farmers to the crop.

More than 20 years after Bond had discovered the first CMS in *Vicia faba*, no hybrid cultivar is grown on any farm. The Plant Breeding Institute, Cambridge, United Kingdom, has stopped working on male sterility. The other country which has a significant program, France, is also questioning the chances of achieving hybrid cultivars in a reasonable time. The chief, and we believe the only reason, is CMS instability.

If a workable CMS, not necessarily a stable one, becomes available, work could start again with, of course, still a lot of questions to be addressed. These would include aspects, as already discussed by Bond (8b) such as the percentage of non-hybrid plants which could be tolerated in commercial seed, and the best techniqu s for hybrid seed production.

Work done with *Vicia faba* CMS is very limited compared to what has been done on maize CMS. Many potential CMS sources were discarded in maize breeding before almost all the USA hybrids were produced, using the Texas CMS. And then Texas was discarded because of susceptibility to a specific disease. By contrast, we have tested only two CMS sources with only one or two thousand genotypes.

A lot of work is under way to determine the nature of CMS at the molecular level, its hereditary support, how it works, and so on. Some research on these aspects is also being done with *Vicia faba*. We can expect to make use of the results of this research in breeding programs in the future.

Also, legumes are very underdeveloped at the level of sophisticated techniques such as tissue culture, protoplast fusion, plant regeneration and so on.

Perhaps the chief reason for *Vicia faba* cultivars not being on the market is that breeders started too early to use a model which worked well in other species.

# REFERENCES

1. Barnes, D.K., Garboucheva 1973 - Intra-plant variation for pollen production in male sterile and fertile alfafa. Crop Sci. 13, 456-460.
2. Benevent, E., Le Guen, J. 1976. Problemes poses par l'utilisation de la sterilite male cytoplasmique en production de feverole hybride. Memoire ENSA de Rennes.
3. Berthelem, P. 1970. Rapport d'activite - Station d' Amelioration des Plantes INRA Rennes - 1968-1970.
4. Berthelem, P., Le Guen, J. 1975. Rapport d'activite - Station d' Amelioration des Plantes INRA Rennes - 1971-1974.
5. Berthelem, P., Formall 1973 . Memoire fin d' etude - B.T.A.
6. Berthelem, P., Le Guen, J., Bourgeois, F. 1977. Recherches sur les methodes de selection en vue de l' exploitation maximale de la vigueur hybride chez la feverole. Utilisation pratique de la sterilite male pour la creation de varietes hybrides. Contrat CEE n° 408 - Compte rendu 1° semestre 1977 mimeographed 16 p.
7. Berthelem, P., Le Guen, J., Bourgeois, F. 1978. Recherches sur les methodes de selection en vue de l'exploitation maximale de la vigueur hybride chez la feverole. Utilisation pratique de la sterilite male pour la creation de varietes hybrides. Contrat CEE n° 408 - Compte rendu de fin de contrat mimeographed 31 p.
8. Bond, D.A., Drayner, J.M., Fyfe, J.L., Toynbee-Clarke, G. 1964:
   a. A male sterile bean (*V. faba* L.) inherited as a Mendellian recessive character. J. Agri. Sc. 63, 229-234.
   b. Yield trials of $F_1$ hybrid winter beans produced with the aid of male sterility. J. Agric. Sc. 63, 235-243.
9. Bond, D.A., Fyfe, J.L., Toynbee-Clarke, G. 1966:
   a. Male sterility with a cytoplasmic type of inheritance in field beans. J. Agric. Sc. 66, 359-367.
   b. Use of CMS in production of $F_1$ hybrids and their performances in trials. J. Agric. Sc. 66. 369-377.
10. Bond, D.A. 1966. Yields and components of yield in diallel crosses between inbred lines of winter beans. (*V. faba* L.) J. Agric. Sc. Camb. 67, 325-326.
11. Bond, D.A. 1974. Die zuchtung von hybrid und synthetischen sorten Ackerbohnen in Cambridge. Gottingen pflanzen zuchter seminar 2, 39-62.

12. Bonnet, A. 1975. Introduction et utilisation d'une S.M.C. dans des varietes precoces europeennes de Radis. *Raphanus sativus* L. Ann. Amel. Plantes 25(4) 381-397.

13. Bouverat-Bernier, J.P., Le Guen, J. 1980. Contribution a l'etude de l'evolution de la sterilite male cytoplasmique chez la Feverole. Memoire ENSA Rennes.

14. Burton, G.W. 1977. Fertile sterility maintainer mutants in CMS. Pearl Millet. Crop. Sci. 17, (14), 635-637.

15. Bougeois-Rousselle, F. 1978. Rapport d'activite C.E.E.

16. Clement, W.M. 1975. Plasmon mutations in CMS pearl millet (*Pennisetum* hybrids) Genetics 79, 583-588.

17. Duc, G. 1978. Modalites d'expression et hypotheses explicatives du manque de stabilite de la sterilite male cytoplasmique chez la feverole (*Vicia faba* L.). These de Docteur-Ingenieur - Universite Paris Sud.

18. Erichsen, A.W., Ross, J.G. 1963. Inheritance of colchicine-induced MS in sorghum. Crop. Sci. 3, 335,338.

19. Favret, E.A., Ryan, G.S. 1964. Two cytoplasmic male sterile mutants induced by X-Rays and EMS. Barley Newsletter 8: 42.

20. Fyfe, J.L., Bailey, N.T.J. 1951. Plant breeding studies in leguminous forage crops. II. Further observations on natural cross-breeding in winter beans. J. Agric. Sci. 41. 371-378.

21. Hanelt, 1972. Die Stellung Von *V. faba* in der Gattung *Vicia* L and Betrachtungen zur eltestschung dieser Kultur art. Kultur Pflanze 20.

22. Hogaboam, G.J. 1957. Factors influencing phenotypic expression of CMS in the sugar beet (*Beta vulgaris* L.) J. Am. Soc. Sugar beet Technol. 9: 456-465.

23. Huglo, B., Duc, G. 1980. Effet du genome mainteneur sur le determinisme de l'insta-bilite de la sterilite male cytoplasmique (cyt. 447) chez la Feverole. Memoire ENSA Rennes.

24. Kinoshita, T., Takahashi, M. 1964. Induction of CMS by gamma ray irradiation in sugar beets. Japan J. Breed 19, 445-457.

25. Ladizinsky, 1975. On the origin of the broad bean (*Vicia faba* L.) Isr J. of Bot. 24. 80-88.

26. Laughnan, J.R., Gabay, S.J. 1974. Mutations restaurant la fertilite male du cytoplasme S chez le Maix. Maize genetics corporation 48 : 38-42.

27. Laughnan, J.R., Gabay, S.J. 1975. An episomal basis for instability of S male-sterility in maize and some implications for plant breeding in "Genetics and biogenesis of mito-chondria and chloroplasts". Ed. by C.W. Birky, Jr. P.S. Perlman and T.J. Byers. Ohio State University Press: Columbus.

28. Lechner, L. 1956. Die Pferdebohne (*Vicia faba* L.) Handbuch der Pflanzenzuchtung Band IV Bogen 1-5.

29. Leclercq, P. 1966. Une sterilite male utilisable pour la production d'hybrides simples de Tournesol. Ann. Amel. Pl. 16, 135-144.

30. Leving *et al.* 1980 . Cytoplasmic reversion of CMS - S in maize. Association with a transpositional event. Science, 209, 1021-1023.

31. Owen, F.V. 1945. Cytoplasmically inherited M.S. in sugar beets. J. Agric. Res. 71: 423-440.

32. Picard, J. 1960. Donnees sur l'amelioration de la feverole de printemps *Vicia faba* L. Ann. Amel. Plantes 121-153.

33. Nouy, B., Le Guen, J. 1977 . Contribution a l'etude de l'evolution de la sterilite male cytoplasmique en production de feverole hybride. Memoire ENSA Rennes.

34. Sequier, J. 1975. Etude de la stabilite  de la S.M.C. chez la lignee de MIL penicillaire. Tifton 239 D.A. DEA Universite de Paris-Sud Orsay.

35. Thiellement, H. 1977. De la sterilite male cytoplasmique chez *Vicia faba* L. These Docteur 3eme cycle. Universite Paris -Sud. 59 p.
36. Van Cruchten, C. 1974. Etude de la croissance des tubes polliniques dans des croisements intra et interspecifiques de *Vicia faba* et *Vicia narbonensis* Mimeographed 13 p.
37. Vear, F.A. 1973. Genetical studies of mildew resistance and M.S. in sunflowers. Thesis of the Univ. of Reading.

# 6. POPULATION IMPROVEMENT IN FABA BEANS

ABDULLAH M. NASSIB AND SHAABAN A. KHALIL

*Food Legume Research Section, Field Crops Institute, Agricultural Research Center, Giza, Egypt*

Mass selection, by far the most widely practised form of population improvement, has been used to improve the commercial faba bean varieties in Egypt since the mid-1960s. A continuous program involving single plant selection of a few hundred plants every year and bulking their progenies after screening for uniformity in plant and seed morphology, along with the yield, has led to higher average yields and stability. Giza 2 which has been subjected to this method, has been widely distributed to farmers in Minia and Assiut provinces since late 1960's. (Fig.1a). The decline in average yields in Minia province during the past few years is probably because many farmers have been planting beans early to allow them to grow a following crop of cotton.

Line breeding based on recombinations from biparental crosses did not permit rapid progress in faba bean improvement (8). On the other hand, the high percentage of cross pollination and the fact that mass selection has been conducted primarily in gentically heterogenous populations propagated throughout under natural cross pollination, has led to the belief that a form of population improvement rather than line breeding would result in higher yields and stability.

*Methods of population improvement-General:*

Sprague (11) defined improvement population as a general term to include all operations within a system in which the end product sought is a sexually-propagated improved type; either a random-mating population or pure line. Population improvement may take many forms, but, in each case, the effectiveness of selection practised, is based on the utilization of additive gene effect.

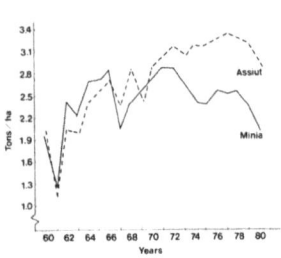

Figure 1 a

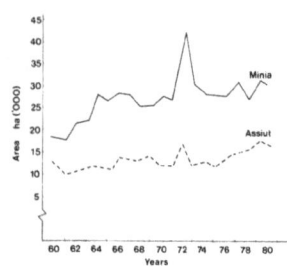

Figure 1 b

*G. Hawtin & C. Webb (Eds.): Faba Bean Improvement*
©1982 ICARDA. ISBN -13:978-94-009-7501-9

A comprehensive breeding system proposed by Eberhart et al. (1) is applicable to both allogamous and autogamous species, but it is certainly most applicable in cases where mass hybridization can be made cheaply. Jensen (6) suggested the "diallel selective mating system" which was designed primarily for autogamous species. With these procedures, the breeding populations of plants are seen as dynamic gene pools (a) to which new sources of germplasm are added whenever feasible, (b) in which the frequencies of favorable alleles are progressively increased via recurrent selection, (c) in which genetic recombination is enhanced by mass hybridization among selected genotypes, and (d) from which cultivars, inbreds or parental lines can be extracted at any stage (2). The breeding population being improved is dynamic in that migration and mutation add new alleles to it; selection increases the frequencies of favourable alleles; opportunity for recombination is maximized, and useful agricultural cultivars can be extracted at any stage in its evolution.

Rachie and Gardner (10) have reviewed selection methods which are practical for partially outcrossing crops. $S_1$ and $S_2$ testing offers better precision for estimating family means than half-sib or full-sib methods with consequent rapid genetic gain if three or four generations can be obtained in a single year. Khan (7) recommended the utilization of the outcrossing potential in pigeonpea in the formation of random mating composites. These composites would serve as the dynamic reservoir of variability and could be used in natural selection, mass selection and recurrent selection. Onim (9), working on an early maturing pigeonpea composite population got, after four cycles of stratified mass selection, an average progress of 2.3% compared with 1.9% over two cycles of mass selection with progeny testing.

Hawtin, et al. (5) suggested an $S_2$ testing procedure in a four-step cycle for population breeding program in pulses. The mainstream of the program should be devoted to the improvement and stabilization of yield and other agronomic characters. Separate sub-populations can be established for developing other characters such as disease resistance and nutritional factors. The selection pressure applied to the sub-population will be for a single character only, and when enough advance has been made in the sub-population, the character can be transferred to the mainstream population through an intermediate back-up population. Random crossing may be achieved through the use of pollinating insects or genetic male sterility. The identification of reliable gametocide would allow the method to be used in crops for which no male sterility has been discovered.

*Population improvement methods in faba beans:*

Gallais (3) proposed a three-part program: (i) introduction of new variability, (ii) cycles of recurrent selection with alternation of selection and crossing, and (iii) utilization of selected units for developing cultivars at every cycle. This work aimed at increasing favourable genes, lessening the risk of genetic drift, use of epistasis, breaking up genetic linkages, and weakening the sensitivity to inbreeding.

Hawtin (4) has outlined simple schemes for population improvement in faba beans. A four-step cycle included recombination, selection in presumed $F_1$ plants and $S_1$ rows and testing $S_2$ lines. Additional $S_3$ testing could be undertaken for special purposes e.g. protein. If recombination could be made off-season, the cycle would take two years; otherwise, it is completed in three years with two generations each where line testing should be done in the main season. Honey bees are used at flowering to induce a high degree of outcrossing. Random mating may be necessary for several generations with low selection pressure before beginning selection cycles.

In Egypt, a one generation a year population improvement program was initiated in 1974 (8). The base populations included a mixture of cultivated landraces, hybrid derived lines selected from yield trials, commercial cultivars and introductions. Using a genetic marker, it was found that 79.9% cross pollination could be reached by introducing honey bees in the population cage during flowering.

In subsequent $S_0$, $S_1$ and $S_2$ generations, very few lines turned out to have desirable agronomic characters, good adaptability or disease resistance. This is probably due to (i) choice of population components, (ii) small size of the population, (iii) low percentage of cross-pollination probably resulting from inefficiency or preference habits of honey bees and difference in flowering time of the population components.

*Recent developments*

The genetic variability retained within and among the best rated $F_3$ or $F_4$ lines in hand-made elite crosses, preferably three way or multiple crosses, could be exploited in compositing the populations for special purposes; for example, disease or *Orobanche* resistance, good adaptability, combining high protein content with elite agronomic background; thus increasing the probability of building up additive gene effects and of breaking up undesired linkages. Later, elite $S_2$ lines of two or more of these populations are hand crossed. Evaluation of their $F_3$ progenies will establish populations for wider breeding objectives (Fig. 2).

Availability of a wide range of genetic variation in the parents of the crosses will ensure better identification of the best cross in which to select among progenies for the base population. Again sufficient number of $F_3$ or $F_4$ lines within an elite cross will increase the probability of including superior genotypes in the base population.

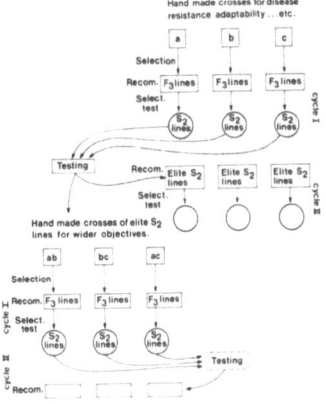

Figure 2.

REFERENCES

1. Eberhart, S.A. Harrison M.N., Ogada F. 1967. A comprehensive breeding system. Der Zuchter 37 : 169-174.
2. Frey, K.J. (n.d). Lectures in plant breeding - Alexandria University (unpublished).
3. Gallais, A. 1970. Evolution de la vigueur des varietes synthetiques diploides au cours des generations de multiplication. 1-en panmixie-influence du nombre de parents et du coefficient de consanguinite de depart - Annales Amel. Plantes 17, No. 3:291-301 In Picard J. 1979 Some reflections on problems and prospects in *Vicia faba* breeding in "Some current research on *Vicia faba* in Western Europe". Pup. EEC; Eur 6244 EN pp 23-34.
4. Hawtin G.C. 1978. Introduction to breeding food legumes Technical Manual No. 2 ICARDA, Aleppo, Syria; 109 pp.
5. Hawtin G.C., Rachie K.O., Green J.M. 1977. Breeding strategy for the nutritional improvement of pulses. In Hulse J.H., Rachie, K.O. and Billingsley L.W. Nutritional standards and methods of evaluation for food legume breeders. IDRC Ottawa p 43-51.
6. Jensen, N.F. 1970. A diallel selective mating system for cereal breeding. Crop Sci. 10 : 629-635.
7. Khan, T.N. 1973. A new approach to the breeding of pigeonpea (*Cajanus cajan* Millsp): formation of composites. Euphytica 22 (1973) : 373-377.
8. Nassib, A.M., Ibrahim A.A. and Khalil, S.A. 1979. Methods of population improvement in broad bean breeding in Egypt. In Hawtin G.C. and Chancellor G.J. (ed)." Proceedings of a workshop of food legume improvement and development ", Aleppo, Syria, ICARDA & IDRC p. 176-178.
9. Onim, J.F.M. 1980. Pigeonpea improvement research in Kenya. In International workshop on pigeon peas ICRISAT, Hyderabad, India. pp. 25.
10. Rachie K.O., Gardner C.O. 1975. Increasing efficiency in breeding partially outcrossing grain legumes. In International workshop on grain legumes ICRISAT, Hyderabad, India. 349 pp.
11. Sprague, G. F. 1967. Quantitative genetics in plant improvement. In Frey, K.J. (ed) : Plant Breeding, a symposium held at Iowa State Univ. University Press, Ames. Iowa. p. 315-354.

7. NEED, CONCEPT AND BREEDING STRATEGY FOR WIDER ADAPTABILITY IN *VICIA FABA*.

E. VON KITTLITZ

*Universitat Hohenheim, Postfach 70 05 62, D-7000 Stuttgart, FRG*

The term "adaptability" has been used in a number of contexts, for example, it has been described as the ability of a species to answer different environmental conditions by the development of new ecological races or subspecies. However in the context of this chapter, breeding for wider adaptability may be described as the development of cultivars which satisfy human requirements over a wide range of geographical conditions. Such cultivars would thus show adaptation across a range of diverse environments, clearly exceeding the limits of homologous environments as they are presented e.g. by Simmonds (6).

In the development of wider adaptation in faba beans, there are many good reasons which are likely to preclude the development of cultivars which are adapted to very widely differing environments; it is hard to imagine a cultivar which would be high yielding both in Aleppo and Hohenheim. There is no evidence in the literature to suggest that such a situation might be possible. What then, could be achieved in a breeding program for wider adaptability ? This chapter aims to contribute towards a preliminary answer to this question. A discussion of the matter from the viewpoint of breeding cultivars with a wide adaptability could at least raise some interesting points.

Many components of the proposed breeding strategy have been discussed by other scientists, for example the collection, maintenance and evaluation of genetic variability (3,4). These topics are only dealt with here in respect to the peculiarities of the subject.

*Need and Concept*

In order to have a clear starting position, it seems necessary to first discuss the need of breeding for wider adaptability.

If the best utilization of the genetic variability with respect to the human use of improved cultivars can be made from a breeding program for wider adaptability, then we should concentrate all our efforts towards achieving this goal. In stating this; however, it is apparent that we do not have adequate supporting data from properly designed experiments. Even genetic materials to test the feasability of the approach are lacking, for an

*G. Hawtin & C. Webb (Eds.): Faba Bean Improvement*
©*1982 ICARDA. ISBN -13:978-94-009-7501-9*

adequate experiment would require genotypes which had been effectively selected for wider adaptability !

Table 1 gives the mean seed yield of ten cultivars across nine rather diverse locations. The productivity of any location can be characterized by the mean seed yield of all entries. The differences between the productivity of the various locations are more likely to be due to differences in moisture supply than to edaphic diversity. In the context of wider adaptability it is interesting to note that the cultivar Minica, which produced the greatest mean yield across all locations, was also ranked in the top three cultivars at most individual locations. Although it might be considered a widely adapted cultivar, it should be stressed that Minica was certainly not selected for wider adaptability but was bred under very favourable growing conditions. Can it be concluded from this that breeding for wider adaptability could be successful when selection is made in the most favourable environment ? The answer would obviously have a great bearing on breeding strategies. However, the trial was not specifically designed to address the problem of adaptability and hence locally well adapted cultivars had not been included at all sites. Unfortunately I know of no other faba bean trials which cover the question of wider adaptability.

The concept of broadening adaptability, therefore will be based on the following hypothesis : that the original gene pool on which breeding is to be practised, necessarily has to include the variability contained within the germplasm originating in the entire range of habitats for which the widely adapted cultivars are ultimately intended. This hypothesis is based on the assumption that wider adaptability may be interpreted genetically as an accumulation of chromosomal or cytoplasmic factors which are currently distributed in different landraces, breeding material or cultivars, throughout the whole range of habitats. Other possibilities, such as that indicated in Table 1 for the cultivar Minica, are not excluded but are considered less promising, at least with the present state of knowledge.

If this is the case, then the assembling of germplasm is an important prerequisite. More detailed concepts with regard to the utilization of germplasm collections are given elsewhere e.g. Chang (2), Simmonds (6) or CIMMYT (1). These activities are summarised in Fig. 1.

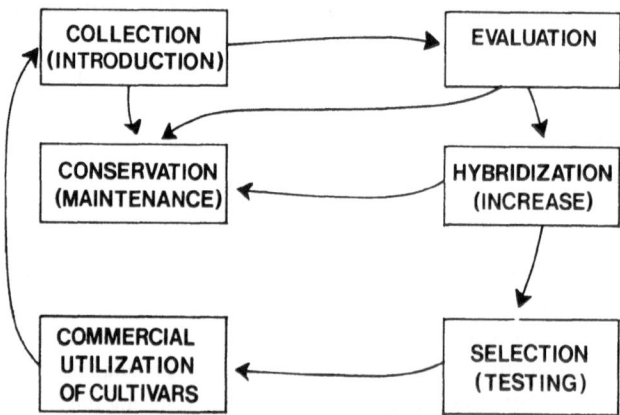

Fig. 1   Features of the utlization and maintenance of a germplasm collection.

Table 1.

Seed yield in a trial conducted across a wide range of environments[++]

| | | A) Frequency of cultivars achieving a specific rank | | | | | | | | | | Mean Yield t /ha |
|---|---|---|---|---|---|---|---|---|---|---|---|---|
| Rank: | | 1 | 2 | 3 | 4 | 5 | 6 | 7 | 8 | 9 | 10 | |
| Cultivars | | | | | | | | | | | | |
| 1 | Dacre | - | 1 | - | 2 | 4 | 2 | 4 | 2 | 1 | - | 3.75 |
| 2 | M. Bead | - | 1 | - | - | 3 | - | 1 | 5 | 4 | 2 | 3.53 |
| 3. | Blaze | 1 | 1 | 3 | 3 | 1 | 1 | 1 | 1 | 2 | 2 | 3.96 |
| 4 | Herra | 1 | 2 | - | 3 | 3 | 1 | 1 | 1 | 2 | 2 | 3.85 |
| 5 | Kristall | 2 | 2 | 1 | 2 | 1 | 6 | 2 | - | - | - | 4.15 |
| 6 | Russian | - | - | - | 1 | - | 1 | 2 | - | 1 | 11 | 2.78 |
| 7 | Felix | 3 | 4 | 2 | 1 | - | - | - | 1 | 5 | - | 4.06 |
| 8 | Minica | 6 | 3 | 4 | - | - | 2 | 1 | - | - | - | 4.64 |
| 9 | Wierboon | 3 | 2 | 6 | 1 | - | 1 | - | 1 | 1 | 1 | 4.23 |
| 10 | Rowena | - | - | - | 3 | 1 | 2 | 5 | 4 | 1 | - | 3.58 |
| LSD for comparison of varietal means $P \leqslant 5\%$ | | | | | | | | | | | | 0.43 |
| $P \leqslant 1\%$ | | | | | | | | | | | | 0.57 |

B) Experimental Mean Yields (t/ha)[+]

| | | |
|---|---|---|
| De Est | (NL) | 6.66 |
| Gottingen | (D) | 4.51 |
| Ihingerhof | (D) | 3.28 |
| Hohenheim | (D) | 3.01 |
| Cambridge | (GB) | 3.30 |
| Aberystwyth | (GB) | 3.50 |
| Dundee | (GB) | 4.74 |
| Dijon | (F) | 4.79 |
| Fuchsenbigl | (A) | 2.52 |

[+]Mean of two trials at each location.
(De Est and Gottingen only one trial)

[++] Unpublished data, obtained from a joint field bean test by G. Dantuma, D.A. Bond, M. Frauen and E.V. Kittlitz

The implications of some aspects of this scheme for breeding strategies will be discussed, but first a general view of the concept will be given. In Fig. 1, a kind of recycling system is indicated, which contains a flow-back of germplasm to the basic gene pool at each step of evaluation, testing and utilization. Genetically speaking this represents a 'world wide' system of recurrent selection for wider adaptability, the implications of which will be discussed in more detail later. The success of such a system clearly depends on the effectiveness of selection for wider adaptability during the various stages.

One of the most important points of this system is the flow-back of germplasm to the basic gene pool. However there may be contradictory requirements of a long-term gene bank and an active breeding program. The proposed strategy for achieving wider adaptability must take into account the final utilization of the selected materials. This in turn is subject to economic and/or political considerations the demands of which could result in an excessive narrowing of the genetic base which in turn would restrict the advances possible in breeding for wider adaptability.

*Breeding Strategy*

Fig. 2 gives an outline of a breeding program. The plan shows two interacting units: regional and central activities, where regional refers to activities in institutions or countries which participate in the overall program. For central activities ICARDA may be an ideal institute. The proposed model is shown through to a second cycle of selection. However, it is very probable that the whole system will only succeed, if at all, with more than two cycles. The figure shows that there is a permanent flow of material (and of course of information) from the regional institutions to the central "headquarters" and back to the regional units. The flow-back of locally selected germplasm to the central gene pool is a vital component of the system. After the first joint tests, the progenies are introduced to a central pool of material with improved adaptability, thus building up a population for reselection in the second cycle. As simple as this system appears, a number of important technical problems are to be expected in its execution, although these will not elaborated on here.

Three main aspects will be discussed further:
- The synthesis of the original gene pool for a selection program for wider adaptability.
- The problem of evaluation and adequate testing
- The maintenance of the original gene pool.

*Synthesis of a gene pool*

As mentioned before, the gene reservoir in a program for wider adaptability should include germplasm originating from the whole range of environments considered within the program. Its synthesis has to start with crosses between parents of only little common heritage. The expected consequence is a decrease in the average yield potential of the progenies. On the other hand, in a crop such as faba beans, heterosis may play an important role in increasing yielding capacity as well as yield stability. Therefore one cannot neglect heterogenity between basic populations for characters responsible for heterosis. The problem could be solved by a gradual improvement of the basic population at each regional unit as shown in Fig. 2.

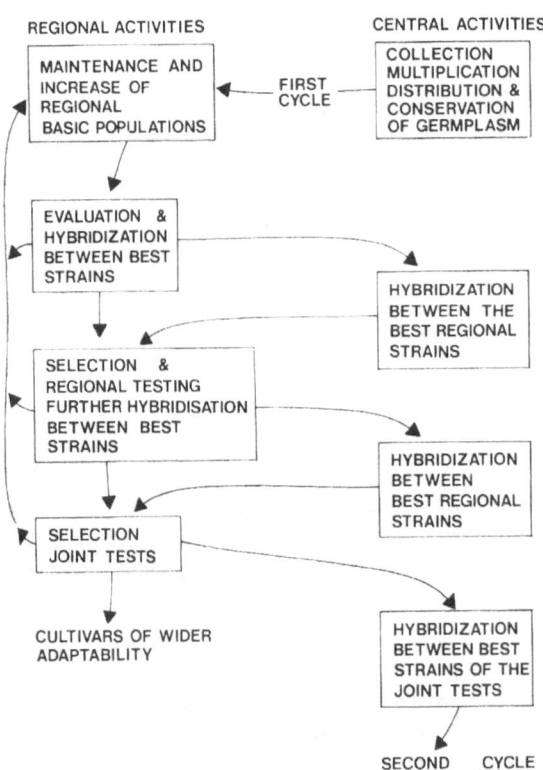

Fig. 2 A possible breeding strategy for the improvement of adaptability.

The essential feature here is an exchange of genes within the whole range of habitats in cycles of selection and hybridization at the regional units. From each step of selection, after regional trials, interesting material is sent to the central unit for hybridization. The hybrids are then returned to the regional units for further selection. This may cause an accumulation of favourable new gene combinations for wider adaptability.

At the same time, genetic diversity between the regional populations could increase within materials of agronomic worth along the lines proposal by Schnell (5) for the improvement of genetic diversity. At the final stage of the procedure, crosses can be made between parents of similar agronomic value, but which may differ in genetic factors responsible for heterosis in yield or other desirable characters.

It is apparent that the program proposed would be very time-consuming. Some steps within each cycle might be replicated two or more times and the regional adaptation of material will take many generations of careful screening under the pressure of natural selection.

*Evaluation and testing problems*

A major problem, at least for the first cycle of the procedure, is that we do not know anything about secondary traits for wider adaptability. So the chance is only small that the first evaluation will provide good material for further selection. In general, characters of resistance or tolerance against diseases, pests or climatic stress conditions, are regarded as factors of wider adaptability. But the value of resistance, for instance, against parasitism by *Orobanche crenata* may be very different for a breeder in central Europe compared to a Syrian faba bean grower. Equally, tolerance against drought in semi-arid climates may be only an escape from summer drought by early ripening. Under the conditions of central Europe, a real tolerance against unforeseen drought during the growing season is wanted.

So, what is the meaning of general climatic tolerance ? Furthermore, when adaptability is measured in terms of yield, a high average yield potential, as confirmed by regional yield trials, may not necessarily be an indication of its wider adaptability. Yield stability in regional trials is likely to be a better measure of wider adaptability. If so, one would have to make crosses in advanced generations not between the highest yielding but the most stable progenies. (Stability here is characterized by little variation in yield in the regional trials).

*The maintenance of the original gene pool*

Techniques for collecting the full spectrum of germplasm in a species are well documented, e.g. Chang (2). There is no reason for not applying them to *Vicia faba*. However, in maintaining a germplasm collection, some special problems have to be considered.

i. Siginficant changes of gene frequencies within the populations may occur due to natural selection.

ii. When material from very different geographical origins is collected, natural selection between populations may cause serious gene erosion in only one season of growing. In progenies of *Vicia faba*, with its high rate of selfing, such influences are much more important than in fully allogamous species where gene erosion is retarded by natural crossing.

iii. When the material is grown for multiplication in bee-proofed cages, one has to expect losses in plant material by autosterility.

Technical aspects which contribute to the maintenance of the original gene pool have been discussed in the chapter by Witcombe. However, it is clear that only very careful observation during the growing season, adequate conservation and an effective control of multiplication can avoid, or at least minimize, losses of germplasm.

*Discussion*

The chance of creating cultivars with wide adaptability in the sense used here is thought to be only very small. This may be true, even if selection for simple or complex characters could be practised effectively. The reason for this is that too many regionally specific factors are important in determining the actual yield. However there are two aspects worthy of further discussion and joint research:

i. The breeding strategy, as described above, may provide a higher level of general adaptability as a basis for further selection on regionally important traits. The effect could be comparable to the improvement of general combining ability by means of recurrent selection e.g. in maize, followed then by selection of favourable specific combinations.

ii. The maintenance and increase of genetic diversity at regional units, and its assimilation in current breeding material, is a very important objective, and it seems possible to achieve this through a strategy of breeding for wider adaptability.

REFERENCES

1. Anon, 1979. CIMMYT Review 1979 Centro International de Mejoramiento de Maiz y Trigo.
2. Chang, T.T. 1979: in Sneep, 1., and A.I.T. Hendriksen (eds.): Plant breeding perspectives. Centre for Agricultural Publishing and Documentation, Wageningen.
3. Frankel, O.H., E. Bennett (eds.). 1976: Genetic resources in plants-their exploration and conservation. Blackwell Scientific Publications, Oxford and Edinburgh.
4. Frankel, O.H., I.C. Hawkes (eds.). 1975: Crop genetic resources for today and tomorrow. Cambridge University Press.
5. Schnell, F.W. 1980: Aspekte der genetischen Diversitat im Problemkreis der Pflanzenzuchtung. In: Robbelen, G. (ed.) : Gottinger Pflanzenzuchter-Seminar 4, 5-17.
6. Simmonds, N.W. 1977: Principles of crop improvement. Longman, London and New York.

# 8. MUTATION BREEDING IN FABA BEANS

MAZHAR M. F. ABDALLA

*Faculty of Agriculture, Cairo University, Giza, Egypt*

One reason why problems associated with the production of faba beans have remained unsolved e.g. susceptibility to pests, yield instability, is because the crop suffers from a relatively narrow useful variability. Consequently no drastic improvement in the production of faba beans has occurred in comparison to other crops in the same area, and acreages have diminished in some countries.

Because of its few chromosomes (n=6) and their relative size, faba bean has been subjected to a number of different genetical and cytological studies. Physical and chemical mutagens have been applied to investigate their effects on growth and cytology. Mostly root tips, but sometimes microsporocytes, have been analysed. The voluminous work in this field has been reported by several authors (9) (24) (10) (25) (21) (13) (15) (23) (22). This chapter focuses on mutation studies where the induced variability may be of immediate use to faba bean breeders.

The most common mutagenic agents which have been used to induce variability in faba beans are X-rays, neutrons, ethyl methane sulphonate (EMS), methyl methane sulphonate (MMS), ethylene imine (EI), and diethyl sulphonate (DES). A combination of certain mutagens has also been applied.

*Induced Variability in Vegetative and Generative Characters.*

Using the cultivar Svalof Primus, Sjodin (30) found that the simple-leafed mutant referred to as the unifoliate mutant which converted the compound leaf into a simple one throughout its life (obligate) or only during the early stages of development (transnormal), was recessive (un). The mutants analysed showed a simple inheritance by a single recessive gene. All mutants except one (un-bc$^{-1}$) proved to be controlled by multiple alleles at the same locus. Gottschalk (14) used a unifoliate recessive mutant to study natural cross pollination. The early flowering mutant (ea) which Sjodin analysed was inherited as a recessive or as an incompletely dominant gene.

*G. Hawtin & C. Webb (Eds.): Faba Bean Improvement*
*©1982 ICARDA. ISBN -13:978-94-009-7501-9*

The terminal inflorescence (ti) mutant, sometimes known as "topless", was found to be controlled by one recessive gene. Flower colour was shown to be affected by genes at four loci, viz sp-a, sp-b, dp-a and dp-b. The first two are independent loci for colour development. The dp-a recessive gave a range from dark brown to violet; one of the alleles produced uniform yellow flowers. The dp-b locus in the recessive state gave yellow spotted wings. The "white-flower" characteristic is reported to be controlled by a recessive gene(s). Other monfactorially inherited mutations which Sjodin reported, included modified pollen grains, late ripening, hooked seed and chlorophyll mutations.

With the cultivar Rebaya 40 and a major type (Cyprian Roumi), Hassanien (17) found that lower gamma ray doses favoured early flowering taller plants and heavier seeds, whereas the higher ones favoured late flowering, stunting growth, fewer seeds/pod and a lower seed index. Mutants which were deficient in chlorophyll, had colourless hila and coloured seeds were also reported.

Hussein and Abdalla (18) found that Giza 2 was relatively more radio-resistant while Rebaya 40 was relatively EMS-sensitive. Combined treatments were most effective, followed by EMS then gamma rays. This was observed by evaluating the frequencies of chlorophyll mutations in the $M_2$. Gamma rays produced no chlorophyll mutations. EMS and combined treatments induced viridis, xantha and albina mutations. Higher EMS concentrations resulted in higher frequencies of chlorophyll mutations, but combined treatments doubled the mutation frequencies compared with EMS. In addition, $M_1$ fertility was affected by different mutagenic treatments.

Treatment of Giza 1 and Giza 2 by El-Hosary (11) with gamma rays, EA and/or DES produced small and abnormal leaves, dwarf and fasciated plants, spotless flowers and colourless hila, later ripening and chocolate spot "resistant" mutations, as well as other changes. Generally, DES was the most effective, followed by gamma rays. The frequency of mutations increased with increasing radiation from 3 to 9kR and from 0.015 to 0.03% DES, but 0.5% EA gave higher mutations than 0.05-0.25% concentrations which did not differ from each other.

After Hassan (16) had treated Giza 2 and Rebaya 40 with gamma rays, the seeds and terminal buds were treated with 0.1-0.7% colchicine. He reported variability in different characters and the occurrence of sterile mutants, colourless hila, dark seeds, long and short stems, albino mutants, and variants with different yielding ability.

Abdalla and Hussein (4) found an interaction of gamma rays and EMS in their effects compared to single mutagenic agents when used on Giza 2 and Rebaya 40 cultivars. The coefficients of variability increased in the populations derived from the mutation treatments. The numbers of tillers increased in both cultivars, but plant height was differentially affected.

Closed flowers characterized five plants out of 15 $F_2$s which Poulsen (28) discovered from a cross between the cultivar Kleine Thuringer and the terminal inflorescence mutant (ti) of Sjodin. The closed flower (cf) is assumed to prevent bee visitation not only because it remains almost closed, but also because of the while flower colour, lack of scent and possibly lack of nectar. The 6% outcrossing in closed flower populations was probably due to distorted flowers on some plants.

The closed flower was found to be inherited as a recessive character controlled by one gene. Unfortunately the flowers containing the mutant are poor yielders. They produce fewer pods and fewer seeds/pod. They require tripping and stigma scarification. If the level of autofertility could be increased in plants having the cf gene, it should be possible to obtain "pure lines" of faba bean which would continue to breed true without significant cross pollination.

Seeds of the closed flower, grown in an insect-free cage at Giza, showed a more prolonged growth and flowering than the other genotypes available. The fact that no single pod was observed on any of more than 100 plants of the closed-flower type, was unique as all other European lines in the same cage produced at least some pods.

Poulsen and Martin (29) referred to the discovery of tetraploid plants whose parent was discovered by Sjodin after X-irradiation, as pollen mutant (po-1). Tetraploidy was suspected because of low fertility and failure of natural cross pollination in the field. The tetraploid plants were reported to be morphologically similar to diploids. While such tetraploids cannot be used commercially before their fertility and yielding ability have been raised to acceptable levels, they may be used for cytogenetic studies.

Low doses of gamma rays applied by El-Kady (12) to Giza 2 and Rebaya 40, resulted in higher values and coefficients of variation for different characters in the $M_2$, but higher doses tended to give an opposite effect.

Nagl (26) isolated early flowering and mutants with different seed weights from plants grown from seed of the minor cultivar Kornberger Kleinkornige which he treated with gamma rays. Variability increased in some traits in the mutant 28/1 which flowers early and has large seeds and which was treated with gamma rays. Comparison between the original and $M_2$ mutant population showed the mutant to have shorter stems and to flower early. Mutants with short stems, higher tillering, determinate growth habit and up to 2-week earlier flowering were isolated in addition to mutants with different leaf sizes and shapes.

Gamma ray treatments increased the variability in different characters of Giza 2, Rebaya 40, Mutant 1011, White-flower, Purple-coat-white-hilum, and Purple-coat-black-hilum (31). Early flowering, taller plants and an increased number of fertile branches were observed in $M_2$ and/or $M_3$ generations. Rhizobium inoculation affected some of the characters positively.

*Induced Variability in Seed and Yield Characteristics.*

Different studies have shown that recessive as well as dominant mutations may occur in seed coat colour. Coloured seed coats were mostly complementary or intermediate over buff, but sometimes the reverse was true particularly when green and red colour seeded plants were crossed to buff ones (30).

Abdalla and Hussein (4) found that gamma rays, EMS and a combination of both mutagens increased $M_2$ seed index in all treatments. The combined treatments also resulted in more pods/plant, more seeds/plant and higher seed yield/plant than the control.

Large-seeded spontaneous mutations have been found in specific stocks (Abdalla, unpublished). Six major-like mutants were discovered in our materials in 1976; the possibility of these having arisen through cross-pollination has been ruled out.

Table 1.

Seed index (g/100 seeds) of spontaneous major-like mutants and their parental stocks (1976).

| Parent stocks: | 27 M | 12 M(102) | 26 M(57) | Y.F.M. | 28 M | 30 M(104) |
|---|---|---|---|---|---|---|
| Parent seed index | 33 | 55 | 49 | 49 | 58 | 65 |
| Mutant seed index | 95 | 115 | 102 | 117 | 125 | 106 |

One of these mutants, denoted 57, was sown in Germany with its parents, in 1977. (Table 2).

Table 2.

Characteristics of Mutant 57 and its parent.

| | mean stem height (cm) | mean no. of tillers | mean lst flowering node | mean no. of flowering nodes | mean no. of fruiting nodes | mean Pods/ plant | mean Seeds/ plant | mean Seed Yield (g) | mean Seed index (g/100 seed) |
|---|---|---|---|---|---|---|---|---|---|
| Parent | 57.0 | 2.9 | 6.6 | 6.4 | 4.4 | 12.3 | 33.6 | 12.4 | 36.4 |
| Mutant | 71.0 | 2.8 | 6.0 | 9.5 | 4.1 | 13.4 | 38.0 | 29.4 | 77.0 |

The progenies of three of the mutants mentioned in Table 1 were sown later in Egypt, (Table 3).

Table 3.

Seed yield and seed index of three spontaneous mutations at Qualyub (north of Cairo, Egypt).

| | Season | Mut. 57 | Mut. 102 | Mut. 104 | Cultivar Giza 2 |
|---|---|---|---|---|---|
| Yield/plant (g) | 1977-78 | 33.0 | 52.5 | 58.0 | 13.0 |
| Yield/plant (g) | 1978-79 | 20.0 | 27.5 | 26.0 | 22.0 |
| Seed index (g/100 seeds) | 1978-79 | 82.0 | 85.0 | 94.0 | 69.0 |

The $M_2$ population of the cultivar Kornberger Kleinkornige had fewer seeds/pod and/plant, and lower yield than the parent stock (26). Mutants with terminal and axillary pods and reduced pod shattering were isolated. High variability was observed in seed size and weight. The interesting discovery of a negative correlation between seed weight and seed coat percentage means that mutations with heavier seeds may be nutritionally more valuable, even when they have similar yielding ability to the parent.

Mutants selected in the $M_2$ generation from Giza 2 by Hussein and Abdalla (19) showed an increase of 6-59% over the yield of the parent. Combined mutagen treatments were particularly efficient in inducing useful change. Yields of $M_2$ mutants of Rebaya 40 were mostly lower than those of their parent.

In $M_4$, Giza 2 mutants gave deviations in seed yield that varied from -31% to +54% of the control whereas in $M_5$, yielding ability was 0-+40% above the control. With only one exception, mutants from Rebaya gave higher yields than the parents in both $M_4$ and $M_5$. The relative weight of seed coat to whole seed was about 12% in both varieties but the percentage varied more in Giza 2 than in Rebaya 40 (20).

Seed yield of mutants from Giza 1 and Giza 2, selected by Abo-Hegazi (6), was -13% to +12% above the parent. Significant increase or decrease in seed yield occurred in different high protein mutants. Tolba (31) obtained increased pod set/plant, seed set/pod and seed yield/plant in $M_2$ and/or $M_3$ generations in all or some of the stocks used.

*Induced Variability in Protein Content*

Sjodin (30) selected within a population (Sv 0720) that was developed from crossing between two mutants for high protein (more than 33%) and low protein (less than 29%). After one year, the mean values were 33.7% for the former; 31.1% for the latter. Mutants analysed from the *equina* variety Giza 2, used by Abo-Hegazi *et al.* (7) were 22-38 % protein in the $M_3$ generation against 27% in the control.

According to Hassan (16), colourless hilum and dark seed mutants from Giza 2 had higher contents of water and salt soluble fractions; the opposite was true in the taller-stemmed mutant.

Hussein and Abdalla (19) analysed the mutants selected for high yield in $M_2$ for protein content in $M_3$ and found that mutants isolated from Giza 2, using single mutagenic treatments, showed deviation in protein content of about -0.5 to +1.05 units from the parent cultivar. However deviations from the combined (EMS + gamma ray) treatment ranged from -3.85 to -1.57% in the flour of the whole seeds, and -3.15 to 0.70% in decoated seeds. Mutants from Rebaya 40 had 4 to 14% higher protein than the parent.

In Rebaya 40, too, there was some increase in the proportion of total essential acids at the expense of non-essential acids. This occurred in the flour of whole seed but the situation was slightly different in decoated seeds; the clear changes being in lysine, histidine, arginine and aspartic and glutamic acids.

Hussein and Abdalla (20) found that the seeds of the mutants from Giza 2 deviated in $M_4$ in crude protein from -1.4 to +4.8% with the seed coat and -1.6 to +7.7% in decoated seeds. Corresponding deviations in $M_5$ were -1.1 to +4.9% and -0.2 to +6.5%. Mutants from Rebaya 40 gave ranges in the whole seed flour from -1.7 to +1.28% in the $M_4$ and -1.1 to +3.5% in $M_5$.

Five mutants from Giza 2 and two from Rebaya 40 gave superior yield of protein/plant than their parents in $M_4$ and $M_5$. Other mutants showed inconsistent results in the two generations. Analyses of three mutants from Giza 2 and two from Rebaya 40 showed a general increase in amino acid content associated with crude protein, but higher increases occurred in arginine, glutamic acid, aspartic acid and leucine relative to the total increase in all amino acids.

High protein mutants, selected by Nagl (26) had a 33% increase in arginine and 10% less in lysine. Similar differences occurred between the low protein mutants. The protein content of mutants which Abo-Hegazi (6) selected from gramma ray-treated Giza 1 and Giza 2 varied from about +9 to +35 % above the control varieties. The heritability estimates for protein content ranged from 32 to 70 % in the different mutants.

In work by Pandey *et al.* (27) following mutagen treatment in the cultivar Diana, three cycles of selection for high and low methionine resulted in a higher methionine (1.66 mg/g against 1.37 mg/g dry matter), higher protein content (36.3 against 33.2% ) and higher methionine in the protein (4.58 against 4.15 mg/g protein). Unfortunately the yield/plant was 16.76 g in higher methionine selections against 22.35 g in low methionine ones. Plant height was also reduced significantly.

*Discussion*

*Yield and protein content:* Nagl  (26) reported that protein content was positively correlated with plant yield, but other workers (5) (20) showed absence of correlation between protein content and seed yield.

Abo-Hegazi (6) found that some, but not all, of the high protein mutants produced equal seed yield to the parents. Some mutants were better yielders than others. Abo-Hegazi *et al.* (7) previously showed that some higher yielding mutants had higher protein percentages than the control.

Significant, but very small, negative correlation coefficients between protein and seed yield/plant were found by Pandey *et al.* (27). This occurred between accessions in a germplasm collection, but generally not in mutants selected for high methionine. The authors reported that selection on the basis of methionine would be superior to selection on the basis of protein content.

As there was no strong and consistent association between protein content and seed yield, it seems logical to select for high yielding stocks without fear of affecting protein yield. However one would hesitate at selecting only on the basis of protein content, unless for specific purposes.

*Early ripening mutants:* Relatively early ripening varieties, useful under certain circumstances, might be isolated through mutation induction or through appropriate crossing. As cotton sowing is finished by the end of March in Egypt, an early faba bean cultivar could be used in the cotton rotation. Early varieties could also be used in Europe in places where growing conditions may not favour late maturing ones, and in places where two crops can be grown in a year.

*The topless mutants*: It is possible that the topless character could be transferred to locally adapted genotypes, and might prove useful under certain local conditions. But experience has shown that one could not recommend in Egypt cultivars having the topless characteristic, but which have been bred for European conditions.

*Future Areas of Mutation Research*

Areas in which mutation research may have a role, include:

*Disease and pest resistance:* In view of the limited variation for disease and pest resistance in faba beans, cooperative research to develop resistant cultivars should be encouraged.

*Plant models:* Factors which need to be considered in the development of plant models include environmental conditions, the balance between vegetative and generative organs, the number of tillers, the type and size of seeds and leaves, and the number and distribution of the inflorescences and flowers.

*Breeding system:* Mutations may be developed that shift the pollinating system towards either self-fertilization or cross-fertilization. This may simplify the breeders' task. Fertility and incompatability phenomena need to be studied.

*Enriching useful variability:* Mutation studies are needed to enrich the useful variability in faba beans.

*Tolerance to environmental conditions:* Although the sensitivity of faba beans to environmental conditions may be attributed largely to factors such as fertility and susceptibility to pests, it is possible that the yield stability can be bred for as an independent character. Mutation breeding might prove useful in developing cultivars with a more stable performance.

## REFERENCES

1.  Abdalla, M.M.F. 1975. Use of mutations in some leguminous crops in crop breeding research. Egypt. J. Genet, Cytol, 4:17-27, Suppl. in Arabic.
2.  Abdalla, M.M.F. 1979a. The origin and evolution in *Vicia faba* Proc. First Mediterrenean Conference of Genetion 28-30 March, Cairo, Egypt : 713-746.
3.  Abdalla, M.M.F, 1979b. A bibliography of field beans (*Vicia faba* L.) research in Egypt, ICARDA. (unpublished).
4.  Abdalla, M.M.F., Hussein, H.A.S. 1977. Effects of single and combined treatments of gamma-rays and EMS on $M_2$ - quantitative variation in *Vicia faba* L. Z.Pflanzenzuchtg. 78:57-64.
5.  Abdalla, M.M.F., Morad, M.M., Roushdi, M. 1976. Some quality characteristics of selections of *Vicia faba* L. and their bearing upon field bean breeding. Z. Pflanzenzuchtg. 77: 72-79.
6.  Abo-Hegazi, A.M.T. 1979. High protein lines in field beans *Vicia faba* from a breeding programme using gamma rays 1. Seed yield and heritability of seed protein. Seed Protein Improvement in Cereals and Grain Legumes IAEA, Vienna, Vol.II. 33-36.
7.  Abo-Hegazi, A.M.T., Shoeb, Z.E., Salama, F.A., Hakam, M. 1973. Breeding for improved protein in pulses using radiation induced mutations. Nuclear Techniques for Seed Protein Improvement, IAEA, Vienna: 265-268.
8.  Bond, D.A. 1979. Breeding work in *Vicia faba* in the U.K. FABIS 1: 5-6
9.  Brewen, J.G. 1964. Studies on the frequencies of chromatid aberrations induced by X rays at different times of the cell cycle of *Vicia faba*. Genetics 50: 101-107.
10. Deufel, J. 1951. Untersuchungen uber den Einfluss von Chemikalien und Rontgenstrahlen auf die Mitose von *Vicia faba*. Chromosoma 4: 239-272.
11. El-Hosary, A.A. 1977. Effect of some chemical and physical mutagens on *Vicia faba*, M.Sc. Thesis, Fac. Agric., Ain-Shams University, Egypt.

12. El-Kady, M.A. 1978. Induced variability of yield and yield components in two Egyptian broad bean cultivars by gamma radiation. Ain Shams University, Fac. Agric. Res. Bull. 820 : 1-12.

13. Galal, H.E., Abd-alla, S.A. 1976. Chromosomal aberration and mitotic inhibition induced by sodium fluoride and diethyl amine in root-tip cells of *Allium cepa, Allium sativum* and *Vicia faba*. Egypt, J. Genet. Cytol. 5: 262-280.

14. Gottschalk, W. 1960. Untersuchungen uber die Befruchtungsverhalfnisse von *Vicia faba* mit Hilfe en einer erkennbaren Mutante. Zuchter: 30:22-27.

15. Hanna, E.M. 1978. Cytological and morphological studies on the effect of $Co^{60}$ gamma radiation on *Vicia faba*. Ph.D. Thesis, Fac. Sci., Cairo University, Egypt.

16. Hassan, H.F. 1977. Mutation studies on *Vicia faba*. M.Sc. Thesis, Fac. Agric. Al-Azhar University, Egypt.

17. Hassanien, A.H. 1973. Effect of radiation on broad beans, *Vicia faba* L. M.Sc. Thesis, Fac. Agric. , Ain Shams University, Egypt.

18. Hussein, H.A.S., Abdalla, M.M.F. 1974. Effects of single and combined treatments of gamma-rays and EMS on the $M_1$- fertility and $M_2$ - chlorophyll mutations in *Vicia faba* L. Egypt, J. Genet. Cytol. 3:246-258.

19. Hussein, H.A.S., Abdalla, M.M.F. 1978. Protein and yield traits of field bean mutants induced with gamma rays, EMS and their combination. Seed Protein Improvement by Nuclear Techniques, IAEA, Vienna, 253-264.

20. Hussein, H.A.S., Abdalla, M.M.F. 1979. Gamma-ray and EMS induced mutations in *Vicia faba* L. Evaluation of yield and protein traits of mutants in the $M_4$ and $M_5$ generations. Seed protein Improvement in Cereals and Grain Legumes. IAEA, Vienna Vol. II, 23-31.

21. Janakiraman, R. Harney, P.M. 1976. Effects of ozone on meiotic chromosomes of *Vicia faba*. Can. J. Genet. Cytol. 18:727-730.

22. Kihlman, B.A., Natarajan, A.T., Anderson, H.C. 1978. Use of the 5-bromodeoxyuridine-labelling technique for exploring mechanisms involved in the formation of chromosomal aberrations. Mut. Res. 52: 181-198.

23. Kihlman, B.A., Sturelid, S. 1978. Effects of caffeine on the frequencies of chromosomal aberrations and sister chromotid exchanges induced by chemical mutagens in root tips of *Vicia faba*. Hereditas, 88: 35-41.

24. McLeish, J. 1953. The action of maleic hydrazide in *Vicia*. Heredity 6 Suppl. 125-147.

25. Miller, M.W., Voorhees, S.M., Carstensen, E.L., Eames, F.A. 1974. An histological study of the effect of ultrasound on growth of *Vicia faba* roots. Rad. Bot. 14.201-205.

26. Nagl, K. 1978. Breeding value of radio-induced mutants of *Vicia faba* var. minor. Seed Protein Improvement by Nuclear Techniques, IAEA, Vienna, 243-252.

27. Pandey, M.P., Frauen, M., Paul, C. 1979. Selection for methionine by GLC after CNBr treatment in a germplasm collection and mutagen-treated population of *Vicia faba* L. Seed Protein Improvement in Cereals and Grain Legumes, IAEA, Vienna, Vol. II. 37-46.

28. Poulsen, M.H. 1977. Obligate autogamy in *Vicia faba* L. J. agric. Sci. Camb. 88:253-256.

29. Poulsen, M.H., Martin, A. 1977. A reproductive tetraploid *Vicia faba* L. Hereditas 87: 123-126.

30. Sjodin, J. 1971. Induced morphological variation in *Vicia faba* L. Hereditas 67:155-180.

31. Tolba, A.M. 1980. Studies on characters of gamma irradiated field bean. M.Sc. Thesis, Fac. Agric. Al-Azhar University, Egypt.

# 9. INTERSPECIFIC HYBRIDIZATION IN VICIA

JOSE I. CUBERO


*Departmento de Genetica, Escuela Tecnica Superior de Ingenieros Agronomos, Apartado 3048, Cordoba, Spain*

Interspecific crosses were used to obtain new forms of horticultural and ornamental species, long before the existence of genetics as a science, or the modern concept of 'species' had been developed. They are widely used in plant breeding as well as in evolutionary studies. Basic and applied aims are, in fact, complementary aspects of the same study. Once the interspecific hybrid has been obtained, the success of each aim depends on the materials used.

The primary and most common objective in plant breeding is to enlarge the genetic variability of a cultivated species. Until now, disease and insect resistance have been the most commonly sought characters in wild relatives of crops, but interest in transferring agronomic characteristics is increasing as classical breeding is exhausting crop genetic resources. The development of new (synthetic) species is also possible by means of duplicating the chromosome complement of the interspecific hybrid, provided that the parental species are phylogene ically distant enough to show a lack of chromosome pairing in the hybrid, so ensuring a good pairing in the alloploid. This process has been undergone naturally in many genera and families e.g. *Triticum, Gossypium* and *Nicotiana.* Triticale is perhaps the best known example of a new Man-made crop.

If we accept the biological concept of species i.e. the limits of a species being defined by its reproductive limits, then interspecific crossing is an impossibility. A morphological concept of species is only acceptable when reproductive limits have not been established. The situation is primarily a question of semantics, coupled with the difficulty of defining reproductive limits. Nature is not clear-cut: the crossability of one group of plants with other groups can range from 100 % to zero. Only groups which cross readily can be considered within the same species; but again the term "readily" is open to different interpretations. Only when a strong barrier to crossing exists, which is not due to factors such as alleles for incompatibility or differences in reproductive periods, can we confidently speak of different species. In fact, a definition of species which is valid for all organisms is not possible. Information on reproductive isolation, and morphological, and ecological characteristics of populations can be used in the following classification, as proposed by Mayr (18), provided problems of reproductive limits are resolved.

*G. Hawtin & C. Webb (Eds.): Faba Bean Improvement*
©1982 ICARDA. ISBN -13:978-94-009-7501-9

| Individuals | Reproductively isolated | |
| --- | --- | --- |
| | No | Yes |
| Identical in morphology physiology and/or ecology: | | |
| sympatric | Same population | Sibling species |
| allopatric | Same subspecies | Sibling species |
| Different in morphology etc. | | |
| sympatric | Same polymorphic population | Separate species |
| allopartic | Separate subspecies | Separate species |

## Cultivated species

A rather different and eclectic approach to the problem, which is of greatest interest for plant breeders, is that of Harlan and De Wet (15). For every cultivated species, these authors proposed three informal categories (gene pools) based on the ease of intercrossing with other populations and/or species.

The primary gene pool corresponds to the traditional concept of biological species; cultivated and wild races are included as subspecies A and subspecies B respectively. There is generally one A subspecies; if there are more, the material has probably not been extensively studied. In some cases, two, or even more, cultivated subspecies can be found. Several authentic B subspecies can often be defined, their relationship with the cultivated one(s) ranging from very close to rather distant, but always within the specific limits: easy intercrossing, hybrids generally fertile, good chromosome pairing, gene segregation normal (i.e. Mendelian) or nearly so, and gene transfer generally simple. Thus, any cross between forms belonging to this first gene pool must be considered intraspecific.

The secondary gene pool includes the other biological species that will cross with the species under consideration. As Harlan and De Wet said, 'gene transfer is possible, but one must struggle with those barriers that separate biological species'. In spite of high sterility levels in the hybrids, poor (if any) chromosome pairing, poor development of hybrids and other problems, both interspecific crosses and gene transfer are possible without any special techniques e.g. embryo culture. The assigning of a specimen to the primary or secondary gene pool can be difficult in some groups. A detailed study of the possibility of intercrossing, together with the chromosome behaviour of hybrids and following generations, and the segregation of known characteristics, for which biochemical markers are very appropriate, will help to resolve the question.

The tertiary gene pool includes species that can cross with the one under study, but only with the assistance of special techniques such as embryo culture, tetraploids, grafting, tissue culture and the use of bridging species.

As an example of these ideas, Fig. 1 shows the gene pools of wheat. The three biological species share the same secondary gene pool (GP-2), which includes *Aegilops, Secale, Hayna-ldia* and some species of *Agropyron.* Logically, they also share the same tertiary gene pool (GP-3) which includes several species of *Agropyron* and *Elymus,* as well as some *Hordeum* species (at least *H. chilense*).

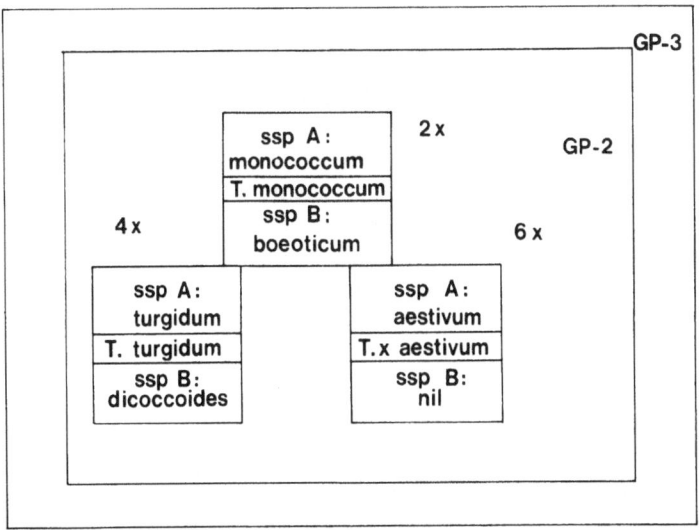

Fig. 1.    The gene pools of wheat (redrawn from Harlan and De Wet, 1971).

Fig. 2 shows the gene pool of *V. faba:* from the point of view of a plant breeder, it is really very sad.

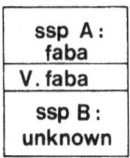

Fig. 2.    The gene pool of faba beans.

*Pre-and Post-zygotic mechanisms of isolation*

Underlying the above problems, is the question of the mechanisms which cause reproductive isolation, the study of which is basic to the creation of authentic interspecific hybrids. There are basically two groups of mechanisms:pre-and post-zygotic. Pre-zygotic mechanisms operate through different ecological requirements (habitat, cycle), reproductive behaviour and allelic incompatibility. A study of such mechanisms provides information to overcome them, thus allowing hybrids to be obtained. Post-zygotic mechanisms such as zygote degeneration, hybrid sterility, and abnormal $F_2$'s, pose a stronger challenge to the plant breeder.

It is worth mentioning that not only intraspecific crosses but also interspecific ones, can depend on the genotype of the lines used as parents: the experience in the *Triticineae* demonstrates this. The direction of the cross is also an important factor, for example hybrids between *Hordeum chilense* and wheat can be obtained using the former as the female parent, but the reciprocal cross has not yet been produced.

*Genetic engineering*

Since 1970, new techniques for handling genes, known as genetic engineering, are increasing in importance and will probably be used in plant breeding within a few years. The description of these techniques is far beyond the scope of this chapter. Basically, a format of DNA of a certain species A, carrying the required information, is 'transplanted' to another species B which lacks such information. The aims of this transfer can be many, for example to obtain a large number of copies of the A gene, to produce the protein coded in the A gene in the B organism, to introduce into B the desired genes present in A, or to substitute some unfavourable genes in B with favourable A genes.

A particular case, that can be defined as 'natural genetic engineering', may help plant breeding in the future. The transfer of plasmids between *Agrobacterium tumefaciens* and its hosts has been demonstrated; thus providing a natural bridge for introducing alien characteristics into any of the many species it parasitizes.

Because of the availability of genetic engineering techniques, a quaternary gene pool of a given species could be defined: it is constituted by the rest of living beings.

*Recombination or alloploidy*

In this section some of the different possibilities resulting from interspecific crosses are discussed. We will only be concerned with crops and crosses under the plant breeder's control, so the general discussion by Mayr (19) will not apply. The direct use of interspecific hybrids will not be considered because this is only possible in cases of asexual reproduction.

It is necessary to keep in mind that main aims of interspecific crossing are:
i.   To introduce into a given species some genetic information (nuclear and/or cytoplasmic) from another species.
ii.  To produce a new chromosome combination in such a way that the characteristics of the parental species are mixed, producing a new artificial species.

These possibilities are strongly related to the degree of genetic relationship between the parents. Although no rule can be given, the closer the phylogenetic relationship, the greater the amount of genetic recombination and, on the other hand, the greater the phylogenetic distance, the greater the probability of a complete lack of homology between the two parental genomes, thus allowing alloploids to be produced provided the duplication of the chromosomes is possible. The alloploid can be used by itself (e.g. Triticale), as a 'bridge' to provide new chromosome combinations, or in other ways.

When we cross different species by standard methods (i.e., not using special techniques such as protoplast fusion) some degree of phylogenetic relationship must exist: otherwise hybridization would be completely unsuccessful. Thus, we will get materials within the continuum 'complete chromosome homology-complete lack of chromosome homology'. As a general rule, the alloploids will be most successful when our material is situated as close as possible to the 'complete lack' extreme. If we duplicate the chromosome complement of a hybrid showing some degree of homology between the parental genomes, we will get a segmental alloploid, characterized in meiosis by multivalent associations, producing unbalanced gametes. Even in extreme cases of lack of homology, some degree of instability may be found in the alloploid as a result of residual homology, an interaction between genes of the two parents, cytoplasm x chromosome interactions and so on.

Nature has produced thousands of alloploids that are chromosomically stable. However this has been reached after a process of diploidisation: natural selection for individuals which produce larger numbers of perfect gametes; the result of higher frequencies of normal meioses. Man-produced alloploids have to be selected for chromosomic stability, and it is possible to achieve excellent results: in 50 years, triticale has been transformed from a classical Man-made alloploid into a new crop.

As a final observation, Nature tried and tested a huge number of interspecific crosses long before humans appeared. So, when thinking of new alloploids, it makes sense to look at related taxa to determine how frequent polyploidy is in them. This information will provide a rough indication of the ease or difficulty we can expect.

Returning to the use of interspecific hybridization for introducing desired characteristics into given species, if there is some degree of homology between the two parental genomes, some genetic recombination can occur. The recombination will generally concern 'blocks' of genes, so the genes in which we are interested may be accompanied by undesirable ones. If the donor species is wild, there is a high probability that these will include many undesirable ones. The recovery of the desired gene alone can be achieved by recurrent backcrossing and/or selecting in large $F_2$ populations.

The above discussion supports the idea of choosing species for crossing which are phylogenetically related. Incompatibility barriers will not be as strong as when the parents are phylogenetically distant. However, occasionally, even sympatric species have developed incompatibility barriers, and allopatric species are known which can cross easily. Gregory et al. (12) have shown a good example of this in the genus Arachys, and we have successfully produced hybrids (and alloploids) between Triticum turgidum (a Mediterranean species) and Hordeum chilense (a South American weed), but crosses between the former and H. vulgare (also belonging to the Mediterranean complex) show a high degree of chromosomal incompatibility. Thus, the need for systematic research must be stressed.

## THE GENUS *VICIA*

The biological concept of species has not been systematically applied to *Vicia:* only the *sativa* group has been widely studied by means of interspecific crosses. Very few attempts have been made on the more than one hundred other species. Most of the work discussed here concerns morphological and/or karyological studies. One of the aims of this chapter is to encourage further systematic research on interspecific crosses in *Vicia.*

Basically, *Vicia* is an Euro-Asiatic genus, but there are representatives in North America (related to some Asiatic species), South America (related to the *Ervum* European types) and Africa (not well known, but more related to the *Euvicia* Mediterranean types). The range of adaptation is wide, from Central Asia to Central Africa and Mediterranean regions: from mountainous to cultivated, deep, rich soils; from humid, shaded forests to open, sunny places.

Morphological features are also very variable. The leaflets can range from two per leaf to more than two dozen, and from very tiny to more than ten centimeters long and five centimeters wide. The biggest leaflets do not correspond to *faba,* but to some specimens of *unijuga, orobus* and some others. Floral racemes can be very long with many flowers (*villosa*), have only 2-4 tiny flowers (*tetrasperma*) can be short with 3-7 flowers (*faba, pannonica*) or are absent (*sativa, narbonesis*). Flowers range from very small (4-5 mm; *nana, pubescens)* to more than 20 mm (*grandiflora, faba*) and, ovules per ovary range from 2-4 (*pubescens, tetrasperma*) to nearly 20 (*grandiflora*). The seeds are generally rounded but are sometimes irregular or flattened.

The basic chromosome number seems to be n=7, from which n=6 and n=5 have originated. Different karyotypes have been described for the same species differing mostly in chromosomal rearrangements rather than in chromosome number. However, polyploidy is generally rare; less than 10 of the species studied have a chromosome number higher than 2n=20.

*Vicia* species are generally annual herbs, but some are perennial. Both allogamy and autogamy (and even functional cleistogamy) are present in the genus, but the relative numbers of each are not known.

### The taxonomic position of V. faba.

This is still a matter for discussion among taxonomists. One extreme position is that *Vicia* and *Faba* are two distinct genera, the other extreme includes *V.faba* as a true *Vicia* species. The former viewpoint originated in the Linnaean period; the genus *Faba* being named by Tournefort. Research data supporting this position is still occasionally reported. I strongly support an intermediate view: faba beans are a *Vicia* species, but should be given a mono-specific sub-genus status.

These three positions can be summarized as:

i.    The genus is different-from, but related-to *Vicia.* Faba beans would named *Faba bona* Medikus (Syn, *Faba vulgaris* Moench).

ii.   Faba beans are a full species of *Vicia,* subgenus *faba. V. narbonesis* (and its related *V. galilea, V. johannis* etc.), and *V. bithynica* would also belong to this sub-genus.

iii.   Faba bean is the only species of the subgenus *faba* of genus *Vicia.*

Most modern taxonomists, e.g. Ball (1), Hanelt *et al.* (14), support position ii, but 1 believe that this is due to insufficient information on morphological, karyological, taxonometric and chemotaxonomic aspects of the genus. The sub-genus *faba* as defined in ii is commonly based on leaflet size, but one wonders why *amoena, unijuga, dumetorum* and others are not included.

In a study by Bueno (2) all the possible Euclidian distances between specimens (species or subspecies) were calculated in a living *Vicia* collection. Faba bean appeared as an isolated point: the greatest distances found were those between *V. faba* and the other species. Thus, this taxonometric study supported the Fedchenko system (10) in which *Vicia* was divided into three sub-genera; *Faba* including only *V. faba: Ervilia* including only *V. ervilia;* and *Vicia* including the rest. Bueno's work showed that the similarity between *faba* and *nabonensis,* for example, is more apparent than real. The only common feature is the apparent leaflet morphology, but the differences in floral morphology (*narbonensis* is clearly closer to *sativa*) and seed and pod morphology, mark these two species as very far apart.

Chooi (3) studied the variation in nuclear DNA in 45 species of *Vicia.* Giving an arbitrary value of 100 to the DNA nuclear content of *Vicia faba,* the closest values were those of *V.melanops* (86) and *peregrina* (71), followed by *sylvatica* (65) and *michauxii* (62). *V. narbonensis* (55) and *bithynica* (34) were not close to *faba.*

The average DNA per chromosome produced a similar result: *V. faba* (16.7) and *melanops* (17.2) had the highest values, followed by *hajastana* (11.2) and *peregrina* (10.2). *V. narbonensis* (7.8) and *bithynica* (4.9) were again very far from faba.

Chooi (4) estimated that about one third of the total DNA was repetitive ('fast') in *faba* and *narbonensis,* but only 15% was repetitive in *melanops.* Hybrid DNA reassociation experiments indicated that there is little nucleotide divergence among species, suggesting DNA duplication as an evolutionary mechanism in *Vicia.* It is worth mentioning that *V. faba* chromosomes are not duplochromosomes, i.e. with double chromatids. Briefly, the DNA duplication is in tandem and not in parallel. It is not well known how this DNA is arranged in the chromatin fiber, or how the duplicated genes are expressed in this kind of 'internal polyploid'. Whatever the situation, these studies again demonstrate the special place of *V. faba* in the genus.

Ladizinsky (16) studied the electrophoretic pattern of *V. faba* and the *narbonensis* group (*narbonensis, galilea, johannis*). The protein profile of *V. faba* was completely different from the other three, which were very similar to each other. Interspecific crosses were successful within the *narbonensis* group, but crosses with *faba* failed completely.

Thus in summary, although faba beans are clearly different from other *Vicia* species, separation of the genus is considered too much. Thus a sensible solution appears to be to assign it to a separate sub-genus.

*The infraspecific taxonomy of V. faba.*

This section provides only a brief summary of the subject which is dealt with in greater depth elsewhere, e.g. Cubero (6) (7) (8). Extensive crosses between cultivars, populations and lines of faba beans have shown that few reproductive barriers exist within the species. If some genotypes cross more easily than others it is because of the existence of incompatibility alleles that very rarely have strong effects such as those in *Trifolium*. The most likely candidate to constitute a subspecies, the Muratovian *paucijuga*, crosses readily with other geographical, as well as botanical groups.

Thus, if we accept the biological species concept, we must admit only one subspecies, including all cultivated forms. The wild forms remain unknown, but they might exist in the region from southern Turkey to northern Iran. Afghan and Indian *paucijuga* and *minor* types seem to be a relict of primitive populations characterized by a remarkable absence of genes producing quantitative variation. The case is similar to that of lentils, for which it was possible to formulate a hypothesis to explain such a fact. With faba beans, the 'companion species' are lacking, and therefore we do not have enough data to explain how the variability of western populations was large enough to respond to selection, while that of the eastern materials was so little, at least from a quantitative point of view.

The result is shown in Fig. 2: a primary gene pool constituted by the cultivated forms included in only one subspecies, with no secondary or tertiary gene pool. The most common type of interspecific cross, cultivated x wild, is not possible in faba beans.

*The Potential Value of Interspecific Crosses in Faba Beans.*

The following characteristics might be transferred to faba beans from other *Vicia* species:
  i.   *Resistance to Orobanche* : Some *Vicia* species studied show true resistance to *Orobanche*, for example we have been able to select some lines of *V. sativa* which are practically immune to broomrape.
  ii.  *Resistance to Botrytis fabae:* Although no systematic study has been made I believe that most *Vicia* species are resistant to chocolate spot. *V. sativa*, for example, is grown in environments which are suitable for the disease but is immune to existing races.
  iii. *Resistance to other diseases such as rust, Ascochyta and viruses:* As with *Botrytis*, no systematic study has been made but many *Vicia* species are expected to be resistant.
  iv.  *Resistance to aphids:* This may be harder to solve because aphids attack most, if not all *Vicia* species. However, characters such as hairy racemes (*V. benghalensis, V. pannonica*) might play a role in resistance. Again it is essential to set up systematic studies.
  v.   *Winter hardiness:* Most *Vicia* species are winter hardy, including those which are cultivated for their seeds e.g. *V. ervilia.*
  vi.  *Adaptation to poor soils:* Many *Vicia* species are well adapted to soils of low fertility, as well as those which have other characteristics.

The above list could be increased, however it should be stressed that interspecific hybridization is of value only when the existing variability within the species has been fully explored. Thus characters such as protein content and quality have not been listed. Other characters such as a high number of ovules per ovary are already being studied in faba beans. Whereas cultivars with 7-9 seeds per pod are not rare, an increase might be possible from interspecific crossing; e.g. *V. grandiflora* has almost 20 ovules per ovary.

*Interspecific Crosses in Vicia*

*Interspecific crosses involving species other than V. faba*

As stated above, very little work has been done on interspecific hybridization in *Vicia*. Recent symposia on this subject (30) (23) contain no references to *Vicia* or even to the *Vicieae* tribe, and very few to other important legumes such as *Phaseolus* or soybeans. Mettin (20) reported success only in very few cases. The lack of success is very discouraging; chlorosis and sterility are the rule even when crossing botanical varieties of the same species, as it happens in *V.pannonica* and *V. narbonensis*. Only some crosses which involve species of the *sativa* group seem to produce fertile progenies (namely *amphicarpa x obovata*, with the best results using *amphicarpa* as the female parent), but even in this group *angustifolia* has never produced hybrids when used as a female. Mettin has stated that the number of chromosomes does not seem to be a barrier because crosses between *sativa* subspecies having different chromosome numbers (*obovata x amphicarpa, obovata x angustifolia*) have produced much better results than crosses between subspecies having the same number (*obovata x angustifolia*).

In one case studied by Mettin (*amphicarpa,* 2n=10, x *obovata*, 2n=12), both pollen fertility and the seed setting of the hybrid were very low. The chromosome numbers in the $F_2$ did not appear in the expected frequencies (in $F_1$, gametes with n=5 and n=6 were produced in roughly the same proportion): plants with 2n=10 and 2n=11 were more or less in the same proportion. Plants with 2n=10 resembled *amphicarpa*, indicating a non-random genetic segregation, an additional difficulty in achieving interspecific recombination.

Other authors have reported similar results. Sterility was the rule in hybrids between *sativa x angustifolia* obtained by Sveschnikova, in Donnelly and Clark (9), even when the $F_1$ showed considerable hybrid vigour; the $F_2$ were mixtures of etiolated, lethal and some normal types. Watanabe and Yamada (25) also reported crosses between *sativa* and *angustifolia,* using the latter as the male parent. The $F_1$ pods resembled those of *angustifolia*, but pollen fertility was less than 5%. The $F_2$ segregated almost exclusively for parental types. Sekizuka *et al.* in Donnelly and Clark (9), crossed *angustifolia* var. *segetalis* with *sativa*: pollen fertility and seed set of the hybrid were very poor.

Yamamoto (26) (27) (28) studied crosses between *sativa, amphicarpa, pilosa* (another species of the *sativa* group) and *macrocarpa*. Chromosome pairing at meiosis was very poor, and as a consequence pollen fertility was low. He reported a stable karyotype produced by the substitution of one chromosome of *sativa* in a *macrocarpa* karyotype derived from the interspecific hybrid. The same happened with one chromosome of *amphicarpa*. This result suggests the possibility of looking for substitution lines instead of hybrids and/or alloploids. Similar results were obtained in the case of *pilosa x angustifolia* and *pilosa x macrocarpa*. Trivalents were of normal occurrence in the meiosis of the hybrid. The existence, not only of new chromosome combinations, but also of genetic recombination, suggests a possible use in plant breeding. These results were supported later by Yamamoto (29), studying the amylase isozyme pattern in the cross *pilosa x macrocarpa*. Recombination between parental genes did occur in the progeny of the hybrid.

Outside the *sativa* group there is little experience and even less success. Cooper (5) reported a cross between *sativa* and *calarata,* and Sekizuka *et al.* in Donnelly and Clark (9), reported a cross between *V. angustifolia* and *V. grandiflora.* Crosses within the *narbonensis* group are possible, but they also depend both on the parents and the direction of the cross. Crosses are possible especially between *narbonensis* and *galilea* (probably conspecific) and also between these and *hyaeniscyamus* (Yamamoto, pers. comm.). *V. serratifolia* also shows some affinity with those species, but there is a greater difficulty in obtaining hybrids (20). The pollen of some lines of *narbonensis* reach the ovules of *V. grandiflora* and *macrocarpa,* as with the *grandiflora* pollen in crosses between *macrocarpa x grandiflora* (24). No success was attained in crosses between *dumentorum, grandiflora, lutea, pannonica, sepium* and *amurensis* with *sativa,* and only a low percentage of success (5% ) between *sativa* and *tetrasperma.* This is surprising, given the systematic position of the parents (20). Donnelly and Clark (9) crossed *angustifolia, atropurpurea, articulata, dasycarpa, ervilia, galeata, grandiflora, lutea, pannonica, sativa, serratifolia* and *villosa,* with only two positive results: *dasycarpa x villosa* and *sativa x angustifolia.* Table 1 summarizes most of the results to date.

Table 1

Interspecific hybridation in *Vicia*

| Cross | Success | Remarks | Author |
|---|---|---|---|
| *angustifolia x amphicarpa* | Yes | no data | Svesnikova 1940(Mettin 1962) |
| *angustifolia x sativa* | no | | Tupikova 1927(Mettin 1962) |
| *angustifolia x sativa* | yes | no data | Sekizuka *et al.* 1958(Mettin 1962) |
| *dumetorum x sativa* | no | | Schelhorn 1940(Mettin 1962) |
| *grandiflora x sativa* | no | | Schelhorn 1940(Mettin 1962) |
| *lutea x sativa* | no | | Schelhorn 1940(Mettin 1962) |
| *macrocarpa x amphicarpa* | yes | no data | Svesnikova 1940(Mettin 1962) |
| *macrocarpa x angustifolia* | yes | no data | Svesnikova 1940(Mettin 1962) |
| *macrocarpa x sativa* | yes | no data | Svesnikova 1940(Mettin 1962) |
| *narbonensis intermedia x narbonensis integrafolia* | 7% | no germination | Schelhorn 1940(Mettin 1962) |
| *id reciprocal* | 20% | chlorosis | Schelhorn 1940(Mettin 1962) |
| *narbonensis x serratifolia* | no | | Tupikova    (Mettin 1962) |
| *pannonica (red) x pannonica (white)* | scarce | no germination | Schelhorn 1940(Mettin 1962) |
| *sativa x amphicarpa* | yes | hybrid vigour, sterility | Svesnikova 1940(Mettin 1962) |
| *sativa x angustifolia* | yes | hybrid vigour, sterility | Svesnikova 1940(Mettin 1962) |

| | | | |
|---|---|---|---|
| *sativa x angustifolia* | 14% | hybrid inter-<br>mediate | Yamamoto 1955(Mettin 1962) |
| *sativa x angustifolia* | yes | no data | Shimano *et al.* 1959(Mettin 1962) |
| *sativa x amphicarpa* | yes | hybrid vigour,<br>sterility | Yamamoto 1959 (Mettin 1962) |
| *sativa x dumetorum* | no | | Schelhorn 1940 (Mettin 1962) |
| *sativa x grandiflora* | no | | Schelhorn 1940 (Mettin 1962) |
| *sativa x lutea* | no | | Schelhorn 1940 (Mettin 1962) |
| *sativa x macrocarpa* | no | | Tupikova (Mettin 1962) |
| *sativa x macrocarpa* | yes | no data | Svesnikova 1940(Mettin 1962) |
| *sativa x pannonica* | no | | Schelhorn 1940 (Mettin 1962) |
| *sativa x sativa (wild?)* | scarce | sterility, no<br>seeds | Hirayoshi & Matsumura 1952<br>(Mettin 1962) |
| *sativa x sepium and*<br>  *reciprocal* | no | | Schelhorn 1940 (Mettin 1962) |
| *sativa x tetrasperma* | 5% | 50%fertility<br>$F_1$ similar female | Yamamoto 1954(Mettin 1962) |
| *serratifolia x narbonensis* | no | chlorosis | Tupikova (Mettin 1962) |
| *villosa x dasycarpa* | scarce | chlorosis | Tupikova (Mettin 1962) |
| *pannonica x galeata* | no | | Donnelly & Clark, 1962 |
| *pannonica x sativa and*<br>  *reciprocal* | no | | Donnelly & Clark, 1962 |
| *sativa x angustifolia* | yes | | Donnelly & Clark, 1962 |
| *serratifolia x articulata* | no | | Donnelly & Clark, 1962 |
| *serratifolia x dasycarpa* | no | | Donnelly & Clark, 1962 |
| *serratifolia x sativa* | no | | Donnelly & Clark, 1962 |
| *sativa x macrocarpa* | yes | only karyo-<br>logical data | Yamamoto 1971 |
| *amphicarpa x macrocarpa* | yes | id. | Yamamoto 1971 |
| *pilosa x angustifolia* | yes | id. | Yamamoto 1974 a |
| *pilosa x macrocarpa* | yes | id. | Yamamoto 1974 a |
| *narbonensis x galilea* | yes | only karyo-<br>logical data | Ladizinsky, 1975 |
| *narbonensis x johannis* | yes | | Ladizinsky, 1975 |
| *narbonensis x grandiflora* | ? | Pollen<br>reachs ovules | Van Cruchten, 1974 |
| *macrocarpa x grandiflora* | ? | Pollen<br>reachs ovules | Van Cruchten, 1974 |
| *sativa x angustifolia* | yes | poor pollen<br>fertility | Watanabe & Yamada, 1958 |
| *angustifolia segetalis x sativa* | yes | " | Moriya (Donnelly & Clark, 1962) |
| *angustifolia segetalis x sativa* | yes | " | Sekizuka *et al.* (Donnelly & Clark |
| *angustifolia x grandiflora* | yes | very poor<br>seed set | Sekizuka " " "1962) |

| | | | |
|---|---|---|---|
| sativa x calcerata | yes | no data | Cooper(Donnelly & Clark 1962) |
| sativa obovata x lutea and reciprocal | 30% | Abortion of seeds | Mettin, 1962 |
| sativa obovata x amurensis | no | | "        " |
| pannonica x sativa angustifolia | no | | "        " |
| sativa obovata x pannonica | no | | "        " |
| sativa angustifolia x lutea | no | | "        " |
| sativa obovata x sativa angustifolia | 9% | | "        " |
| sativa angustifolia x sativa obovata | no | | "        " |
| sativa obovata x sativa amphicarpa | 50-63% | | "        " |
| sativa amphicarpa x sativa obovata | 92-100% | | "        " |
| sativa angustifolia x sativa amphicarpa | no | | "        " |
| sativa amphicarpa x sativa angustifolia | 75% | | Mettin, 1962 |
| atropurpurea x sativa | no | | Donnelly & Clark, 1962 |
| atropurpurea x serratifolia and reciprocal | no | | "        "        " |
| dasycarpa x villosa | yes | | "        "        " |
| dasycarpa x atropurpurea | no | | "        "        " |
| grandiflora x angustifolia | no | | "        "        " |
| grandiflora x articulata | no | | "        "        " |
| grandiflora x galeata and reciprocal | no | | "        "        " |
| grandiflora x lutea and reciprocal | no | | "        "        " |
| grandiflora x pannonica and reciprocal | no | | "        "        " |
| grandiflora x sativa and reciprocal | no | | "        "        " |
| grandiflora x serratifolia and reciprocal | no | | "        "        " |
| lutea x angustifolia | no | | "        "        " |
| lutea x articulata | no | | "        "        " |
| lutea x ervilia | no | | "        "        " |
| lutea x pannonica | no | | "        "        " |
| lutea x sativa and reciprocal | no | | "        "        " |
| faba x narbonensis | no | | Tupikova (Mettin, 1962) |

| | | | |
|---|---|---|---|
| faba x narbonensis | no | also using 4x | Rousselle (pers. comm) |
| faba x galilea | no | | "            "            " |
| faba x hyaeniscyamus | no | | "            "            " |
| faba x johannis | no | | "            "            " |
| faba x procumbens | no | | "            "            " |
| faba x narbonensis | no | pollen can reach ovules | Van Cruchten, 1974 |
| faba x macrocarpa and reciprocal | no | pollen can reach ovules | "      "              " |
| faba x grandiflora and reciprocal | no | pollen can reach ovules(?) | "      "              " |
| faba x ervilia | no | | "      "              " |
| faba x narbonensis | no | genotipo collapses | Yamamoto (pers.comm.) |
| faba x narbonensis | no | faba 4x | Martin and Cubero |
| faba x galilea | no | faba 4x | "            " |
| faba x ervilia | no | faba 4x | "            " |

*Interspecific crosses involving V. faba*

Only a few attempts to cross *V. faba* with other species have been published. Van Cruchten (24) reported the results of several crosses between two lines of *V. narbonensis* and six lines of *Vicia faba*. The faba pollen germinated on the stigma of one of the *narbonensis* lines, but the pollen tubes did not reach the stylar bend, a few millimeters under the stigma. In the other *narbonensis* line, the pollen tubes reached the ovules. The reciprocal crosses produced the same results: only the pollen of the second *narbonensis* line was able to reach the *faba* ovules. Interspecific crosses strongly depend on the genotypes of the parents as this study again confirms. The pollen tube growth was much slower than that recorded for intraspecific crosses. Van Cruchten also reported that *faba* pollen tubes could reach *macrocarpa* ovules.

Yamamoto (pers. comm.) crossed *narbonensis x faba* using embryo culture to avoid the poor development, or lack, of the hybrid endosperm, but embryos lived only about one week. Rousselle (pers. comm.) has also failed to produce these hybrids. Thus the differences between *narbonensis* and *faba,* indicated above, are reflected in these results.

There are several references in the literature on *Pisum sativum x Vicia faba* crosses, most of which were reviewed by McComb (17). He discussed in some detail the work of some Russian researchers who described *Pisum x Vicia* (mainly *faba*) hybrids, especially, the work of Sobolev *et al.* Sobolev described the morphology of some hybrids, as well as the segregation among the progeny. Cytological analysis supported the assumption that the hybrids were true. The possibility that the hybrids were derived from apomictic development of the seeds, which is known to occur in *Pisum,* was discounted by Sobolev on the basis of chromosome number. The authentic hybrids generally have 2n=14, with a range of 2n=12 to 16. The cytological behaviour of these hybrids was very unusual: the non-homologous chromosomes of peas and faba beans formed bivalents but two sets of chromosomes were formed, each one corresponding to a parental set. To increase the seed set of the $F_1$ (hybrid), Sobolev and other Russian workers irradiated seeds of the parental lines and/or pollen of the male line, with gamma rays.

In other studies in Russia, Gritton and Wierzbicka (13) reported successful crosses between *Pisum sativum* and *V.faba, V. villosa* and *V. sativa* from which high yielding and very early ripening forms of peas were obtained. However, these results have not been reproduced elsewhere. When *P. sativum* was used as the female parent, the percentage of *Pisum* ovules fertilized by *V.faba* pollen ranged from 2 to 8, which must be compared with parellel results attained in *Pisum x Pisum* crosses of 25.5 - 48 %. *Vicia* pollen grew more slowly than that of *Pisum* on *Pisum* stigmas, supporting the data of Van Cruchten. A great portion of the lack of fertilization must be due to the slow growth of the alien pollen tubes.

The rate of development of the intergeneric embryo was much slower than that of the intraspecific one. Four days were required to reach two-celled proembryo stage compared to only two days. Five days after pollination, the intergeneric embryo was in a four-cell stage, whereas the intraspecific one had reached the multicellular stage. Collapse of the former was evident from the sixth day, and was complete by the eighth day. The intergeneric endosperm cells showed many chromosomal abnormalities and formed scattered masses of densely stained nuclei. Such abnormalities have occasionally been reported in pure *Pisum* and *V. faba* embryos.

This tendency is not 'counter-balanced' in the intergeneric hybrid and the result is the complete failure of the hybrid embryo development. The exact causes of endosperm failure are not well known; a likely reason is the lack of translocation between the maternal tissue and the endosperm, but this has yet to be demonstrated.

Underlying this physiological process is the genetic fact that in an interspecific embryo, two different patterns of development coexist which, if they are compatible, will result in the embryo developing more or less easily. If they produce 'contradictory orders', the embryo will collapse. There has been no consideration up to now of the importance of differences in the DNA content per genome and per chromosome, to explain this collapse. *V. faba* has much more DNA than any other *Vicia* species (with the probable exception of *V. melanops*) which means that its DNA reproductive cycle must be different. 'Contradictory orders' could thus simply be the lack of co-adaptation of two different DNA sets provoking, for example, bridges, fragments, or laggards, and not necessarily contradictory genetic information. In somatic hybrids, obtained by protoplast fusion, the chromosomes can follow independent reproductive cycles and in true hybrids between some *Hordeum* and *Triticum* species with *H. bulbosum,* a true embryo is formed which loses the *bulbosum* chromosomes very quickly.

If this hypothesis is true, the possibility of obtaining hybrids between *V.faba* and any other species is very poor. It would be interesting to try *V.melanops* because of its high DNA content.

*Towards A Solution*

Much work remains before admitting the total failure of genic transfer to *Vicia faba*. Even then both protoplast fusion i.e. somatic hybrids, and genetic engineering remain as possibilities for the future. The relative lack of such research on faba beans in the past may be due to the crop being considered of insufficient importance to justify the expense entailed.

Considering the classic approach to interspecific hybridization, a systematic approach is required in *Vicia*. Most crosses attempted have been within the *sativa* group, and between these and a few other *Vicia* species. Even the *narbonensis* group is not as well studied as the *sativa* group. Several other species clusters can exist, e.g. *pannonica*, *villosa* and *pubescens*. Knowledge of the reproductive barriers within and between these groups is not only of academic interest, but also of practical importance. Genes will move through barriers once the barriers are known and understood. Techniques for interspecific crossing need to be studied. In this respect, factors found to be important in other groups, such as the *Triticineae*, should be considered:

i.    Suitable lighting conditions, as well as temperature and other environmental factors are basic in obtaining hybrid seeds.

ii.   Hormonal treatments may be necessary. The most widely used hormone is giberellic acid (GA) applied, for example, as a spray on the whole plant or ovaries, or to stem cuttings placed in a GA-supplemented nutrient solution. We have used GA when crossing our tetraploid *Vicia faba* with other *Vicia* species (*narbonensis*, *galilea*, *ervilia*), but this work is far from complete and systematic. Many more species and treatments must be tried.

iii.  Embryo culture has been tried unsuccessfully, but very difficult cases in cereals have been solved. It may be only a question of time and research input.

iv.   A more extensive use of tetraploid strains should be tried. We have only used *Vicia faba* as a tetraploid parent but to break incompatibility barriers, it may be necessary for both parents to be tetraploid.

v.    Miscellaneous techniques, such as grafting and decapitating styles, can also be tried in an attempt to break the barriers. However in the case of *Vicia faba*, the incompatibility is not in the stigma and/or style, but in the zygote. Although some crosses fail because the pollen tubes do not reach the ovules, the strong barriers seem to be post-zygotic.

*Introgression and/or Alloploidy*

We have seen that even when genetic segregation is not random in *Vicia* interspecific hybrids, there is some degree of recombination. Sekizuka *et al.* in Donnelly and Clark (9) reported the selection of *Vicia sativa* lines resistant to *Collectotrichum* derived from an *angustifolia x sativa* hybrid. More recently, Yamamoto (29) has confirmed by electrophoretic techniques the existence of re-combination.

On the question of alloploids, it must be pointed out that polyploidy is not frequent in *Leguminosae*. Goldblatt (11) has calculated that only about 20% of legume species have a polyploid origin. In *Vicia*, Mettin and Hanelt (21) have reported the existence of 13 species of polyploid origin out of a total of 128 examined (5 with 2n=24, 6 with 2=28, one with 2n=36 and one with 2n=42). It is not possible, at present, to decide on the relative importance of auto-and allopolypoid in *Vicia*, but neither is common.

Thus producing artificial alloploids may be a very difficult task. Confirming this, Mettin (20) found very few polyploid plants in the progeny of an *amphicarpa x obovata* cross: only two highly sterile triploids were found among 190 plants. In the progeny of one of these, a 2n=22 individual was found; probably an alloploid. Sixty per cent of the pollen grains were fertile and seed set was poor. The only other reference to alloploidy in the literature is Shimano *et al.*, in Mettin (20), the alloploid having been produced in a cross between *sativa* and *angustifolia*.

A chromosomically stable alloploid would not be *Vicia faba* but a new crop. Any new crop, especially in the legume family would be welcomed, but here we are dealing with the use of interspecific crosses in faba bean improvement. In this context, if any alloploid involving *V. faba* is obtained, it has to be considered as an intermediate step in transferring useful characters to this crop. This task could be achieved in many ways, depending on the constitution of the new materials obtained. The steps below outline the most common way of handling alloploids in plant breeding:

i.   The alloploid must be backcrossed to *Vicia faba*. From this hybrid, plants having the 12 *faba* chromosomes plus one or more chromosomes of the other parent species, will be obtained.

ii.  From these plants, after adequated cytological identification, chromosome addition lines will be derived by backcrossing these plants with *faba*. These lines will be either monosomic (i.e., 12 *faba* plus one alien chromosome) or disomic (i.e. 12 *faba* plus a pair of alien homologus chromosomes). They must be identified cytologically and phenotypically. Additional lines are not likely to be suitable for direct use, even if they show regular meiosis, but the alien characteristics would be easily identified.

iii. Homology between the chromosomes of *faba* and those of any other *Vicia* species is not expected. So recombination between *faba* and alien chromosomes seems unlikely to occur. To transfer the desirable characteristics from the alien to the *faba* chromosomes, 'chromosome grafting' seems the most convenient technique. Through irradiating such *faba* plants (seeds, flowers, buds etc.) fragments of the alien chromosomes may be translocated to any of the *faba* genome. The progenies of such irradiated plants whether backcrossed again to *faba* or not, must be carefully checked to identify those plants having 12 chromosomes and showing the alien characteristic sought. The smaller the chromosome fragment translocated, the better the result is likely to be.

iv.  The alloploid can also be used as a 'bridge' to introduce characteristics from other *Vicia* species. Because of the presence of two or more different genomes in their genetic constitution, alloploids are 'buffered' and may admit others genomes. Thus, a third species which does not cross directly with faba beans could be used as a source of useful genes through the 'bridge' formed by the alloploid. The production of an alloploid of *Vicia faba* would enhance the possibility of transferring genes from alien species to the cultivated one.

## ACKNOWLEDGEMENT

I am indebted to Drs. Picard, Bond, Martin, Yamamoto and Rousselle for communicating unpublished research results and for suggesting certain ideas for this article.

# REFERENCES

1. Ball, P.W. 1968. Flora europaea 2: 128-136. Ed. by T.G. Tutin *et al.* Cambridge Univ. Press, London.
2. Bueno, M.A. 1975. Taxonometria y cariologia en el genero *Vicia*. Ph.D. Thesis, Facultad de Ciencias, Universidad Complutense, Madrid.
3. Chooi, W.Y. (1971 a). Variation in nuclear DNA content in the genus *Vicia*. Genetics 68: 195-211.
4. Chooi, W.Y. (1971 b). Comparison of the DNA of six *Vicia* species by the method of DNA-DNA hybridization. Genetics 68: 213-230.
5. Cooper, R.L. (1959). Hybridization in *Vicia*. Agronomy Abstracts. Ann. Soc. Agronomy, p. 56.
6. Cubero, J.I. 1973, Evolutionary trends in *Vicia faba*. Theoret. Appl. Genetics, 43: 59-65.
7. Cubero, J.I. 1974. On the evolution of *Vicia faba*. Theoret. Appl. Genetics 45: 47-51.
8. Cubero, J.I. 1980. Modern and primitive forms in *Vicia faba*. Z.Kulturplf, (in press).
9. Donnelly, E.D., Clark, E.M. 1962. Hybridization in the Genus *Vicia*. Crop Science 2: 141-145.
10. Fedchenko, B.A. 1948. *Vicia*. In Flora of the USSR, ed. by V.L. Komorov, Moscow.
11. Goldblatt, P. (1978). Cytology and the Phylogene of Leguminosae. I. International Congress on Legumes, Kew (Handout).
12. Gregory, W.C., Kaprovickas, A., Pfluge-Gregory, M. 1980. Structure, variation, evolution and classification in *Arachys*, in Advances in Legume Science. pp. 469-481.
13. Gritton, E.T., Wierzbicka, B. 1974. An embryological study of a *Pisum sativum x Vicia faba* cross. Euphytica 24: 277-284.
14. Hanelt, P., Schafer, H., Schultze-Motel, J. 1972. Die stellung von *Vicia faba* L. und Botrachtungen zur Entstehung diescr Kulturart. Kulturpflanze 20: 263-275.
15. Harlan, J.R., De Wet, J.M.J. 1971. Toward a rational classification of cultivated plants. Taxon 20: 509-517.
16. Ladizinsky, G. 1975. Seed protein electrophoresis of the wild and cultivated species of section *faba* of *Vicia*. Euphytica 24: 785-788.
17. Mac Comb, J.A. 1974. Is intergeneric hybridization in the leguminosae possible ? Euphytica 24: 497-502.
18. Mayr, E. 1948. The bearing of the new systematics on genetical problems. The nature of species. Advances in Genetics 2: 205-237.
19. Mayr, E. 1963. Animal species and evolution. Belknap Press of Harvard University Press, Cambridge (Mass.).
20. Mettin, D. 1962. Bastardierungs versuche in der fattung *Vicia*, Der Zuchter 32: 146-155.
21. Mettin, D., Hanelt, P. 1968. Bemerkungen zur Karyologie und Systematik einiger Sippen der Gratting *Vicia* L. Feddes Repertorium 27: 11-30.
22. Mettin, D., Hanelt, P. 1973. Uber Speziationsvorgange in der fattung *Vicia* L. Kulturpflanze 21: 25-54.
23. Sanchez-Monge, E., Garcia-Olmedo, F. (eds.). 1978. Interspecific hybridization in plant breeding. ETSIA. Madrid.
24. Van Cruchten, C. 1974. Etude de la croissance des tubes polliniques dans des croissements intra et interspecifiques de *Vicia faba* and *Vicia narbonensis*. Station d'Amelioration des Plantes, rapport.

25. Watanabe, K., Yamada, T. 1958. (Studies on the interspecific hybridization of common vetch. *Vicia sativa*, and Yahazuendo, *V. angustifolia* var. *segetalis*) (Eng. summary). Bull. Nat. Inst. Agric. Sciences, Japan, series G, 15: 109-145.
26. Yamamoto, K. 1971. On the secondary balanced karyotypes of $M\text{-}m_4 + s_1\ M\text{-}m_4 + s_1$ and $M\text{-}m_4 + a_4\ M\text{-}m_4 + a_4$. Tec. Bull. Fac. Agr. Kagawa U. 22: 1-8.
27. Yamamoto, K. 1974 a. On the interspecific hybrids between *Vicia pilosa* and *V. angustifolia*, *V. pilosa* and *V. macrocarpa*. Tec. Bull. Fac. Agr. Kagawa U. 25: 179-190.
28. Yamamoto, K. 1974 b. Hybrid plants between two races of *Vicia amphicarpa* having $2n=14$ chromosomes. Japan J. Breed. 24: 15-22.
29. Yamamoto, K. 1975. Estimation of genetic homogeneity by isozymes from interspecific hybrids of *Vicia*. Japan J. Breed 25: 59-64.
30. Zeven, A.C., Van Harten, A.M. (eds.) 1979. Broadening the genetic base of crops. PUDOC, Wageningen.

## 10. FABA BEAN AGRONOMY IN EGYPT

ALI A. IBRAHIM, ABDULLAH M. NASSIB, MOHAMED H. EL-SHERBEENY

*Field Crops Research Institute, Agric. Res. Centre, Giza, Egypt*

One of the major constraints to production of faba bean in Egypt is that it is often grown under sub-optimal production systems. Hence considerable research emphasis is being placed on determining optimal agronomic criteria under different agroecological conditions covering sowing dates, plant population and distribution, fertilizer requirements, *Rhizobium* inoculation and water requirements. Studies are also being conducted to determine the effect of tillage and no-tillage systems of farming on the performance of faba beans as well as the behaviour of the crop under intercropping.

*Date of sowing :* Ali (3) indicated that highest yields of seeds/plant and per feddan at Giza were obtained from planting between November 3 and 23.

Salem (20) pointed out that late sowing delayed flowering and reduced the number of flowers, pods, seed yield and 100-seed weight. He concluded that seed and straw yields progressively decreased with delaying sowing date from the last week of October until the third week of December at Sharkia province in Lower Egypt.

Mohamed (17) stated that better yields of faba bean were obtained from plantings during the first through the third week of November at Monofia province in the Delta. Delayed planting decreased the number of tillers, number of pods/plant, number of seeds/pod and 100-seed weight.

Hakam and Ibrahim (8) have reported that the optimum planting dates were November 1 for Giza 1 in Lower Egypt, and from October 15 to November 1 for Giza 2 and Rebaya 40 in Middle and Upper Egypt. Delaying sowing until November 15 is recommended for the north of the Delta to control, to a certain extent, chocolate spot *(Botrytis fabae)* and rust *(Uromyces fabae)* diseases which usually prevail in this region.

In Lower Egypt, experiments in 1977-78 and 1978-79 (Table 1) showed a highly significant effect of planting dates on the average yield of Giza 1 and Giza 3 with no significant interaction between cultivars and sowing dates. Planting late in December resulted in a significant decrease in plant height, number of pods and seeds per plants, and seed yield (Table 2).

In Middle Egypt, in 1978-79 and 1979-80, there was a significant effect of sowing dates on the average yield of Giza 2 and Giza 4, while the interaction between cultivars and sowing dates was not significant. In 1978-79, planting on October 15 gave the highest yield. However, in 1979-80, when planting dates of October 1 and October 15 were not

*G. Hawtin & C. Webb (Eds.): Faba Bean Improvement*
©1982 ICARDA. ISBN -13:978-94-009-7501-9

carried out due to delay in land preparation, the earliest sowing date in the experiment i.e. November 1, gave the highest yield.

Table 1. Effect of sowing date on seed yield (kg/ha) at Sakha (Lower Egypt), Sids (Middle Egypt) and Shandoweel (Upper Egypt) Research Stations.

| Station : | Sakha [1] | | Sids [2] | | Shandoweel [3] | |
|---|---|---|---|---|---|---|
| Season : | 1977-78 | 1978-79 | 1978-79 | 1979-80 | 1978-79 | 1979-80 |
| Sowing date | | | | | | |
| Oct. 1 | — | 609 | 3033 | — | 1311 | |
| Oct. 15 | 3353 | 892 | 4218 | — | 1171 | 1950 |
| Nov. 1 | 4848 | 1041 | 3582 | 2430 | 1634 | 2222 |
| Nov. 15 | 2909 | 727 | 3031 | 1718 | 1738 | 2432 |
| Dec. 1 | 2725 | 178 | 1212 | 816 | 965 | 1929 |
| Mean | 3459 | 689 | 3015 | 1655 | 1363 | 2133 |
| L.S.D. 5 % | 855 | 199 | 516 | 304 | 409 | N.S. |

[1] / mean of cvs. Giza 1 and Giza 3
[2] / mean of cvs. Giza 2 and Giza 4
[3] / mean of cvs. Giza 4 and R 40

Table 2. Effect of sowing date on seed yield and some agronomic characters of faba bean at Sakha Research Station, 1977-78.

| Sowing date | Characters | | | | | | |
|---|---|---|---|---|---|---|---|
| | Plant height cm | branches/ plant | No. of | | seeds/ plant | 100-seed weight g | seed yield g/plant |
| | | | pods/ plant | seeds/ plant | | | |
| Oct. 1 | 92.1 | 2.4 | 14.5 | 29.3 | 2.2 | 74.6 | 22.1 |
| Oct. 15 | 88.9 | 2.4 | 14.2 | 29.8 | 2.2 | 73.6 | 21.8 |
| Nov. 1 | 72.8 | 2.2 | 12.7 | 27.9 | 2.3 | 72.2 | 20.0 |
| Nov. 15 | 72.1 | 2.1 | 10.9 | 22.5 | 2.2 | 76.9 | 17.3 |
| Mean | 81.5 | 2.3 | 13.1 | 27.4 | 2.2 | 74.3 | 20.3 |
| L.S.D. 5 % | 7.9 | N.S. | 2.0 | 4.5 | N.S. | N.S. | 3.4 |

In Upper Egypt, a significant effect of planting dates on the average yield of Rebaya 40 and Giza 4 was found only in 1978-79, where planting on November 1 through November 15 recorded the highest yields of seeds/ha.

Generally, it may be concluded that planting faba bean on November 1 is recommended for Lower Egypt; from October 15 to November 1 for Middle Egypt, and early in November for Upper Egypt. These recommendations are in general agreement with those of Hakam and Ibrahim (8).

*Population density and distribution :* Although seeding methods have little effect on faba bean yields (21) (1), most of the crop in Egypt is planted on ridges for ease of manual planting, weeding, furrow irrigation and other agronomic operations.

Hakam and Ibrahim (8) reported that seeding rates between 95 and 240 kg/ha had no clear effect on the seed yield. They also reported that row spacing is the major factor affecting seed yield regardless of hill spacing (15, 20, 25 cm) and number of plants/hill (1 or 2 plants). Thie highest seed yield/ha was obtained when sowing was done on both sides of ridges 60 cm apart in 2-seeded hills 20 cm apart. However, decreased densities lead to high tillering, high number of pods and high seed yield/plant.

Rizk (19), pointed out that increasing density through close hill spacing or increased number of plants per hill resulted in fewer seeds/plant and lower yield of seeds. El-Morshidy *et al.* (6) recommended sowing on both sides of ridges, 60 cm apart and 10 to 15 cm between hills.

In experiments on the effect of row and hill spacing and number of seeds per hill on the new cultivars Giza 3 and Giza 4 in 1979-80, at Sakha, the lowest seed yield resulted when planting was at a medium density (33 plants/$m^2$) but with high plant competition in one row/ridge, while planting at low density (17 and 25 plants/$m^2$) but with better plant distribution on three rows/ridge yielded significantly higher than the previous combination. However, high plant densities (42 and 50 plants/$m^2$) had no advantage over the medium one (33 plants/$m^2$) in four well distributed treatments on two or three rows/ridge. Out of these four treatments, double-seed spacing at 20 cm apart on two rows/ridge may be recommended for Giza 3.

Table 3. Effect of population density and plant distribution on seed yield (kg/ha) of faba beans at Sakha, Bahteem and Sids Research Stations, 1980.

| No. of rows/ridge | No. of Plts/hill | Seed spacing cm. | Plt/$m^2$ | G.3 Sakha | G.4 Bahteem | G.4 Sids | Mean yield (kg/ha) |
|---|---|---|---|---|---|---|---|
| 3 | 1 | 30 | 17 | 2814 | 3784 | 1197 | 2598 |
| 3 | 1 | 20 | 25 | 2704 | 4179 | 1142 | 2675 |
| 1 | 2 | 10 | 33 | 2329 | 3833 | 1566 | 2576 |
| 2 | 1 | 10 | 33 | 3006 | 4308 | 1237 | 2850 |
| 2 | 2 | 20 | 33 | 2907 | 4152 | 1482 | 2847 |
| 3 | 1 | 15 | 33 | 2844 | 4505 | 1781 | 3043 |
| 3 | 2 | 30 | 33 | 2966 | 4340 | 1630 | 2978 |
| 3 | 1 | 12 | 42 | 2881 | 4269 | 1534 | 2894 |
| 3 | 2 | 20 | 50 | 2870 | 4257 | 1730 | 2952 |
| Mean | | | | 2813 | 4180 | 1477 | 2823 |
| L.S.D. 5 % | | | | 235 | 396 | N.S. | |

At Bahteem, the lowest density (17 plants/m$^2$) but well distributed plants on three rows/ridge, yielded nearly the same as the medium one (33 plants/m$^2$) with high plant competition in one row/ridge. Increasing plant population from 17 to 25 plants/m$^2$ by decreasing hill spacing from 30 to 20 cm resulted in higher yield. Again, high plant densities (42 and 50 plants/m$^2$) had no advantage over the medium density in four well distributed treatments on two or three rows/ridge. Out of these four, single seed spacing at 15 cm apart may be recommended for Giza 4 at Bahteem.

At Shandaweel, planting three rows/ridge in one-seeded hills 15 cm apart tended to give a higher yield, although there were no significant differences between treatments.

*Fertilization and Rhizobium inoculation*

Hamissa (11) summarized the results of 31 fertilizer trials carried out between 1967 and 1973 in the major faba bean areas. It was concluded that :

i.  In general the soils are slightly alkaline, are rich in calcium carbonate, poor in organic matter and nitrogen, have moderate levels of available phosphorus and are fairly rich in exchangeable and soluble potassium.

ii.  Faba bean responded significantly to nitrogen. An average increase of 9.4% and 10.5% over the control in seed yield/ha was recorded at the rates of 18 and 36 kg/ha respectively.

iii. The phosphorus response was rather high. Seed yield increased gradually and significantly as the rate of P increased up to 72 kg $P_2O_5$/ha. An average increase of 9.8 and 15.7% in seed yield over the control was recorded with 36 and 72 kg $P_2O_5$/ha.

iv.  The maximum yields were produced when nitrogen and phosphorus were applied together, and the optimum economic rates were 36 kg N + 72 kg $P_2O_5$/ha.

Again, Hamissa *et al.* (9) indicated that seed yield of faba beans increased after nitrogen application (10 to 12% higher than the control treatment) up to 18 kg N/ha. Response to P was also high (12 to 20% ) and maximum yield was obtained from a combination of 36 kg N + 72 kg $P_2O_5$/ha. About 80% of the field trials responded positively to N and P whereas 20% gave no response. They all demonstrated no positive response due to K.

Hamissa (10) concluded from later work that, in general, the response of the new cultivars Giza 3, Giza 4 and F 402 was not so marked. In most cases, the highest yield was from the recommended rate of 36 kg N + 72 kg $P_2O_5$/ha. No significant response to K application was detected.

Previous investigations indicated that symbiotic nitrogen fixation decreased, and fertilizer absorption increased, with additional available nitrogen. The lower rates of inorganic nitrogen stimulated both plant growth or seed production but had a tendency to replace the fixation process (5) (10). Evidence suggests that effectively nodulated faba bean plants gain little or no benefit from N fertilizer (1) (14) (12) (5) (10); indicating that the crop could obtain its nitrogen requirements from the soil nitrogen and atmospheric N-fixation. Hakam and Ibrahim (8) found that faba beans responded to *Rhizobium* inoculation, indicating the importance of seed inoculation to ensure the presence of highly effective bacteria in Egyptian soil.

The response of faba bean cultivars to *Rhizobium* inoculation and N fertilizer was further studied in 1978-79 at four locations in Lower, Middle and Upper Egypt. Contrary to the results of Hakam and Ibrahim (8) no response to inoculation was observed, indicating that native *Rhizobium* levels were adequate. Significant responses to N were only recorded at Sids and Shandoweel. From the overall data it may be concluded that a small dose of nitrogen (approx. 18 kg/ha) applied 21 days after planting can be recommended as an economic practice.

In the following season a trial was conducted on three farmers' fields in Minia Province (Middle Egypt) and at Gemmeza Research Station in the Delta. Two *Rhizobium* cultivars (Local and Aleppo) and different levels of N and P were tested. Significant results were only obtained from two locations, and these are given in Table 4.

Table 4. Effect of nitrogen and phosphorus application and *Rhizobium* inoculation on seed yield (kg/ha) and number of nodules, 1980.

| Treatment | Gemmeza Res. Sta. yield (kg/ha) | nodule no. | Site 1 Minia yield (kg/ha) | nodule no. |
|---|---|---|---|---|
| 1- Control | 842 | 14 | 3171 | 41 |
| 2- Local inoc. | 1236 | 231 | 3308 | 123 |
| 3- Aleppo inoc. | 1089 | 25 | 3279 | 103 |
| 4- Aleppo inoc. + 36 $P_2O_5$ (kg/ha) | 990 | 27 | 3316 | 90 |
| 5- Aleppo inoc. + 72 $P_2O_5$ (kg/ha) | 1087 | 41 | 3446 | 61 |
| 6- Aleppo inoc. + 18 N + 72 $P_2O_5$ (kg/ha) | 1249 | 26 | 3466 | 84 |
| 7- Aleppo inoc. + 36 N + 72 $P_2O_5$ (kg/ha) | 1088 | 49 | 3735 | 80 |
| 8- 120 N equally split + 72 $P_2O_5$ (kg/ha) | 1072 | 11 | 3349 | 70 |
| LSD 5 % | 120 | 37 | 315 | — |

At Gemmeza, Giza 3 responded positively to both *Rhizobium* cultures and N + P. *Rhizobium* inoculation significantly increased seed yield; the local inoculum producing a better seed yield and a greater number of nodules than the introduced one. With Aleppo culture, the application of 18 kg N + 72 kg $P_2O_5$ resulted in the highest seed yield. At the site in Minia , *Rhizobium* inoculation with both the local or introduced inoculum resulted in a slight, but statistically non-significant increase in seed yield. The number of nodules was significantly higher in both treatments. With Aleppo culture, increasing the level of nitrogen application from 18 to 36 kg/ha did not result in any significant increase in seed yield. However, the application of 36 kg N + 72 kg $P_2O_5$, along with *Rhizobium* inoculation, significantly increased the seed yield compared with the two inoculated treatments without N.

*Irrigation :* Previous investigations indicated that it was necessary to ensure that the plants were not exposed to soil moisture stress, particularly during the critical pod-filling stage.

El- Gibali *et al.* (7) found that six irrigations at 20-day intervals resulted in highest seed yield from Giza 2 at Mallawi (Middle Egypt), while at Mataana (Upper Egypt) irrigations at 15-20 day intervals gave the best yield. Tawadros *et al.* (22) reported that the highest seed yield was obtained at Sids (Middle Egypt) when irrigation took place at 50-60% depletion of available soil moisture (about 18-day intervals). Irrigation at 80-90% depletion (about 26-day intervals) resulted in the lowest yield of seeds. Tawadros *et al.* (22) also found that both frequent light irrigations and heavy applications (4098 to 5288 $m^3$/ha) at five irrigations, produced the highest seed yield. Thus, from the economic stand-point, frequent intervals with light applications, which save about 1,200 m$^3$ water per hectare, are recommended. Tawadros *et al.* (23) pointed out that the peak water use by faba beans was found to be during February and March because of rising temperatures and vigorous growth of the plants.

Ahmed (2) has pointed out that increasing the number of irrigations from 1 to 2, 3 and 4 increased the net assimilation rate, dry weight, tillers, pods and seed yield per plant.

Miseha *et al.* (16) found that the water requirements of faba beans at Sids (Middle Egypt) were 4517, 3113 and 2856 m$^3$/ha for 15, 30, 45 days intervals. They showed frequent intervals (15 days) to be the best treatment (4517 m$^3$/ha) and that one ardab (155 kg) of faba beans was produced by 189 m$^3$ of water. Metwally (15) at Giza, and El-Maghraby (4) at Moshtohor (South Delta), showed that five irrigations gave the best yield of faba beans, and that increasing the depletion of the available soil moisture before irrigations decreased the seed yield. Hakam and Ibrahim (8) detected no significant differences due to the number and time of treatments in the North Delta, probably because of the higher rainfall of this region. But in Middle and Upper Egypt, four irrigations at 30-day intervals or three irrigations given at 30, 60, 120 days after sowing, gave significantly higher yields than all other treatments.

Recently, Tawadros (24) concluded that seed yield did not respond significantly to different soil moisture levels at Sakha (North Delta) because of the high rainfall. However, at Sids (Middle Egypt) and Mataana (Upper Egypt) soil moisture stress badly affected the yield of faba bean plants. Maximum seed yield was obtained from the treatment in which irrigation water was applied at 60% depletion. It has been found also that seed yield of faba beans responded to an increasing number of irrigations during the pod-development stage; treatments having three irrigations during this period yielded about 0.22 ton/ha more than those having only two irrigations.

*Tillage :* Corn, cotton and rice are the three major summer crops preceding faba beans in the rotation in Egypt.

A survey in Minia Province during September 1979 indicated that about 57% of the bean acreage was preceded by corn and 24% by cotton.

About 50% of the farmers did not till their soil and used corn and cotton ridges for bean sowing.

In the Delta, where rice is the major summer crop before faba beans, tillage is not common, and the farmers plant beans in untilled soils after rice.

A study was carried out at Sakha in 1979-80 on tilled and non-tilled soil following rice. 25 cultivars and breeding lines were evaluated under the two systems. With the exception of line 129/1813/76 all entries produced lower yields when tillage was not carried out. The mean yield of the 25 entries in tilled soil was 3.12 t/ha compared to only 2.32 t/ha in non-tilled soil.

A similar trial at Gemmeza Research Station, on 36 cultivars and breeding lines sown either into tilled soil or directly into the ridges of the preceeding cotton crop, produced the opposite result. The mean yield of the entries planted into non-tilled soil (2.583 t/ha) was significantly higher than in tilled soil (2.374 t/ha). A significant cultivar x tillage interaction was found; 16 entries responded positively to no-tillage, eight responded negatively and 12 showed no effect of tillage. Giza 2 yielded higher in untilled soil whereas F402 did not respond. A similar result was obtained at Sids Research Station.

*Intercropping :* Although considerable increase in average yield of faba beans has been achieved through the improvement of cultivars and agronomic practices, the area has decreased progressively because of competition with other winter crops, mainly Berseem clover and wheat.

It was thought that this reduction in acreage might be compensated by intercropping faba bean with other crops which could minimise the gap between national production and consumption, as well as adding to farmers' incomes.

Intercropping faba bean with autumn planted sugar cane (16) gave high yields, averaging 2.176 t/ha, while at the same time, no significant differences were detected between yields of solid and intercropped sugar cane.

Recent studies at the Sugar Crops Research Section, Agricultural Research Center, indicated that increasing the rate of fertilizer for the intercropped plantings from 214 kg N + 35 kg $P_2O_5$/ha to 275 kg N + 71 kg $P_2O_5$ + 67 k/ha resulted in significant increases of 16% and 12% for sugar cane and faba bean respectively.

Intercropping faba bean with sugar cane is now a common practice in the cane growing areas of Middle and Upper Egypt.

## REFERENCES

1.  Abdallah Hussein, M., 1966. Effect of methods and rates of seeding on broad bean *(Vicia faba* L.) yield and other characters. Bull. Fac. Agric., Cairo Univ., 17, 11-23.

2.  Ahmed, M.M.E., 1973. Growth and yield response of horse bean to plant population and nitrogenous fertilizer. M.Sc. Thesis, Fac. Agric., Ain Shams Univ.

3.  Ali, A.M., 1969. Influence of seeding date and harvest time on yield and its components in beans *(Vicia faba* L.) M.Sc. Thesis, Fac. Agric., Cairo Univ.

4.  El-Maghraby, S., 1980. Effect of water regime, nitrogen and phosphatic fertilizers on growth and yield of broad bean. M.Sc. Thesis, Dept. of Agronomy and Agricultural Engineering, Fac. of Agric. Sc., Moshtohor, Zagazig Univ. Egypt.

5.  El-Sherbeeny, M.H., 1977. Symbiotic variation in *Vicia faba.* Ph.D. Thesis, University College of Wales, Aberystwyth, UK.

6.  El-Morshidy, M.A., Khalifa, M.A. and Hassaballa, E.A., 1977. Seed yield of broad beans *(Vicia faba* L.) as affected by plant density. Contribution from Agron. Dept. Assuit Univ. (Assuit J. Agric. Sci. in press).

7.  Gibali, A.A., Shenouda, N., Badawi, A.Y and Mansoor, S.F. 1968. Irrigation requirements, frequency and its effect on yield and quality of horse bean grains in Middle Egypt. Agric. Res. Rev., 46, 91-98.

8.  Hakam, M.M., Ibrahim, A.A., 1973. Cultural practices of grain legumes in A.R.E. Improv. and prod. Field Food Crops. First FAO/STDA Semi. for plant scientists from Africa and Near East, Cairo, Sept.

9.  Hamissa, M.R., Abdel-Samie, M.E., El-Banna, E.A. Abdel-Aziz, M.S., Abdel-Bary, S. and Abdel-Moneim, M.I., 1975. Response of bean plant *(Vicia faba)* to N, P and K. Agric. Res. Rev. 53, 155-166.

10. Hamissa, M.R., 1980. Annual Coordination Meeting, 1979-80, Cairo, Egypt, August 25 to 27.

11. Hamissa, M.R. 1973. Fertilizer requirements for broad beans and lentils. Improvement and Production of Field Food Crops, First FAO/SIDA Seminar for Plant Scientists from Africa and Near East. Cairo, Egypt. Sept.

12. Hassaballa, E.A., El-Morshidy, M.A., Khalifa, N.A. and Abdel-Rahman, K.A., 1977. Seed yield of broad bean *(Vicia faba* L.) as affected by different fertilizer rates. Contribution from Dept. Agron., Assuit Univ. (in press).

13. Ibrahim, A.A., Nassib, A.M., Khalil, S.A. and El-Sherbeeny, M.H., 1980. Faba bean breeding and agronomy, Annual Coordination Meeting, 1979-80 Cairo, Egypt, August 25 to 27.

14. Kassem, E.S., Hassaballa, E.A., El-Morshidy, M.A. and Khalifa, M.A., 1977. Flowering, pod formation and seed yield and its components of broad beans *(Vicia faba* L.) as affected by different levels of N-P-K combinations. Contribution from Agron. Dept. Assiut Univ. (in press).

14. Metwally, M.A. 1973. Study of the effect of irrigation and fertilization on yield and technological properties in field bean *(Vicia faba* L.). M.Sc. Thesis, Fac. Agric. Al-Azhar University.

16. Miseha, W.I., Hassanen, M.A., Tawadros, H.W. Abdel Rasool, S.F., 1971. Effect of irrigation regime and nitrogen fertilizer on the yield and yield component of field beans Fertilization Conf. Ain Shams Univ. (From Deptl. Abstr.).

17. Mohamed, L.K., 1972. Physiological response of the field bean plants to planting dates and population density. Ph.D. Thesis, Fac. Agric. Ain Shams Univ.

18. Nour, A.H., Ghafaga, E.E., Elsamia, M.A., Eladawi, M.K., Goher, M.A., 1971. Intercropping bean *(Vicia faba)* with sugar cane in U.A.R. Sugar Journ. Vol. 3 No. II, April, U.S.A.

19. Rizk, M.A.M., 1973. The effect of density and distribution of plants on the yield and its components in two varieties of field bean *(Vicia faba* L.). M.Sc. Thesis, Fac. Agric., Cairo Univ.

20. Salem, A.H., 1969. Fruit setting and yield of bean plant *(Vicia faba* L.) as affected by some agronomic factors. M.Sc. Thesis. Fac. Agric., Ain Shams Univ.

21. Shami, A., 1958. Effect of some cultural treatments on the growth and yield of *Vicia faba* L. M.Sc. Thesis, Fac. Agric. Ain Shams Univ.

22. Tawadros, H.W., Miseha, W.I., El-Serogy, S. 1969. Influence of irrigation practices on the yield of horse beans. First Arab Conference of Physiological Sciences, Cairo.

23. Tawadros, H.W., Miseha, W.I., Hassanen, M.A. and Abdel Rasool, S.F., 1971. Effect of irrigation regime and nitrogen fertilizer on the water consumptive use by field beans. Fertilization Conf. Ain Shams Univ. (From Deptl. Abstr.).

24, Tawadros, H.W., 1980. Water management. Annual Coordination Meeting, 1979-80, Cairo, Egypt, 25 to 27.

25. Unpublished data, Sugar Crops Res. Section, Field Crops Res. Institute, Agric. Res. Center, Giza, Egypt.

# 11. FABA BEAN AGRONOMY IN THE SUDAN

OSMAN A.A. AGEEB

*Gezira Research Station, P.O. Box 126, Wad Medani, Sudan*

## Sowing time

The crop growing season is short as it is limited by high temperatures and disease stress at both beginning and end. Also, the high day temperatures during the season tend to set an upper limit to potential maximum yields. Under such conditions, timely sowing of the crop is of paramount importance.

Last and Nour (23) found that the optimum date of sowing beans in northern Sudan was affected by pests and diseases. In experiments at Shambat, crops sown in early November gave the largest yields. Earlier sowing tended to be seriously damaged by insects, especially *Helicothrips* spp. At Zeidab, 175 miles north of Khartoum, where the winter begins earlier, high yields were given by crops sown from mid-October to early November. Yields drop very rapidly when sowing is delayed after mid-November, as early-sown crops develop most of their pods before powdery mildew becomes widespread.

Heipko (19) believed that the negative effects of powdery mildew could be reduced or avoided by early sowing in the second half of October as other husbandry practices e.g. spacing and irrigation practices, seem to offer comparatively little prospect for overcoming the yield reductions which this disease causes.

Abu Salih *et al.* (1) showed that delaying the sowing of faba beans after October greatly lowered the yield of seed as a result of an increased infestation by aphids, and incidence of Sudanese broad bean mosaic virus (SBBMV). The mean seed yield in the two seasons for plots sown in October, November and December was 1387, 494 and 50 kg/ha respectively. Seed yield was found to be positively correlated with number of pods per plant ($r = 0.96^{**}$) but negatively correlated with percentage infection with SBBMV ($r = -0.95^{**}$).

The scientists considered that the optimum period for sowing at Hudeiba Research Station was around mid-October. Sowing earlier in 1971-72 resulted in low yield, probably because of wilt and root-rot diseases attributed to *Fusarium oxysporum* and *F. solani* F. sp. *Fabae* respectively (20). In recent years, both diseases have become the major factors affecting crop yield of plants sown in early October (16) (17).

Ageeb (4) showed a highly significant effect of sowing date, watering interval and their interaction on grain yield (Table 1). The optimum sowing date varied with irrigation interval, being in the second half of October and the first week of November when the plots were watered weekly or 14 and 21 days intervals respectively. This was because irrigation

*G. Hawtin & C. Webb (Eds.). Faba Bean Improvement*
©*1982 ICARDA. ISBN -13:978-94-009-7501-9*

Table 1. The effects of sowing date and watering interval on seed yield of faba beans (kg/ha) (Ageeb, 1977).

| Watering interval (days) | October | | | | November | | | | | December | | | | Mean |
|---|---|---|---|---|---|---|---|---|---|---|---|---|---|---|
| | 4 | 11 | 18 | 25 | 1 | 8 | 15 | 22 | 29 | 6 | 13 | 20 | 27 | |
| 7 | 1308 | 2081 | 2386 | 2209 | 2043 | 2150 | 1820 | 1653 | 1377 | 1250 | 1000 | 787 | 547 | 1585 |
| 14 | 333 | 927 | 1338 | 1379 | 1459 | 1578 | 1344 | 1289 | 847 | 928 | 763 | 572 | 413 | 1013 |
| 21 | 231 | 512 | 889 | 1017 | 1033 | 1083 | 921 | 896 | 695 | 476 | 449 | 344 | 282 | 679 |
| S.E. | | | | | | | ±81.5 | | | | | | | ±110.7 |
| Mean | 624 | 1173 | 1538 | 1535 | 1512 | 1604 | 1362 | 1279 | 973 | 885 | 737 | 568 | 414 | |
| S.E. | | | | | | | ±25.6 | | | | | | | |

at weekly intervals during October, when the temperatures are relatively high, creates a micro-climate which is less favourable for wilt and root rot development and more favourable for plant growth and survival (17). Longer watering intervals in October lead to significant reductions in plant stand.

The number of pods per plant was highest (34 pods/plant) for the early October-sown plants due to the low plant stand and the consequent high degree of branching, but it decreased with delayed sowing, reaching very low levels with the late December sowings (approx. 5 pods/plant). Signficantly more pods were found on the frequently watered plants.

Seed size decreased with both delayed sowings and with moisture stress but the effect of sowing date was more pronounced. Delaying the sowing until after the optimum date exposed the plants to disease and heat stress at the end of the season, thereby limiting their photosynthetic capacities and grain filling. Plant height was also decreased by delaying the watering from 7 to 21 days intervals.

Since it is difficult to convince farmers to water at shorter intervals than two to three weeks, the optimum sowing date for faba beans should be moved to late October- early November.

Freigoun (17) found similar results after a three-year study (1977-1979) on the effect of sowing date and watering interval on incidence of root-rot and wilt disease. He and Salih have found some promising lines (Freigoun, pers. comm.) which, with time, may form a good base for the solution of this disease problem and allow early planting to be practised.

*Water management*

Since the energy crisis of the mid-1970's, lift irrigation by diesel pumps has become expensive. In addition, petrol is sometimes not available at the right time and/or in adequate quantities. Moreover the recent expansion of perennial horticultural crops, mainly citrus and mangoes, has put more strain on the existing irrigation facilities which are often old and in need of frequent repairs. So, whatever optimum conditions are put forward, the farmer, being influenced with the above factors, will decide his own course of irrigation policy. Nevertheless efforts shall continue to improve our knowledge of responses of crops to irrigation and the economics of water use.

An irrigation trial was conducted at Hudeiba Research Station in which three irrigation intervals (7, 14 and 21 days) were applied either throughout the season or interchanged between the vegetative growth phase and the reproductive phase from pod setting to maturity. The average number of irrigations during the season ranged from 14.7 to 4.7 for the wettest and dryest treatments respectively.

Water stress during the first phase had a small effect on grain yield during the first season, but increased during the following two seasons due to a build up of root-rot and wilt diseases and consequent reduction in plant stand.

The results (Table 2) clearly indicated that the second phase of development was more sensitive to water stress. This was because more than 50% of the total number of flowers and most of the pods were formed during this phase. Water stress was found to greatly reduce the formation of fruiting bodies and hence the total number of pods that developed to maturity. These results are in agreement with those of other scientists e.g. Salter and Goode (30) El Nadi (12) and Farah (15). Ali *et al.* (5) reported that weekly irrigations increased seed yield by 42% compared to fortnightly irrigations while Babiker (8) found a

92% increase for the same treatments. Ayoub (7) found that 8-day intervals gave yield increases of 22, 63 and 108% over 13, 18 and 23 day treatments respectively.

Table 2. Effect of watering interval at two growth phases of faba beans on seed yield (kg/ha).

| Watering intervals (days) | | Seasons | | | |
|---|---|---|---|---|---|
| Phase I (vegetative) | Phase II (reproductive) | 1974-75 | 1975-76 | 1976-77 | Mean |
| 7 | 7 | 2528 | 1910 | 2698 | 2379 |
| 14 | 7 | 2527 | 1427 | 2002 | 1985 |
| 21 | 7 | 2313 | 1417 | 1458 | 1729 |
| 7 | 14 | 2032 | 1205 | 1520 | 1586 |
| 14 | 14 | 2042 | 1226 | 1307 | 1525 |
| 21 | 14 | 1782 | 1167 | 1183 | 1377 |
| 7 | 21 | 1601 | 879 | 1108 | 1196 |
| 14 | 21 | 1462 | 792 | 1166 | 1140 |
| 21 | 21 | 1361 | 865 | 999 | 1075 |
| S.E. | | ±74 | ±101 | ±89 | |

Salih (25) showed that the crop greatly benefited from watering until maturity. Early termination of irrigation significantly reduced the yield.

In 1979-80, Fadl (14) showed that a crop watered every 7 days and consequently no suffering from visual symptoms of water stress used 5960 cubic meters of water/ha during the growing season (November to February) to give a grain yield of 1342 kg/ha.

The mean daily ET for this crop which was initially 1.3 mm, reached a maximum of 11.3 mm by mid-January, then gradually decreased to 7.4 mm in early February. It dropped sharply to 2.0 mm at the end of the month. On the other hand, the ratio ET/EO increased from an initial of 0.21 to a peak of 1.75 throughout January 1 to 20, then it dropped to 1.42 at the end of January, 1.10 at the beginning of February and 0.26 at the end of the month.

However when a similar crop was watered every 14 days, it used 4240 cubic meters of water/ha to give a grain yield of 659 kg/ha. Mean daily ET reached a maximum of 6.6 mm in mid-January, and the ET/EO ratio reached a peak of 1.05 at the beginning of January.

## Seed rate and plant spacing

Seed rate : Last and Nour (23) found a 48 and 18% increase in the grain yield of Beladi (B.F.M.) and an introduced cultivar (R. 34) respectively when the seed rate was increased from 78.6 to 157.2 kg/ha. They believed that the actual seed rate used by farmers, at least for the local variety was sub-optimum.

On the other hand, Heipko and Dafalla (18) found no significant difference in the grain yield of B.F.M. and R 40 when seed rates of 67, 100 and 133 kg/ha were used.

El Saeed (13) found that the yield of Beladi increased with an increase in the seed rate from 38 to 302 kg/ha but the amount of increased yield diminished beyond 151 kg/ha.

*Plant spacing :* In a trial at Hudeiba in 1976 and 1977, Ageeb (3) showed that variation in row spacing (60, 40 and 20 cm), plant spacing (20, 10 and 5 cm) and number of plants per hole (1 or 2) had no significant effect on the grain yield of the Hudeiba 72. (Table 3). The results reflected the plastic behaviour of H. 72 which makes it insensitive to wide variations in plant population as a result of the compensatory effect of the variation in number of pods/plant as neither seeds per pod nor seed size was significantly affected by the treatments used.

Table 3. The effect of plant and row spacing on the seed yield (kg/ha) of faba beans at Hudeiba.

| Row spacing (cm) | Season | | Average |
|---|---|---|---|
| | 1975-76 | 1976-77 | |
| 20 | 2236 | 2254 | 2245 |
| 40 | 2452 | 2438 | 2445 |
| 60 | 2394 | 2387 | 2395 |
| S.E. | ±69 | ±55 | |
| Plant spacing (cm) | | | |
| 5 | 2294 | 2311 | 2303 |
| 10 | 2395 | 2356 | 2376 |
| 20 | 2393 | 2413 | 2403 |
| S.E. | ±69 | ±55 | |
| Number of plants per hole | | | |
| 1 | 2349 | 2350 | 2350 |
| 2 | 2373 | 2370 | 2372 |
| S.E. | ±52 | ±45 | |

Crop plasticity, to large variations in plant population, was also evident in the work of Ishag (22) who did not find any significant reduction in grain yield when plant spacing was varied between 10 and 30 cm or when 2 or 3 plants/hole were used, but yield was significantly decreased with 40 cm or single plants/hole. Salih (27) also did not find any significant effect on the grain yield of four cultivars when he compared the planting of one row or two rows per ridge.

However in the marginal agroclimatic zone of Gezira Research Station, the crop responded differently to variation in plant spacing as decreasing plant spacing from 20 to 10 cm, and increasing the number of plants/hole from 1 to 2 or 3 significantly increased grain yield (Table 4).

It is quite clear that the branching indeterminate cultivars adaptable to the growing conditions of the Northern Region have a high degree of plasticity to wide variation in plant stand . The recommendation which came out of Hudeiba to plant the crop at a spacing of 60 x 20 cm with 2 to 3 plants/hole, or to use 70-120 kg of seed/ha is adequate for obtaining a good seed yield.

Table 4. The effect of plant population on seed yield of faba beans at the Gezira Research Station.

| Treatments | |
|---|---|
| Plant spacing (cm) | Seed yield (kg/ha) |
| 10 | 1016 |
| 14 | 903 |
| 20 | 878 |
| S.E. | ±34.1 |
| Plants per hole | |
| 1 | 800 |
| 2 | 980 |
| 3 | 1017 |
| S.E. | ±34.1 |

## Fertilizers

Nitrogen : Soils of the Sudan are deficient in nitrogen, but most of the indigenous leguminous crops including faba beans nodulate well with the naturally occurring Rhizobia, and do not respond to fertilizer nitrogen.

Ishag (21) showed that maximum nodulation occurred at 9 weeks from sowing, and that inoculation with Sudanese or French strains helped to increase nodulation and persistance of nodules in plants (Table 5). However, this was not reflected in significant increases in grain yield. As expected, addition of fertilizer nitrogen significantly depressed nodulation; more so in the presence of the French strains.

Table 5. Effect of inoculation and nitrogen on nodulation and seed yield (Ishag, 1970-71).

| Treatments | Number of nodules per plant at | | | Dry wt. of nodules per plant at 12 weeks (mg) | Grain yield (kg/ha) |
|---|---|---|---|---|---|
| | 6 weeks | 9 weeks | 12 weeks | | |
| Control | 61.8 | 162.5 | 63.4 | 148 | 2052 |
| Nitrogen | 47.8 | 104.0 | 54.9 | 154 | 2195 |
| Rhizobium (F) | 59.0 | 143.0 | 95.8 | 198 | 2307 |
| Nitrogen + (F) | 46.3 | 93.0 | 29.0 | 78 | 1979 |
| Rhizobium (S) | 72.0 | 168.8 | 90.0 | 295 | 2164 |
| Nitrogen + (S) | 45.5 | 146.5 | 55.0 | 131 | 2169 |
| S.E. | ±6.5 | ±7.7 | ±13.6 | ±37.0 | ±171 |

F and S denote French and Sudanese strains respectively.

Other workers (6) (9) (24) did not get significant increases from nitrogen. However Salih (24) indicated that application of nitrogen after one or two months from planting might increase grain yield.

*Potassium and phosphorus :* As the clays of the Sudan are known to be rich in potassium faba beans are expected not to respond to the application of potassium fertilizer (6) (9). The situation for phosphorus is complex. As the soil is alkaline (ph $>$ 8.0) a very small amount of phosphorus is available to plants because of chemical fixation. In such a situation, placement of phosphorus near to the plant's absorbing system is important. No response was found at Shambat to phosphorus when it was broadcasted on top of the ridges, but a positive response occurred when a similar amount was placed in the holes with the seeds.

*Soil salinity :*
Faba beans are traditionally grown in the fertile "Gureir" soils which are silty loams deposited in a narrow strip along the banks of the Nile. Due to the scarcity of good soils and the high demand for faba beans, coupled with its ever increasing prices, new lands of inferior quality such as "kuru" and "terrace" which are affected to various degrees of alkalinity and salinity, are coming into production. Faba beans are not expected to do as well in these soils.

On saline alkaline (ph 8.7) soils at Soba Research Station, Karouri (1979) showed that the salinity level at which 50% reduction in bean yield occurred was 9 mmho/cm. The negative correlation between yield and ECe (r = −0.883) together with field observations, made him believe that soil salinity may be responsible for the different growth and yield of faba bean fields in the salt-affected areas of the Northern Region.

*Seed quality*

*Seed size :* Salih and Salih (29) have clearly indicated that grading faba beans seed would be of no economic value.

*Seed age :* Salih and Salih (28) found that storing of faba beans seed under ordinary laboratory conditions had little effect on germination percentages (92-87%) but, for longer periods, the percentage of germination declined more rapidly reaching 20% after 8 years. Grain yield changed gradually from 1738 to 1105 kg/ha when 1 and 5-year-old seeds were used respectively. The drop in yield was then more rapid until it was only 105 kg/ha when 8-year-old seeds were sown (Table 6).

Table 6. Effect of seed age on germination and yield of faba beans (Salih and Salih, 1975-1976).

| Seed age (years) | Germination (%) | Field emergence (%) | Seed yield (kg/ha) |
|---|---|---|---|
| 1 | 92.3 | 92.1 | 1738 |
| 2 | 91.3 | 86.1 | 1567 |
| 3 | 93.5 | 75.1 | 1410 |
| 4 | 94.8 | 71.3 | 1238 |
| 5 | 87.7 | 48.7 | 1105 |
| 6 | 44.0 | 40.8 | 210 |
| 7 | 34.6 | N.A. | N.A. |
| 8 | 19.9 | 13.7 | 105 |

N.A. = Not available

*Time of harvest :*

Ageeb (3) showed clearly (Table 7) the optimum time for harvest of H. 72 was 110 to 120 days from sowing, as harvesting 10 days earlier or later than that will result in significant reductions in grain yield. Grain yield increased at a linear rate between 90 and 110 days (from 559 to 2186 kg/ha) as a result of increase in grain size, then levelled at 110 to 120 days after which it significantly dropped because of pod shedding. The percentage of hard seed, which greatly affects cooking quality, was at its lowest after full maturity.

In view of the sharp increase in the price of faba beans, especially during the last four years, great and collective efforts have to be made in many aspects of production to improve the commodity situation, partly in the intensification of research efforts for short-season better adapted cultivars with built-in tolerance for heat and disease stress and the perfection of optimum and economic packages of cultural practices for farmers' fields.

The encouraging results that are already coming out of the limited faba bean trials at Gezira Research Station could form the basis for a coordinated research project to look into the possibility of extending the cultivation of faba beans to the irrigated schemes in the northern part of central Sudan.

Table 7. Effect of time of harvest on seed yield and other characteristics at Hudeiba (Ageeb 1976-77).

| Time of harvest (days after sowing) | Grain yield (kg/ha) | Pods per plant | 100 seed wt (g) | Seed hardness (%) |
|---|---|---|---|---|
| 90 | 559 | 29 | 20 | 19 |
| 100 | 1377 | 30 | 29 | 15 |
| 110 | 2186 | 28 | 33 | 10 |
| 120 | 2149 | 25 | 34 | 11 |
| 130 | 1793 | 24 | 34 | 6 |
| 140 | 1627 | 20 | 35 | 8 |
| S.E. | ±94.4 | ±3.0 | ±0.7 | ±1.4 |

REFERENCES

1. Abu Salih, H.S., Ishag, H.M., Siddig, S.A. 1973. Effect of sowing date on the incidence of Sudanese broad bean mosaic virus in, and yield of *Vicia faba*. Ann. Appl. Biol. 74, 371-378.

2. Ageeb, O.A.A. 1975-77. The effect of watering interval at different stages of growth on the grain yield of two varieties of faba beans. Annual Reports, Hudeiba Research Station, Ed-Damer, Sudan.

3. ---------------- 1976-77. Effects of plant and row spacings on the grain yield of faba beans. Annual Reports, HRS, Sudan.

4. ---------------- 1977. The effect of watering interval and sowing date on plant survival and grain yield of faba beans. Annual Reports, HRS, Sudan.

5. Ali, M.A., Khalifa, O., Baghdadi, A.M. 1965-67. Effect of cultural practices on powdery mildew and yield of faba beans. Annual Reports, HRS, Sudan.

6. Ayoub, A.T. 1971. Foul masri (faba beans) fertilizer experiment. Annual Report, HRS Sudan.

7. ---------------- 1971-73. Effect of watering interval on grain yield and quality of faba beans. Annual Reports, HRS, Sudan.

8. Babiker, I.A. 1975. Foul masri (broad bean) nitrogen, phosphate application and watering interval. Annual Report, HRS, Sudan.

9. ---------------- 1975-77. Broad bean NP fertilizer and time of application. Annual Rept. HRS, Sudan.

10. El Karouri, A.M.O. 1979. A review of literature on research carried out on broad bean *(Vicia faba* L.) in the Sudan. Report prepared on the request of and presented to ICARDA.

11. El Karouri, M.O.H. 1979. Effects of soil salinity on broad bean *(Vicia faba* L.) in the Sudan. Expl. Agric. 15, 59-63.

12. El Nadi, A.H. 1970. Effects of differential irrigation on yield and seed size of broad beans. Expl. Agric. 6, 107-111.

13. El-Saeed, E.A.K. 1968. Agronomic aspects of broad beans *(Vicia faba* L.) grown in the Sudan. Expl. Agric. 4, 151-159.

14. Fadl, O.A.A. 1981. Water requirements of irrigated faba beans *(Vicia faba* L.) in Sudan Gezira. FABIS

15. Farah, S.M. 1979. An examination of the effects of water stress on leaf growth of crops of field beans *Vicia faba* L. Ph.D. Thesis, University of Reading.

16. Freigoun, S.O. 1975. Annual Report, HRS, Sudan.

17. ---------------- 1980. Effect of sowing date and watering interval on the incidence of wilt and root rot diseases in faba beans FABIS 2, 41.

18. Heipko, G.H., Dafalla, D.A. 1961-62. Annual Report, HRS, Sudan.

19. Heipko, G.H. 1966. Agronomy of traditional Northern Province crops. Annual Report, HRS, Sudan.

20. Ibrahim, G., Hussein, M.M. 1975. A new record of root rot of broad bean *Vicia faba* from the Sudan. J. Agric. Sci. 83, 381.

21. Ishag, H.M. 1970-71. Effect of seed inoculation on plant growth and yield of field beans. Annual Report, HRS, Sudan.

22. ---------------- 1970-71. Effect of spacing between holes and number of seeds per hole on yield and yield components of field beans. Annual Report, HRS, Sudan.

23. Last, F.T., Nour, M.A. 1961. Cultivation of *Vicia faba* L. in Northern Sudan. Empire Journ. of Exper. Agric. 29, 60-72.

24. Salih, F.A. 1976-78. Effect of level and time of nitrogen application on the yield of faba beans. Annual Reports, HRS, Sudan.

25. ---------------- 1977-78. Water closure experiment. Annual Reports, HRS, Sudan.

26. ---------------- 1977-79. The effect of time of harvest on the grain yield and quality of broad bean. Annual Reports, HRS, Sudan.

27. ---------------- 1978-79. Effect of different methods of sowing on the grain yield of five varieties of faba beans. Annual Reports, HRS, Sudan.

28. Salih, S.H., Salih, F.A. 1975-76. The effect of age of seed on viability and field performance of field bean. Annual Report, HRS, Sudan.

29. ---------------------- 1980. Influence of seed size on yield and yield components of broad bean *(Vicia faba* L.). Seed Sci. and Technol. 8, 175-81.

30. Salter, P.J., Goode, J.E. 1967. Crop responses to water at different stages of growth. Commonwealth Agricultural Bureaux, Farnham Royal.

## 12. SYMBIOTIC NITROGEN FIXATION IN FABA BEANS

YOUSEF A. HAMDI

*Department of Microbiology, Institute of Soil and Water Research, Agricultural Research Centre, Giza, Egypt*

Nitrogen fixation in faba beans is associated with symbiosis with nodule bacteria, *Rhizobium leguminosarum* Frank. This species of *Rhizobium* is also associated with plants in the genera *Pisum, Vicia* and *Lathyrus* and infects clovers, forming ineffective nodules.

*Nodule Structure*

The first drawings of infection threads in legumes were done with *Vicia faba* by Marshal Ward in 1887 (Nutman, 1958). He thought that infection threads were fungal hyphae. It was shown later that the infection thread is laid down by the host and is composed externally of cellulose, and internally of hemicellulose substances embedding the nodule bacteria.

Histologically, the faba bean nodules are composed of cortex, non-bacteroidal tissue, meristmatic tissue, bacteroidal tissue and vascular bundles (15). Table 1 shows that the percentage of bacteroidal tissue increases with time to reach 54% at 40 days of growth.

Table 1. Structure of Nodules of *Vicia faba* cv. Giza 1 (El-Mokadem, 1974).

| Time days | Total nodule area $mm^2$ | Cortex $mm^2$ | Non-bacteroidal tissue $mm^2$ | Meristematic tissue $mm^2$ | Bacteriodal tissue $mm^2$ | Percentage of bacteriodal tissue of total area of nodule |
|---|---|---|---|---|---|---|
|    | 0.78± | 0.224 | 0.209 | 0.035 | 0.298 | 39.02 |
| 20 | 0.0589 | ±0.038 | ±0.054 | ± 0.005 | ± 0.02 | |
|    | 0.73 | 0.191 | 0.185 | 0.079 | 0.289 | 39.01 |
| 30 | ±0.04 | ±0.022 | ±0.025 | ± 0.017 | ± 0.025 | |
|    | 1.186 | 0.253 | 0.169 | 0.112 | 0.643 | 54.29 |
| 40 | ±0.061 | ±0.035 | ±0.015 | ± 0.02 | ± 0.039 | |

*G. Hawtin & C. Webb (Eds.): Faba Bean Improvement*
©1982 ICARDA. ISBN -13:978-94-009-7501-9

*Population Density of Rhizobium leguminosarum in Soils*

Although usually associated with legume root nodules, the *Rhizobia* are capable of an in-dependent saprophytic existence in soil which does not contain legumes. They can survive in this condition for several years but become drastically reduced in numbers or will disap-pear in the continued absence of an appropriate host. Nutman and Ross (36) surveyed the populations of rhizobia in differently manured plots of the Park Grass experiment at Rothamsted which, for more than a century, has been meadow land cut for hay. *R. leg-uminosarum* counted by using *V. faba* were present in most plots, especially limed plots. In a permanent pasture in Kent (sown before 1956) on a surface-dried clay soil of pH 6.7 and containing white clover, more than $10^4$ to $10^5$ *R. leguminosarum* cells/g were counted (De Escuder, 1972). In soils of Yugoslavia (Vijinoric 1970), log counts of *R. legumino-sarum* were : chernozemes; 1.8 to 3.4; somitza, 3.8 to 4.8; brown forest, 3.4 to 4.4; allu-vial, 2.8 to 3.4, and pseudogley, 1.0 to 3.4. Candlish and Clerk (1975) indicated that nod-ulation of faba bean planted in Manitoba soils without inoculation, was either spotty or lacking.

Population density of *R. leguminosarum* varies in the soils of Egypt according to lo-cation, rotation, drying and waterlogging. Average number of nodules/plant in non-inocu-lated fields in Upper Egypt ranged between 13 and 170 with an average of 79 nodules/plant, weighing 0.36 g/plant. In the Delta, the number of nodules ranged between 30 and 137 with an average of 94 nodules/plant weighing 13 g/plant (23). The fluctuation in pop-ulation of *R. leguminosarum* was affected by crop rotation and time of the year (31).

Drying soil (Sharaqi practice) i.e. leaving soils fallow for extended periods, caused a significant decrease in numbers of *R. leguminosarum* in soils (31) (Table 2). The decrease amounted to 63, 84, 95 and 98% of the density of population at the start after periods of 45, 60, 120 and 150 days.

Table 2. Effect of soil drying(Sharaqi)on survival of *R. leguminosarum* (Loutfi *et al.,* 1980).

| | DAYS | | | | |
|---|---|---|---|---|---|
| | Start | 45 | 60 | 120 | 150 |
| Counts/g | $16.8 \times 10^3$ | $62.4 \times 10^2$ | $27 \times 10^2$ | $8.4 \times 10^2$ | $3.7 \times 10^2$ |
| Survival percentage | 100 | 37 | 16 | 5 | 2 |

*Survival of R. leguminosarum on the seeds*

Date (12) showed that the form of inoculum, temperature, pelleting materials and host species and *Rhizobium* strains were major factors controlling the survival of rhizobia on seeds. *R. leguminosarum* population decreases when inoculated on seeds of different species or inoculated with different methods (Table 3).

Table 3. Number of *R. leguminosarum* surviving on host seeds out of 10,000, after 9 and 20 days in the Laboratory at $25^\circ C$ or in the field, in acid soil. (Date, 1969).

| | *Rhizobia* per seed | | | |
| --- | --- | --- | --- | --- |
| | Laboratory | | Field | |
| Inoculation method | 9 days | 20 days | 9 days | 20 days |
| Peat slurry | 420 | 34 | 330 | 54 |
| Gum arabic - no coating | 600 | 35 | 560 | 9 |
| Gum arabic - lime coating | 650 | 32 | 91 | 2 |

*Response to Inoculation*

Candlish and Clark (18) found that all inoculated seeds of Diana and Ackerperle cultivars of faba bean *(Vicia faba,* v. *minor)* in the greenhouse and field (Table 4) produced well nodulated plants. The total amount of nitrogen fixed was 840 mg/plant. The faba bean continued to fix nitrogen after forming pods.

Table 4. Growth, ethylene production and nitrogen content of inoculated Diana and Ackerperle faba beans* (Candlish and Clark, 1975).

| Cultivar Age, wks. | Plant wt., g | Ethylene production μ moles/plant/hr | Nitrogen percentage | g/plant |
| --- | --- | --- | --- | --- |
| | | Diana (Experiment 1) | | |
| 4 | 0.92±0.07 | 0.304±0.09 | -- | -- |
| 6 | 2.63±0.59 | 0.147±0.04 | -- | -- |
| 8 | 3.75± 0.35 | 0.137±0.01 | 4.3 ±0.05 | 0.05 ±0.00 |
| 11 | 6.62±1.03 | 0.137±0.04 | 3.57±0.71 | 0.08 ±0.00 |
| 15 | 18.87±3.79 | 0.032±0.00 | 4.44±0.53 | 0.28 ±0.06 |
| | | Ackerperle (Experiment 2) | | |
| 3 | 4.03±0.41** | 0.18 ±0.06 | -- | -- |
| 5 | 5.53±0.95** | 0.27 ± 0.09 | -- | -- |
| 7 | 34.00±4.10** | 16.47 ± 0.87 | -- | -- |
| 7 | 2.17±0.29 | -- | 4.55± 0.03 | 0.10 ±0.01 |
| 10 | 7.23±2.02 | 14.44 ± 6.5 | 3.80 ±0.63 | 0.27 ±0.05 |
| 13 | 24.23± 2.36 | 38.62 ± 20.6 | 2.82 ±0.57 | 0.68± 0.02 |
| 13*** | | 38.84 ± 11.0 | | |

*Plants inoculated with Nitrogen inoculum.
**Fresh weight of plants, in all other instances dry wt. of plants was recorded.
***Three plants for nitrogen assay.

Studies in Egypt on the increase of crop yield of faba bean through inoculation are inconclusive. Taha *et al* (48), showed that calcium nitrate increased the seed yield of uninoculated plots of faba bean. Inoculation and application of 117 kg N/ha showed a significant increase in yield of seeds. Further increase of nitrogen fertilizers reduced the yield of seeds but increased the dry weight of straw.

Application of phosphorus to the soil, without inoculation, generally gave a marked increase in yield of faba bean (Table 5). Inoculation increased the yield but when combined with phosphorus, gave significantly higher yields than inoculation alone. The best economic rate of superphosphate was 117 - 233 kg/ha for faba bean.

Table 5. Effect of inoculation in presence of superphosphate on faba bean yield and total nitrogen content of the seeds. (Taha *et al.* 1967).

| Phosphate | Mean yield | | Total nitrogen of seeds (mg/g) | |
|---|---|---|---|---|
| kg/ha | Uninoculated | Inoculated | Uninoculated | Inoculated |
| 0 | 1310 | 1550 | 62.9 | 65.4 |
| 117 | 1572 | 1716 | 64.3 | 68.5 |
| 233 | 1716 | 1856 | 64.9 | 72.4 |
| 350 | 1786 | 1856 | 65.2 | 70.7 |
| 466 | 1642 | 1893 | 65.4 | 73.4 |

Loutfi *et al.* (30) showed that two of four experiments in different locations gave a significant increase due to inoculation.

In field experiments at 10 locations of Upper Egypt, significant responses to inoculation in terms of nodule number and weight was observed at 77 - 91 days after planting at all locations. However, a significant seed yield response was only detected in 3 out of the 10 locations, and no significant effect on the yield of straw was obtained (23) (Table 6).

*Nitrogenase Activity of Faba bean Nodules*

Maximum nitrogenase activity of nodules of *V. faba* was detected between 20 and 30°C (11). Sprent and Bradford (45) found that the seasonal pattern of nitrogen fixing (acetylene reducing) activity was markedly affected by population density. At 7 plants/m$^2$, a marked peak of 42 $\mu$ moles $C_2H_4$/plant/hr was observed at the end of June when pods were setting. At the opposite extreme of 200 plants/m$^2$, no such peak was evident. Activity fluctuated about a mean value of about 6 $\mu$ moles/plant/hr. Low activity was always associated with the stress of waterlogging.

Isotopic studies revealed that photosynthates were rapidly metabolized after arrival in the nodules (within 30 min. of feeding the shoot $C^{14}$) (29). The importance of glutamate, glutamine and aspartate as early products of the metabolism of fixed nitrogen, was confirmed by the distribution of radioactivity.

Cooper *et al* (10) using $N^{15}$ in pot trials, showed that the rate of $NO_3$ uptake was relatively constant throughout the experiment, but nitrogen fixation showed a sharp increase after early fruit set. Nitrogen fixation accounted for 80%. Quantitative redistribution of nitrogen in the plants made little contribution to seed nitrogen, in contrast to soybean.

Dean (13) showed that acetylene reduction by faba bean was 37% less in high nitrate soil than in low nitrate soil. At high nitrate locations, yield decreased with inoculation suggesting that more abundant nodules diverted carbohydrates from developing pods.

Table 6. Response of faba bean to inoculation in the field at 10 locations (Hamdi et al., 1980).

| Locations | No. of nodules | | Wt. of Nodules g | | Seeds t/ha | | Straw t/ha | |
|---|---|---|---|---|---|---|---|---|
| | Inoc. | Non Inoc. | Inoc. | Non. Inoc. | Inoc. | Non Inoc. | Inoc. | Non Inoc. |
| Etsa | 426** | 180 | 1.67** | 0.50 | 3.04 | 2.68 | 8.22 | 6.40 |
| Samaloot | 212** | 126 | 0.70** | 0.42 | 3.64* | 2.68 | 11.28 | 9.22 |
| Der Samaloot | 364** | 136 | 1.70** | 0.66 | 2.76 | 2.21 | 14.11 | 12.47 |
| Tiba | 332** | 160 | 1.24** | 0.76 | 3.719 | 4.18 | 4.53 | 6.08 |
| Beni Sharawi | 406** | 186 | 1.96** | 1.05 | 3.77 | 3.82 | 9.27 | 9.56 |
| Beni Abaid -S | 302** | 122 | 1.07** | 0.52 | 3.91 | 3.73 | 11.0 | 10.3 |
| Beni Abaid -M | 416** | 170 | 1.19** | 0.34 | 3.11 | 3.75 | 7.9 | 6.7 |
| Al Fekriah | 479** | 285 | 1.76** | 1.36 | 3.6 | 3.3 | 9.4 | 9.2 |
| Etlidam | 325** | 178 | 0.76** | 0.09 | 3.1* | 2.4 | 6.0 | 4.8 |
| Dermwnas | 406** | 189 | 1.1** | 0.5 | 4.2* | 3.4 | – | – |
| Mean | 367 | 173 | 1.3 | 0.61 | 3.2 | 2.9 | 8.11 | 7.2 |
| L.S.D. 1 % | 69 | | 0.26 | | – | | – | |
| 5 % | 51 | | 0.19 | | 0.58 | | – | |

**inoculation treatment significantly different from uninoculated $P \leq 0.01$.

*Amount of nitrogen fixed*

With reference to the amount of nitrogen fixed by faba beans, 45 to 552 kg/ha are reported with an average of 210 kg/ha (35). In Egypt, figures of 277 kg N/ha (40) and 136 kg/ha (38) have been estimated.

Bond reported that *V. faba* fixed 691 mg N/g dry nodule weight and nodule dry weight was 4% of the whole plant.

Richards and Soper (37) showed that the nitrogen requirements were fully met from soil and symbiotic fixation. The faba beans fixed 87.1% of their total nitrogen content, and substantial amounts were fixed after pod-fill.

*Factors affecting responses to inoculation*

*Genetic factors* : Little is known about the genetics of *R. leguminosarum.* Bose and Venkatoramen (7) reported conjugation among different strains. Ineffective or non-nodulating auxotrophic mutants, requiring adenine and thiamine, were reported by Schwinghamer (42).

El-Zawahry (16) used radiation obtained mutants which were more effective and resulted in an increase in total N in the seeds. Infectivity of EMS- and UV-treated *R. legum-*

*inosarum* was decreased in the cultivar Giza 1 (4), however the effectiveness of the nodules was increased in many cases.

Mytton *et al.* (34) indicated that improvement of symbiotic N-fixation in faba beans was most likely to occur through simultaneous selection of both symbionts.

*Temperature :* Temperature affects all stages of symbiosis by influencing bacterial growth and survival, plant growth, nodule formation and nodule functioning. The critical night temperature for nodulation of *Pisum* sp. and *Vicia faba* is $2$-$6^{0}C$ (46).

*Moisture :* Nodulated plants of *Vicia faba* exhibit a high degree of correlation between $N_2$ fixing ($C_2H_2$ reducing) activity and soil water content. Activity is maximal at field capacity and is suppressed when the soil dries out and wilting of the lower leaves begins (43). In a different study, faba bean plants on excess water regime grew better and fixed more nitrogen than those on drought stress. The nodules showed a rapid response to fluctuations in water supply (20).

Abdel Wahab and Zahran (2) reported the depressing effect of moisture stress on faba beans. After one week, plants growing at less than one third of field capacity showed a marked reduction in nitrogen fixation. A progressive decline occurred and fixation nearly ceased after two weeks, and was completely inhibited in the third week of stress. At two-thirds of field capacity only a slight reduction in activity was observed. Subsequent watering restored activity to normal after about two weeks in all but the severest stress treatment (one sixth of field capacity for three weeks) in which only partial recovery occurred.

*Salinity :* Khadr (28) showed that different sodium salts at 25 and 35 meq/100 g soil completely inhibited the growth of faba beans. Nodule formation, dry weight and nitrogen contents of the plants decreased with increasing of salt concentration. Nodule numbers were reduced even at the lowest concentration of 5 meq/100 g soil. In general NaCl and $Na_2CO_3$ had a greater effect on nodule number than $Na_2SO_4$.

*Biotic factors :* Biotic factors include bacteria, actinomycetes, fungi, mycorrhiza, phages, *B dellavibrio*, protozoa, nematodes and insects. These different organisms may inhibit, stimulate, predate or parasitize rhizobia, hence affecting symbiosis. The interaction between these factors and faba bean-rhizobia symbiosis has not been extensively defined.

Ibrahim (25) showed that filtrates of cultures of *Streptomyces* sp. $R_2$ and $R_8$, *Aspergillus* sp. $R_9$ and $R_{14}$, and *Bacillus subtilus* sp. $R_3$ and $S_{10}$ inhibited the growth of *R. leguminosarum* by 60 to 89%, compared to the control. The effect was variable with strain.

Chhanker and Subba Rao (9) isolated several fungal species associated with the nodules of nine common legumes : *Cephalosporium Alternaria, Aspergillus, Penicillium, Rhizopus, Acrothecium, Fusarium, Rhizoctonia, Curvalaria, Pythium* and *Trichoderma.* Antibiotic activity was detected in culture filtrates of these fungi against different *Rhizobia* among which *R. leguminosarum* was more sensitive than *R. meliloti* and cowpea *Rhizobia.*

*Sitona* larvae mature within nodules of *Vicia faba*, peas and sometimes lupins in Europe (21).

*Competition among inoculated and natural R. leguminosarum strains :* Little is known about competition between inoculated and natural populations of *R. leguminosarum* strains. A preliminary study (23) to estimate the competive ability of inoculants against natural rhizobia in Egyptian soil, (Table 7) showed :

i.   An antigenic relationship between inoculated strains of *R. leguminosarum* (leg 1, 635, 636, and BB) and the natural population.

ii.  A variable competitive ability of strains inoculated singly on different hosts : Strain leg 1 gave 90, 60 and 80% nodules on Giza 3, Giza 4 and F 402, respectively. Strain 635 gave 70, 80, and 80% of nodules on Giza 3, Giza 4 and F. 402, respectively. Strain 636 gave 80, 70 and 100% of nodules on Giza 3, Giza 4 and F 402 respectively. BB strain was present in 60, 80 and 60% on hosts of Giza 3, Giza 4 and F 402.

iii. When a mixture of strains leg 1, 635 and 636 was used, different strains showed different percentages on different hosts : leg 1 gave 50, 60 and 40% of nodules of Giza 3, Giza 4 and F 402 respectively. Strain 635 gave 30, 10 and 20% of nodules on Giza 3, Giza 4 and F 402 respectively. It is also apparent that the natural population was excluded when the mixture of *Rhizobium* was used.

Table 7. Number and percentage of nodules showing cross reactions with different antisera (Determined by agglutination) (Hamdi *et al.* 1980).

| Strains | Antigen Variety | No. of Nodules | Leg 1 No. | Leg 1 % | 635 No. | 635 % | 636 No. | 636 % | BB No. | BB % | Natural* Population No. | Natural* Population % |
|---|---|---|---|---|---|---|---|---|---|---|---|---|
| Control | Giza 3 | 10 | 1 | 10 | | | 1 | 10 | | | 8 | 80 |
| | Giza 4 | 10 | 1 | 10 | | | 1 | 10 | | | 8 | 80 |
| | 402 | 5 | 1 | 20 | | | 1 | 20 | | | 3 | 60 |
| Leg 1 | Giza 3 | 10 | 9 | 90 | | | | | | | 1 | 10 |
| | Giza 4 | 5 | 3 | 60 | | | | | | | 2 | 40 |
| | 402 | 5 | 4 | 80 | | | 1 | 20 | | | 2 | 20 |
| 635 | Giza 3 | 10 | 1 | 10 | 7 | 70 | 1 | 10 | | | 1 | 10 |
| | Giza 4 | 5 | | | 4 | 80 | | | | | 1 | 20 |
| | 402 | 5 | 1 | 20 | 4 | 80 | | | | | | |
| 636 | Giza 3 | 10 | 2 | 20 | | | 8 | 80 | | | | |
| | Giza 4 | 10 | | | | | 7 | 70 | | | 3 | 30 |
| | 402 | 5 | | | | | 5 | 100 | | | | |
| Mixture of | Giza 3 | 10 | 5 | 50 | 3 | 30 | 2 | 20 | | | | |
| Leg 1 + 635 | Giza 4 | 10 | 6 | 60 | 1 | 10 | 3 | 30 | | | | |
| + 636 | 402 | 5 | 2 | 40 | 1 | 20 | 2 | 40 | | | | |
| BB | Giza 3 | 15 | 1 | 8 | 1 | 8 | 1 | 8 | 8 | 60 | 4 | 30 |
| | Giza 4 | 5 | | | | | 1 | 20 | 4 | 80 | | |
| | 402 | 5 | 1 | 10 | 1 | 10 | | | 3 | 60 | | |
| Control + | Giza 3 | 10 | 2 | 20 | | | 1 | 10 | | | 7 | 70 |
| 60 kg N | Giza 4 | 5 | | | | | | | | | 5 | 100 |
| | 402 | 5 | 1 | 20 | 1 | 20 | 1 | 20 | | | 2 | 40 |

*Estimated by subtraction.

*Fungicides :* The toxicity of various fungicides against four strains of *R. leguminosarum* has been studied (17). The fungicides Karathane and Spergen were found to be the least toxic followed by Dithane M 45 and Rhizoctol. Thiram and Cerodon were most toxic. Differences in toxicity between the four strains were observed.

*Herbicides :* roots were deformed and stems were dwarfed upon planting faba beans immediately after the application of trifluralin herbicide at different rates (19). Application of trifluralin 21 days after sowing stimulated plant growth and total nitrogen fixed when used at rates between 1.19 and 4.76 l/ha. The highest rate, 11.9 l/ha resulted in a large decrease in both dry weight and total nitrogen.

Gafar (18) found that of six herbicides tested, Preforan had the least effect on growth and nodulation (Table 8). Other herbicides showed different degrees of inhibition depending on the chemical and rate.

Salama *et al.* (39) showed that Diptrex up to 35.7 kg/ha had no harmful effect on nodulation of broad bean.

In another study, Taha *et al* (47) fount that Diptrex, Andrin and Sevin stimulated the nodulation, dry weight and total nitrogen of plants. The herbicides CIPC inhibited nodulation. The fungicide Cu Oxychloride was more inhibitory than Dithane 78. (Table 9).

Monib *et al* (32) showed that Temik applied alone seven days before sowing and inoculation with *Rhizobia,* had a significant stimulatory effect. No significant effects were observed when Temik was applied at sowing time to soil which had been inoculated seven days earlier or at the time of inoculation. The herbicide Stomp when applied at sowing or inoculation, completely inhibited the formation of nodules; few nodules were formed, without any change in size, when the herbicide was applied seven day before sowing and inoculation. When Temik and Stomp were applied together, the dry weight and nitrogen content of nodules and plants increased compared to when Stomp was applied alone.

The ability of *Rhizobia* to degrade pesticides has been reported. Hamdi and Tewfik (22) showed that the strains of *R. leguminosarum* tolerated 250 ppm of the herbicide 3,5 dinitro 0-cresol (DNOC) in culture media. A prominant degration product was identified as 3-amino-5-nitro-0-cresol.

Mostafa *et al* (33) showed that the degradation of P 32 malathion by *R. leguminosarum* resulted in the production of inorganic sulphate, thiophosphate and S-hydrolytic metabolites.

### Associated growth of plants with faba bean

Ibrahim and Kabish (26) did a lysimeter experiment to study the effect of the associated growth of faba beans and wheat. The increased yield was 11% for the beans, and 20% for wheat in mixture compared to the pure stands.

Table 8. Effect of different levels of herbicides on certain characteristics of faba beans (Gafar, 1976).

| Rates/ feddan (0.42 ha) | Nodules Weight g | gain/loss % | Nitrogen Content mg | gain/loss % | Leghaemoglobin μg | gain/loss % | Whole Plant Weight g | gain/loss % | Nitrogen content mg | gain/loss % |
|---|---|---|---|---|---|---|---|---|---|---|
| Control | 0 | | | | | | | | | |
| | 0.60 | – | 19.1 | – | 4355 | – | 8.25 | – | 322.3 | – |
| Preforan 2 l | 0.61 | + 1.7 | 12.7 | - 33.5 | 4011 | - 7.9 | 9.34 | + 12.8 | 383.5 | + 19.0 |
| 4 l | 0.56 | - 6.7 | 10.8 | - 43.7 | 4011 | - 7.9 | 9.02 | + 8.9 | 333.7 | + 3.6 |
| 6 l | 0.57 | - 5.0 | 14.2 | - 25.9 | 4355 | – | 8.89 | + 7.4 | 334.6 | + 3.9 |
| Linuron 0.375 kg | 0.26 | - 56.7 | 12.5 | - 34.6 | 4355 | – | 4.72 | - 43.0 | 181.2 | - 43.8 |
| 0.750 kg | 0.33 | - 45.0 | 10.4 | - 45.7 | 4985 | + 14.4 | 5.03 | - 39.3 | 189.7 | - 41.1 |
| 1.125 kg | 0.98 | - 53.3 | 5.8 | - 71.8 | 3725 | - 14.4 | 5.23 | - 36.8 | 199.4 | - 38.1 |
| V.C.S. 5.0 kg | 0.33 | - 45.0 | 9.0 | - 53.2 | 8097 | + 85.0 | 7.32 | - 11.6 | 273.9 | - 15.0 |
| 10.0 kg | 0.43 | - 25.0 | 7.2 | - 62.6 | 6532 | + 50.0 | 9.26 | - 11.8 | 328.7 | + 11.0 |
| 15.0 kg | 0.53 | - 10.0 | 6.0 | - 68.8 | 5272 | + 21.0 | 8.52 | + 2.9 | 344.3 | - 6.9 |
| Terbutol 5.0 kg | 0.33 | - 45.0 | 11.2 | - 41.4 | 3724 | - 14.4 | 3.52 | - 9.2 | 247.4 | - 14.8 |
| 10.0 kg | 0.43 | - 25.0 | 6.7 | - 65.0 | 4011 | - 7.9 | 6.36 | - 23.2 | 235.8 | - 26.8 |
| 15.0 kg | 0.54 | - 10.0 | 13.0 | - 32.1 | 3553 | - 18.4 | 6.58 | - 20.5 | 266.3 | - 29.8 |
| Dalapon 5.0 kg | 0.45 | - 25.0 | 20.5 | + 7.1 | 3896 | - 10.5 | 6.73 | - 19.2 | 266.5 | - 17.3 |
| 10.0 kg | 0.55 | - 8.3 | 11.0 | - 42.5 | 4011 | - 7.9 | 7.89 | - 4.2 | 284.7 | - 11.3 |
| 15.0 kg | 0.25 | - 58.3 | 8.6 | - 55.1 | 3553 | - 18.4 | 5.70 | - 31.2 | 219.9 | - 31.7 |
| M 15 6.0 kg | 0.33 | - 45.0 | 11.6 | - 39.6 | 4641 | + 6.7 | 5.84 | - 27.1 | 241.5 | - 25.0 |
| 12.0 kg | 0.20 | - 66.7 | 2.5 | - 86.8 | 4813 | + 10.5 | 5.55 | - 30.4 | 204.8 | - 36.5 |
| 18.0 kg | 0.23 | - 61.7 | 2.4 | - 87.6 | 3094 | - 28.9 | 5.39 | - 33.3 | 177.3 | - 45.0 |
| L.S.D. 5 % | 0.27 | | 13.64 | | – | | 2.32 | | 87.30 | |
| 1 % | – | | 18.25 | | – | | 3.11 | | 116.6 | |

136

Table 9. Effect of several pesticides on nitrogen fixation of faba beans (Taha *et al.*, 1969).

| | Sterile Soil Inoculated | | | Non-sterile Soil | | | | | |
|---|---|---|---|---|---|---|---|---|---|
| | No. of Nodules | Dry Wt. of plants (mg) | Total N mg | No. of Nodules | | Dry Wt. of plants (mg) | | Total N (mg) | |
| | | | | Inoc. | Non-Inoc. | Inoc. | Non-Inoc. | Inoc. | Non-Inoc. |
| Control | 74 | 725 | 34 | 52 | 22 | 714 | 685 | 30 | 24 |
| Insecticides : | | | | | | | | | |
| Diptrex (2.98 kg/ha) | 102 | 850 | 41 | 39 | 7 | 706 | 612 | 30 | 21 |
| Andrin (4.76 l/ha) | 85 | 795 | 35 | 58 | 32 | 790 | 670 | 33 | 25 |
| Sevin (3.75 kg/ha) | 158 | 975 | 46 | 68 | 58 | 858 | 770 | 34 | 33 |
| Herbicides : | | | | | | | | | |
| CIPC (17.86 l/ha) | – | 186 | 10 | | | 158 | 131 | 9 | 8 |
| Fungicides : | | | | | | | | | |
| Copper Oxychloride (o.6 %) | 90 | 170 | 29 | 28 | 4 | 591 | 467 | 25 | 20 |
| Dithane 78 (0.5 %) | 58 | 810 | 26 | 46 | 20 | 765 | 660 | 34 | 26 |

REFERENCES

1. Abdalla, M.M.F. 1969. The origin and evolution in *Vicia faba* L. 1st Mid. Conf. Genetics, Cairo 28-30 March 1979.
2. Abdul Wahab, A.M., Zahran, H.M. 1979. The effect of water stress on $N_2$ ($C_2 H_2$) fixation and growth of four legumes. Proc. Soc. Appl. Microbiol. Ann. Meet. Cairo pp. 261-275.
3. Afifi, N.M., Moharram, A.A., Hamdi, Y.A., Abdel Malek, Y. 1969. Sensitivity of Rhizobia species to certain fungicides. Arch. Microbiol. 66: 121-128.
4. Amin, M.R.A. 1977. The cytochemical effect of Rhizobia on *Vicia faba* MSc Thesis Faculty of Agric., Zagazig University.
5. Berg, E.H.R. Van den 1977. The effectiveness of the symbiosis of *R. leguminosarum* on pea and broad bean. Plant and Soil 48 (3): 629-639.
6. Bond, Y. 1958. Symbiotic nitrogen fixation by non-legumes. In. E.G. Hallsworth Nutrition of the Legumes. Academic Press pp. 216-231.
7. Bose, P.D., Venkatarman, G.S. 1969. Recombination in *Rhizobium leguminosarum* Experimentia 25, 772.
8. Candlish, E., Clark, K.W. 1975. Preliminary assessment of small faba beans grown in Manitoba. Can. Plant Sci. 55 : 89-93.
9. Chhanker, P.K., Subba Rao, N.S. 1965. Fungi associated with legume root nodules, and their effect on rhizobia. Can. J. Microbiol. 12: 1253-1261.
10. Cooper, D.R., Hill-Cottingham, D.G., Lloyd Jones, C.P. 1976. Absorption and redistribution of nitrogen during growth and development of field bean, *Vicia faba*. Physiologica Plantarum 38 : 313-318.
11. Dart, P.J., Day, J.M. 1971. Effects of incubation temperature and oxygen tension on nitrogenase activity of legume root nodules. Plant and Soil Special Vol. 1971 : 167-184.

12. Date, R.A. 1968. Rhizobial survival on the inoculated legume seed. 9th Intern. Congr. Soil Sci. Trans. Vol. 11: 75-83.

13. Dean, J.R. 1977. Inoculation, acetylene reduction and yield of faba beans as affected by inoculum concentration and soil nitrate level. Cong. J. Plant Sci. 57(4) : 1055-1061.

14. De Escuder, A.H.Q. 1972. A survey of Rhizobia in farm soils at Wye College Kent. J. Appl. Bacteriol. 35 : 109-118.

15. El-Mokadem, M.T. 1974. Effect of some pesticides on nodule bacteria. Ph.D. Thesis. Ain Shams Univ., Cairo, Egypt.

16. El-Zawahry, Y.A.M. 1976. Studies on the effect of gamma-radiation on growth and activity of *Rhizobium leguminosarum*. Ph.D. Thesis, Faculty of Sci., Univ. Cairo, Egypt.

17. Fawaz, Kareman, M., Abdel Gaaffar, El Gabaly, M.M. 1972. Sensitivity of root nodule bacteria to different seed protectants. Symp. Biol. Hung. 11: 417-422.

18. Gafar, E.M. 1976. Effect of some herbicides on nitrogen fixation in some legumes. M. Sc. Thesis, Fac. of Agric., Ain Shams Univ., Cairo, Egypt.

19. Gafar, Zenab, A., Hamdi, Y.A., Tewfik, M.S. 1969. Effect of trifluralin on the nitrogen fixed in broad beans. 1st Arabic Confr. Physiol. Sci., Feb. 22 to Feb. 25, Cairo, Egypt.

20. Gallacher, A.E., Sprent, J.I. 1978. The effect of different water regimes on growth and nodule development of greenhouse grown *Vicia faba*. J. Exper. Bot. 29 (109): 413-423.

21. Gibson, A. 1977. The influence of the environment and managerial practices on the legume *Rhizobium* symbiosis. In. R.W.F. Hardy and A.H. Gibson. A Treatise on Dinotrogen Fixation Section IV : Agronomy and Ecology. Wiley-Intersci., Pub., pp. 393-450.

22. Hamdi, Y.A., Tewfik, M.S. 1970. Degradation of 3,5 Dinitro-Cresol by *Rhizobium* and *Azotobacter* sp. Soil Biol. Biochem. 2 : 163-166.

23. Hamdi *et al*. 1970. Nitrogen fixation by faba bean. Annual coordination meeting 1979-1980. Cairo, Egypt 25 to 27 Aug. 1980.

24. Henson, I.E., Wheeler, C.I. 1976. Hormones in plants bearing nitrogen fixing root nodules; The distribution of Cytokinins in *Vicia faba* . New Phytol. 76, 433-439.

25. Ibrahim, A.N. 1970. Growth inhibition of Rhizobia by certain antagonistic organisms Acta. Bot. Polon. Sec. 39 (2) : 333-338.

26. Ibrahim, E.N., Kabesh, M. 1971. Effect of associated growth on the yield and nutrition of legume and grass plants; 1) Wheat and horse beans mixed for grain production. UAR J. Soil Sci. 11 (2) : 271-283.

27. Islam, R., Afendi, F. 1980. Effect of several herbicides on the nodulation and yield of faba bean *(Vicia faba)* FABIS 2 : 36-37.

28. Khadr, M.S. 1969. Effect of salinity and alkalinity on nitrogen fixation by some leguminous crops. M.Sc. Thesis, Fac. of Agric., Al-Azhar University.

29. Lawrie, A.C., Wheeler, C.T. 1975. Nitrogen fixation in the root nodules of *Vicia faba* in relation to the assimilation of carbon. I. Plant Growth and metabolism of photosynthetic assimilates. New. Phytol. 74 (3) 429-436.

30. Loutfi, M., El Sherbini, M.F., Ibrahim, A.N., Casdy, A. 1966. Effect of inoculation of legumes by nodule bacteria on yield. 1. Horse bean. J. Microbiol. UAR 1 (2) 161-168.

31. Loutfi, M., Rizk, S.G., Hamdi, Y.A. 1980. Final report of PL 480 Project 127. Presented to USDA.

32. Monib, M., Belol, M.H., Abdel-Baky, Afaf, Hassan, M.M., Hegazi, N.A. 1980. Effect of some pesticides on nodulation of broad bean. 4th Conf. Microbiol. Cairo 24 -28, 1980.

33. Nostafa, I.Y., Fakhr, I.M.I., Bahig, M.R.E., El Zawahry, Y. 1972. Metabolism of organophosphorus insecticides XIII Degradation of malathion by *Rhizobium* sp. Ark. Mikrobiol. 86 : 221-224.

138

34. Mytton, L.R., El Sherbeeny, M.H., Lawes, D.A. 1977. Symbiotic variability in *Vicia faba* . 3. Genetic effect of host plant *Rhizobium* strain and host X strain interaction. Eyphytica 26 (3) : 785 - 791.

35. Nutman, P.S. 1958. The physiology of nodule formation. In. E.Q. Hallsworth, Nutrition of Legumes. Academic Press pp. 82-107.

36. Nutman, P.S., Ross, J.S. 1969. *Rhizobium* in the soils of the Rothamsted and Woburn farms. Rep. Rothamsted Exp. Sta. 1969. 2 : 148-167.

37. Richards, J.E., Soper, R.J. 1979. Effect of N-fertilizer on yield protein content and symbiotic N. fixation in faba beans. Agron. J. 71 (5): 807-811.

38. Rizk, G.S. 1966. Atmospheric nitrogen fixation by legumes under Egyptian conditions II. Grain Legumes. J. Microbiol. UAR. 1 (1) : 33-45.

39. Salama, A.M., Mostafa, I.Y., El-Zwahry, Y.A. 1974. Insecticides and soil micro-organisms II. Effect of diptrex on nodule formation in broad bean and clover plants under different manural treatments. Acta. Biol. Acad. Sci. Hung. 25. (4) 239-246.

40. Salem, S.H. 1969. Microbiological studies on nodule bacteria. Ph.D. Thesis, Fac. of Agric., Ain Shams Univ.

41. Salem, K.G., Mahmoud, S.A.Z., El-Mokadem, M.T. 1970. Effect of dieldrin and lindane on the growth and nodulation of *V. faba.* Plant and Soil 33: 325-329.

42. Schwinghamer, E.A. 1967. Effectiveness of *Rhizobium* as modified by mutation for resistance to antibiotics. Antonie Van Leeuwenhock. J. Microbiol. Ser. 33 : 121-136.

43. Sprent, J.I. 1972. The effects of water stress on nitrogen fixing root nodules. IV . Effect on whole plants of *V. faba* and G, max. New Phytologist, 71: 608-611.

44. Sprent, J.I., Bradford, A.M. 1977. Nitrogen fixation in field beans *V. faba* as affected by population density, shading and its relationship with soil moisture. J. Agric. Sci. Cam. 88 : 303-310.

45. Sprent, J.I., Bradford, A.M., Norton, C. 1977. Seasonal growth patterns in field beans *(V. faba)* as affected by population density, shading and its relationship with soil moisture. J. Agric. Sci. Cam. 88: 293-301.

46. Stadler, L. 1952. Uber dispostions Verschieburgen Bei Der Bidung Von Wurzelknoll chem. Phytopathol. 2.18:236.

47. Taha, S., Mahmoud, S.A.Z., Slama, A. 1969. Effect of certain pesticides on nodules of clover and broad bean. Vth Arabic Congress, Cairo, Egypt.

48. Taha, S., Mahmoud, S.A.Z., Salem, S.H. 1976. Effect of inoculation with Rhizobia on some leguminous plants in UAR. 1-Phosphorus manuring. J. Microbiol. UAR 2 (1) : 17–29.

49. Taha, S., Mahmoud, S.A.Z., Salem, S.H. 1976. Effect of inoculation with Rhizobia on some leguminous plants in UAR. 2-Nitrogenous fertilizers. J. Microbiol. UAR 2 (1) : 31-41.

50. Vojinovic, Z.A. 1976. Some studies on the necessity of legume inocuation in Serbia. In. P.S. Nutman, Symbiotic Nitrogen Fixation in Plants, pp. 191-198, IBP 7, Cambridge Univ. Press.

## 13. SYMBIOTIC NITROGEN FIXATION IN FABA BEANS IN SUDAN

M.M. MUSA

*Agricultural research Corporation, Wad Medani, Sudan*

In Sudan *Vicia faba* is an irrigated winter crop grown mostly north of Khartoum. Recent trials as far south as Sennar (300 km south of Khartoum) have indicated that the potential exists for expanding the area where land and water are not as limiting (5) (2) (1). With this in mind work on nodulation and nitrogen fixation was initiated in 1973-74 at the Gezira Research Station at Wad Medani. Earlier observations at Shambat and Hudeiba Research Stations have shown that at those locations faba beans were invariably well nodulated. No record of the nature and extent of nodulation has been documented in other areas of production.

A survey of 21 small farms between Wad Medani and Khartoum, was conducted in 1974. Of these, only four were growing small areas of faba beans. Top soil was collected from each site for soil analysis and was tested for the presence of natural *Rhizobia* in pots planted with faba beans. Ten sites were studied in detail (Table 1). Nodules were only formed in the soil from five of the sites, and these were all from farms which had previously grown the crop. There was no apparent relationship between nodulation and other soil characteristics such as clay content, salt concentration or alkalinity.

Table 1. Some soil characteristics of selected sites between Medani and Khartoum, 1973-1974.

| Site of farm | Mechanical analysis | | pH 1:5 | % salt content | Nodulation score (0-5 scale) |
|---|---|---|---|---|---|
| | % sand | % clay | | | |
| 1. Gezirat El Fil | 54 | 35 | 8.8 | 0.120 | 1.0 |
| 2. Wad El Magdoub | 43 | 42 | 8.2 | 0.051 | — |
| 3. Fadasi | 35 | 53 | 8.4 | 0.066 | — |
| 4. Arbagi | 58 | 29 | 8.2 | 0.048 | 1.5 |
| 5. El Masid A | 22 | 66 | 8.8 | 0.255 | — |
| 6. El Gadid | 60 | 31 | 8.3 | 0.057 | — |
| 7. Bageir | 66 | 21 | 8.3 | 0.045 | — |
| 8. Butri | 63 | 28 | 8.6 | 0.057 | 2.0 |
| 9. Soba Scheme | 32 | 48 | 8.9 | 0.240 | 1.0 |
| 10. Shambat Research Farm | 22 | 58 | 8.2 | 0.048 | 3.0 |

*G. Hawtin & C. Webb (Eds.): Faba Bean Improvement*
©1982 ICARDA. ISBN -13:978-94-009-7501-9

About 100 isolates of *Rhizobium* were made from the different soils. These were tested in vitro for their reaction to sodium chloride concentrations (Table 2).

Table 2. Salt reaction of various faba bean rhizobia isolates in vitro culture.

| Concentration of sodium chloride in liquid medium % | % isolates showing salt tolerance | Remarks |
|---|---|---|
| 0.5 (conductivity 2.0 mmoh/cm) | 100 | |
| 1.0 (2.7 mmoh/cm) | 90 | |
| 1.5 (3.6 mmoh/cm) | 42 | Isolates from ligh non-saline sites most affected. |
| 2.5 (5.9 mmoh/cm) | 10 | Isolates from Soba and Masid A were least affected. |
| 5.0 (9.5 mmoh/cm) | 2 | |

Some local *Vicia faba Rhizobium* isolates (coded 1, 2 and 5) collected from the Gezira Research farm and identified as having good nodulating ability and N fixation were used on four cultivars, BF 2, Hudeiba 72, Giza and Silaim, in 1974-75. The trial was conducted in a field where native *Rhizobia* levels in the soil were low. The trend for all four cultivars was similar to that of Hudeiba 72 (Table 3). They responded significantly to inoculation with the three cultures. The ratio of dry matter per plant for all the cultivars relative to that of the uninoculated control ranged between 3.0 and 3.5. These small differences could not be regarded as evidence of cultivar-strain specificity in nitrogen fixation.

Table 3. Dry weight of nodules, dry weight of tops and yield of faba bean cv. Hudeiba 72 with three local isolates of *Rhizobium*.

| *Rhizobium* isolate | Dry weight of nodules mg/plant | Dry weight of tops g/plant | Dry pod yield t/ha |
|---|---|---|---|
| Local 1 | 202 | 28 | 1.99 |
| Local 2 | 188 | 26 | 1.84 |
| Local 5 | 162 | 24 | 1.15 |
| Uninoc. Control | — — | 8 | 7.8 |
| S.E. of mean ± | 12 | 1.6 | 0.073 |

In 1974-75 Isolate 2, considered the best local isolate, was compared with TA 101 of Australian origin, 3 Hog 3 from the U.S.D.A. collection at Beltsville and Nitragin inoculant. Hudeiba 72 was used throughout at sowings on October 15, November 1 and 21 in the field.

The plants were generally small, the October planting showed yellowing and stunting. However both early and late November plantings were showing lush growth. Nitragin inoculant was markedly superior to both local, TA 101 and 3 Hog 3. However the local isolate was comparatively better in nodulation and dry matter production, followed by TA 101 and 3 Hog 3, at all three sowing dates (Table 4).

Table 4. Nodulation and plant performance with different inoculant sources at three sowing dates.

| Date of sowing | Rhizobium Strain or isolate | Nodule dry matter mg/plant | Dry matter of top. g/plant | Dry pod yield tons/ha |
|---|---|---|---|---|
| | Local isolate | 60 | 12 | – |
| | Nitragin | 96 | 16 | – |
| Oct. 10 | TA 101 | 54 | 10 | – |
| | 3 Hog 3 | 42 | 10.5 | – |
| | Control | – | 3.6 | – |
| | Local isolate | 164 | 28 | 1.90 |
| | Nitragin | 224 | 40 | 2.40 |
| Nov. 1 | TA 101 | 148 | 26 | 1.79 |
| | 3 Hog 3 | 108 | 26 | 1.61 |
| | Control | 20 | 8.0 | 1.11 |
| | Local isolate | 185 | 28.8 | 1.84 |
| | Nitragin | 222 | 36.8 | 2.25 |
| Nov. 21 | TA 101 | 155 | 29.2 | 1.82 |
| | 3 Hog 3 | 148 | 28 | 1.56 |
| | Control | 24 | 10.0 | 0.92 |

In experiments in 1976-77 to test reaction of nodulation to fertilizers, the lowest level of 20 kg N/ha stimulated nodulation but higher levels dimished it, and nodulation was virtually absent at 80 kg N/ha. However with P application, some nodulation was maintained at the levels of 40 and 80 kg N/ha. P alone had a negligible effect on nodulation. However, inoculation, in the absence of N and P, raised the level of nodulation about 5 or 6 times (Table 5).

Plant dry matter and yield of pods showed a similar trend. Yield of dry matter increased from a very low value to about 4 times with increasing rates of N. Inoculation alone gave dry matter and yields comparable to 40-60 kg N/ha. P helped to sustain a much higher plant dry matter and pod yield in inoculated than in uninoculated treatments. Generally pod yields were lower than normal because of a heavy attack of powdery mildew and wilt.

The amount of inoculant needed for effective nodulation is known to be a function of a number of ecological conditions during planting and subsequent crop development. This was investigated on the Gezira Research Farm during 1977-78. Rates of Nitragin and $V_2$ inoculant as low as $1/20$ the normal rate of 3 g/kg seed gave some nodulation. There was a negligible difference between the normal and $1/10$ normal rate. Although the highest rate produced heavier nodulation, it was not reflected in the dry matter production of plants after 8 weeks. However, germination was delayed by both the 20 and 5 times rate in the case of the silt-carried V2 isolate, but not with Nitragin peat base inoculant (Table 6).

Although faba beans are alien to the Gezira, some nodulation has been observed in non-inoculated plots. In certain cases nodulation, regarded as satisfactory for adequate nitrogen fixation, has been observed. In general the nodulation in these plots started comparatively late in the season and nodules were diffuse in the main and lateral roots. Nodulation occurred on the tap roots only in very few cases.

Table 5. Effects of inoculation, nitrogen and phosphorus application on nodulation, plant dry matter and pod yield.

| Treatment | | | Nodule dry matter mg/plant | Plant dry matter g/plant | Dry pod yield t/ha |
|---|---|---|---|---|---|
| Uninoculated control | | | 24 | 12 | 0.72 |
| " | " | + ½ N* | 28 | 25 | 1.02 |
| " | " | + 1N | 21 | 33 | 1.16 |
| " | " | + 1½ N | – | 41 | 1.33 |
| " | " | + 2N | – | 42 | 1.53 |
| " | " | + 1P-ON** | 25 | 38 | 0.77 |
| " | " | + 1P-1N | 29 | 40 | 1.28 |
| " | " | + 1P-2N | 14 | 46 | 1.40 |
| Inoculated | | | 130 | 41 | 1.21 |
| " | | + ½ N | 125 | 50 | 1.17 |
| " | | + 1N | 142 | 54 | 1.48 |
| " | | + 1½ N | 81 | 60 | 1.81 |
| " | | + 2N | 61 | 54 | 1.64 |
| " | | + 1P-ON | 68 | 48 | 1.27 |
| " | | + 1P+1N | 53 | 62 | 1.77 |
| " | | + 1P+2N | 48 | 64 | 1.75 |
| S.E. of mean | | | ±10.5 | ± 5.9 | ± 0.82 |

*1N = 40 kg N/ha
**1P = 36 kg $P_2O_5$/ha

Table 6. Nodule and plant dry matter as affected by rate of inoculation.

| Rate of inoculation | Dry matter of nodules of plant (mg/plant) | Dry matter of plants (g/plant) | Nodule wt. as % of plant dry weight |
|---|---|---|---|
| $1/20$ normal | 18 | 56 | 0.32 |
| $1/10$ normal | 40 | 61 | 0.65 |
| Normal (3 g/kg seed) | 41 | 65 | 0.63 |
| 5 normal | 48 | 68 | 0.70 |
| 20 normal | 56 | 66 | 0.85 |
| Uninoc. - control | –– | 28 | –– |
| S.E. of Mean | ± 9 | ± 2.0 | –– |

In situations where such base nodulation exists improvement of the crop productivity by introducing proven local or introduced isolates was not very much evident. This indicates the competitive nature of this native *Rhizobia* and the need for very rigid selection in order to introduce highly adapted efficient isolates. The results of nodulation tests on Gezira soils of different cropping history are shown in Table 7. Fallow generally decreased the chances of survival of the native *Rhizobia* whereas cropping with groundnuts and phillipesara tended to maintain and increase them. Both crops require the promiscuous cowpea group of *Rhizobia*, which is so predominant in the Gezira clays. This work will be continued using plots of known long term cropping history such as the Intensified Crop Rotations an experiment which began in 1931. Current work is also underway in our laboratory to investigate the cross-inoculation tendencies of faba beans with various *Rhizobia* isolates from different origins.

Table 7. Nodulation of faba beans in some plots of the Gezira Research Farm.

| Previous plot history | % plots showing some nodulation | % plots showing highly effective nodulation |
|---|---|---|
| One fallow | 36 | 12 |
| Two fallows | 24 | 6 |
| Permanent fallow (141) GRF | Nil | Nil |
| Cotton | 32 | 14 |
| Groundnuts | 52 | 22 |
| Wheat | 30 | 8 |
| Phillipesara | 60 | 35 |

## REFERENCES

1. Ageeb, O.A.A. 1980. Annual Report Gezira Research Station 1979-80.
2. Ishag, H.M. 1977. Annual Report Gezia Research Station 1977-78.
3. Jewett, T.N. 1955. Gezira Sore, Ministry of Agriculture, Bulletin.
4. Musa, M.M. 1974. Annual Report Gezira Research Station 1973-74.
5. Musa, M.M. 1976. Annual Report Gezira Research Station 1975-76.
6. -------------- 1977. Annual Report Gezira Research Station 1976-77.
7. -------------- 1978. Annual Report Gezira Research Station 1977-78.
8. Saxena, M.C. *et al.* 1980. Faba bean Fertility - cum – inoculation trial 1979-80. ICARDA unpublished.

# 14. PHYSIOLOGICAL ASPECTS OF ADAPTION

MOHAN C. SAXENA


*Food Legume Improvement Program, ICARDA, P.O.Box 5466, Aleppo, Syria*

*Vicia faba* L. is widely grown as a winter season crop in sub-tropical regions with mild winters and at high elevations (above 1200 m) under tropical conditions,whereas in temperate areas it is generally grown as a spring season crop thus avoiding the period of severe frost (26). The range of its cultivation extends from about 9°N to more than 40°N and from near Sea Level to more than 2000 m (26) (47), thus exposing the crop to varied thermal and photoperiodic regimes. The present day landraces in different geographical regions of faba bean cultivation can, therefore, be well considered to have arisen from the natural selection based on the adaptability to local environment with some human selection for specific agroeconomic traits. The landraces should thus contain a wide range of variability in their adaptation as the seasonal and total variations in day length and temperature in these ranges of latitudes and altitudes are enormous.

Variation in photoperiod and maximum and minimum temperatures become progressively more conspicuous away from the Equator. Temperatures are also greatly affected by change in altitude. These variations have been well illustrated by Sinha (47) and Summerfield and Wien (50) for several latitude and altitude combinations.

In practical agriculture, the time available for growth, reproductive development and build-up of yield of a crop species is often limited by factors of environment and, in several cropping systems, even by economic considerations involving competition for use of land by other crops in the rotation. Hence successful adaptation of a crop to its environment, necessitates that it has the ability to make most efficient use of the available time, ensuring an adequate balance between the vegetative growth and reproductive development (46). While aiming for heavier yields, breeders always make a conscious or unconscious effort to produce genotypes that make more productive use of the environmental conditions for which they are being developed. Therefore information on the effect of thermal and photoperiodic regimes and their interactions on growth, development and yield of the diverse genotypes of a crop species can be very valuable in breeding for better adaptation (50).

*G. Hawtin & C. Webb (Eds.): Faba Bean Improvement*
©1982 ICARDA. ISBN -13:978-94-009-7501-9

*Broad Ecological Categorisation of Faba Bean Types*

The so-called Mediterranean types are well adapted to winter planting in sub-tropical conditions including low altitudes in the Mediterranean region. They may need an average temperature between 18-27°C for their optimum growth (26). Such genotypes respond favourably to early planting in winter, and show considerable decline in the productivity as the planting is delayed because of high temperatures during the reproductive period (5) (42) (30). The other two commonly recognized ecotypes are the European 'spring' and 'winter' types which are adapted to spring and winter plantings respectively in temperate regions such as the U.K. The optimum temperature conditions for the spring type appear to be more or less similar to those of the Mediterranean type and therefore they perform well under early winter planting in the Mediterranean region. However they show much greater reduction in the yield as the planting is delayed (30) (43). The European winter-type cultivars are suitable for winter planting in temperate areas which experience a relatively mild winter; average temperature around 2°C without prolonged or severe frost, followed by a warm drier spring and summer (26). Such types are not adapted to the low altitude Mediterranean environment and they yield very poorly , if at all, even when sown in the early winter in this environment. Some may even fail to flower because of their adaptive requirement for a low temperature induction and presence of long photoperiods for successful reproductive growth and yield development (30) (43). The types adapted to the other geographical regions will generally fall into one or the other of these three eco-types although there could be exceptions e.g. cultivar Hudeiba 72 from Sudan.

Most workers have made this kind of a categorisation of cultivars in studies on the response of faba beans to temperature and photoperiod conditions.

*Response to Temperature*

The physical, bio-physical and physicochemical processes involved in germination of seed are all sensitive to temperature. Said *et al.* (40) studied the effect of increasing night temperature from 5°C to 20°C with a constant day temperature of 20°C, on the germination of faba beans in a growth cabinet. The germination rate was markedly increased as the night temperature was raised. Field emergence, which is affected by rate of germination as well as the rate of epicotyl growth, shows a conspicuous negative relationship with the mean ambient temperature up to 20°C. Field emergence data from ICARDA's farm in North Syria for a large number of the Mediterranean types (mostly *equina* and *major*, represented by cvs. Giza-2 and Aquadulce), for European spring type (cv. Express) and for Sudanese (cv. Hudeiba 72) type faba beans are shown in Fig. 1. Temperatures below 10°C resulted in a conspicuous increase in days to emergence. Cultivar Hudeiba 72 was the fastest to emerge. This may be attributed partly to its small seed, as the cv. is of the *minor* group.

Temperature affects both mitosis and cytokinesis in the meristematic tissues. Both these processes are reported to proceed very slowly in faba bean root meristems at temperatures below 3°C, and most rapidly at about 25°C (49).

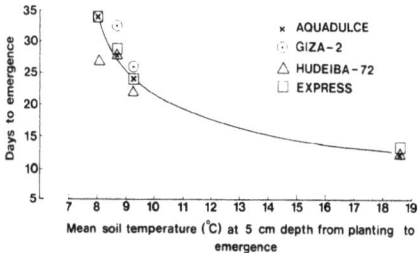

Fig. 1   Effect of mean soil temperature at 5 cm depth, from planting to emergence. on the days to complete field emergance of various cultivars of faba bean at Tel Hadya, northern Syria.

Evans (15) reported heteroblastic change in the leaflet number in many spring and winter stocks of faba bean; the rate of change being affected by temperature. The rate of increase in the number of leaflets rose sharply as the temperature in the controlled chambers was reduced successively by $3^{\circ}C$ intervals from $23^{\circ}C$ to $7^{\circ}C$ for the winter cv. Bullocks. But the total number of leaves formed increased with increasing temperature. Bull (7) reported that the relative rate of increase in leaf area per plant of spring beans in the early stages of growth (until LAI reached 3) was closely related to daily maximum air temperature. The regression equation showed a $Q_{10}$ of 2.1, and it could be predicted that a temperature difference of $2.5^{\circ}C$ in spring would change the time taken to reach LAI of 2 by about 4 days. Dennett (12) and Dennett et al.(13) have also shown that the leaf expansion in faba beans is highly sensitive to temperature changes. Studies at the Rothamsted Experimental Station, U.K., on potted plants of faba beans in a controlled environment have shown that the photosynthesis was a maximum at $25^{\circ}C$ when temperature was raised from $15^{\circ}C$ to $30^{\circ}C$ (29). At $30^{\circ}C$, the photosynthesis declined by about 20%. The dark respiration and compensation points both increased as the temperature was increased.

Controlled environmental studies on the effect of various night temperatures (5, 10, 15 and $20^{\circ}C$) on the growth and dry matter production of the Egyptian cultivar Giza 2 by Said et al. (40) have shown that the optimum value changes with the stage of growth. Up to the 2-4 leaf stage, the stem elongation was fastest at the highest night temperature of $20^{\circ}C$. The day length was kept at 12 h, and the day temperature at $20^{\circ}C$. The optimum night temperature fell as the plants advanced. Highest seed yield was obtained with $10^{\circ}C$ night temperature. Abdalla and Fischbeck (1), also using a controlled environment, studied the temperature response of five stocks of faba beans, including one paucijuga, two minor (Diana and $C_5$), one equina (Giza 2) and one major (Hedosa) type. Three thermal regimes consisting of day/night temperatures of 15/10, 20/15 and $30/23^{\circ}C$ were used with a photoperiod of 10 hours. The plant height and plant dry weight were adversely affected as the thermal regime increased for these spring-type faba bean stocks.

Reports on the effect of thermal regimes on the initiation of flowering and further floral differentiation are conflicting (15). Much of this conflict can be ascribed to the differences in the low temperature inductive requirements of genotypes used, differences in the criteria used for flower initiation, differences in the photoperiods employed, and in other conditions of the experiments. Vernalization response is discussed in detail later. Said *et al.* (39) and Abdalla and Fischbeck (1) suggest that a cycle of 20°C day temperature and 15 to 10°C night temperature results in the earliest flowering, and increase or decrease in the thermal regime delays the flowering. These observations are consistent with the earlier observations of Evans (15) on several spring and winter faba beans which showed inhibition of flowering at continuous temperatures above 23°C. Jaquiery *et al.* (24) also observed that very high temperatures in 1976 stopped the appearance of flowering in spring faba beans in Switzerland.

The reduction in the yield of faba bean genotypes with delay in planting date in the Mediterranean conditions due to exposure of the reproductive phase of the later-planted crop to increasing temperatures, is clearly borne out from the data on day degrees accumulated during the reproductive phase and the grain yields of a number of Mediterranean and spring type European cultivars sown on different dates at ICARDA's farm (Table 1). Based on the experience of past seasons, dates of planting later than Jan. 14 were not included in this study as the yields from such late dates were very poor, possibly on account of still higher average air temperatures during the reproductive phase. Three years' studies by Georgiev (17) in Bulgaria also revealed that the reproductive growth of faba beans was inversely related to the air temperatures in April to July.

*Response to vernalization*

A short period of exposure to low temperatures has been reported to hasten flower initiation in a number of economic legumes including faba beans (14) (47) (50). However, the results of earlier studies on faba beans have not been very consistent. Whereas Souza da Camarra (48) with *Vicia faba equina* and Robitzsch (38) with a range of early and late flowering stocks observed no vernalization response, considerable hastening in flower initiation was reported by several other workers (32) (52) (2) (3) (8) (15). More recent work has mostly supported the latter observation (25) (34) (18) (16) (41).

These apparently conflicting results can be due to differences in genotypes, duration and temperature of vernalization treatment, environmental conditions during the maturation of the mother plant from which seed for vernalization studies is derived, temperature and photoperiodic conditions in which the seedlings from vernalized seeds are grown and also the criteria used for such evaluation.

The temperature and duration of vernalization treatments has ranged from 2-3°C for 1, 3 or 5 weeks (16), to 4-10°C for 1 to 16 weeks (15) in different studies. Ballatore (2) reported that November planting of seeds of a Mediterranean type *major* faba bean vernalized for 35 days at 3-4°C resulted in early flowering, precocious pod set and consequently early yield. Vernalization of seeds of Mediterranean landraces from Cyprus, Jordan, Syria and Lebanon, improved cultivars from Egypt (Giza 2, 3 and 4) at 5°C for 38 days hastened flowering when the plants were raised in a heated plastic house at ICARDA's farm (41). Since the flowering occurred even in the untreated plants, these genotypes showed only a quantitative response to vernalization. Flower initiation was accelerated by vernalization in all except the earliest flowering of the six European *V. faba* genotypes (including three spring beans and three winter beans) in a controlled environment study, particularly when the plants were grown in short days at high temperatures (15). The response to low temperatures was more rapid at 10°C than at 4°C; similar to the observation of Tanaka (52) who also found better vernalization at 10°C than at 5°C.

Table 1.

Days from flowering to maturity (DFM), day degrees accumulated from flowering to maturity (DDFM), and dry bean yield t/ha (YD) of some cultivars of faba beans under different dates of planting at Tel Hadya, N. Syria, during 1979-80. Day degrees are the sum of daily mean air temperatures (°C).

| Cultivars | Planting date | | | | | | | | | | | |
|---|---|---|---|---|---|---|---|---|---|---|---|---|
| | 23 Oct. 1979 | | | 11 Nov. 1979 | | | 4 Dec. 1979 | | | 14 Jan. 1980 | | |
| | DFM | DDFM | YD | DFM | DDFM | YD | DFM | DDFM | YD | DFM | DDFM | YD |
| Mediterranean types: | | | | | | | | | | | | |
| Aquadulce | 63 | 990 | 2.9 | 62 | 1038 | 3.1 | 63 | 1070 | 2.5 | 59 | 1103 | 1.2 |
| Syrian Local Large | 68 | 1118 | 3.6 | 62 | 1038 | 3.2 | 63 | 1070 | 2.9 | 55 | 1022 | 0.6 |
| Lebanese Local Small | 61 | 965 | 3.2 | 60 | 992 | 2.6 | 62 | 1024 | 1.8 | 55 | 993 | 0.6 |
| Seville Giant | 59 | 926 | 3.4 | 57 | 993 | 2.6 | 62 | 1024 | 2.5 | 62 | 1153 | 0.9 |
| Giza 2 | 65 | 1084 | 1.3* | 62 | 1038 | 2.1 | 73 | 1309 | 2.0 | 59 | 1103 | 0.7 |
| Giza 3 | 67 | 1131 | 0.6* | 68 | 1143 | 1.8 | 63 | 1070 | 1.8 | 55 | 993 | 0.8 |
| European spring types: | | | | | | | | | | | | |
| New Mammoth | 63 | 990 | 2.6 | 61 | 992 | 2.9 | 63 | 1070 | 2.4 | 52 | 943 | 0.9 |
| Express | 74 | 1306 | 0.9* | 75 | 1323 | 2.1 | 73 | 1309 | 1.7 | 62 | 1153 | 0.9 |
| Hudeiba 72 | ** | ** | ** | 57 | 933 | 0.1* | 57 | 917 | 0.7 | 52 | 913 | 0.5 |

* Partial frost damage        ** Total mortality because of frost

Evans (15) with Bullocks winter beans, clearly showed that an increase in the duration of low temperature exposure from 1 to 4 weeks resulted in linear acceleration of flowering, but further increase in the duration of vernalization delayed flower initiation. Similar effect has been observed in a study on the duration of vernalization with 12 different genotypes (representing various ecotypes) grown under natural day length (10.5 to 9.8h) in a heated plastic house (mean maximum and minimum temperatures being 24.1°C and 15.2°C over the duration) at the ICARDA farm (Table 2). This study shows quite conspicuous differences in the magnitude of response amongst different genotypes. Fuciman (16) also reported significant genotypic differences in vernalization response : cv. Chulmecky was slower in response than the spring cultivar Windsor.

The low temperature response increases as the plants get older (15) but the induction can occur even during embryo development (45) (15) (19). This explains how seed stocks of one vernalization-responsive cultivar maturing in different environmental conditions may show contradictory responses to vernalization treatment. At temperatures above 14°C, and particularly above 23°C, a reaction inhibitory to flower initiation occurs. Extensive studies of Evans (15) on the interaction of low temperature-induced response with high temperature exposures at various day length conditions at different stages of growth, led to the observation that the accelerating influence of periods at temperatures below 14°C and the delaying effect of temperatures above 23°C, may operate at almost any time in the life cycle of the plants, and subsequent treatment with the one does to some extent remove the effects of the other. The processes responsible for inhibitory effects of high temperature, probably occur during the diurnal dark period, and can be mitigated completely if the low temperature processes have reached saturation by the use of optimum temperature and time combinations for vernalization treatment. These observations further highlight that the vernalization responses in faba beans can be completely altered by the environmental conditions during the subsequent growth when the treatment effects are expressed.

The vernalization requirement is an important adaptation of faba bean genotypes to environment. Where winters are severe, such a requirement may permit a genotype to escape frost during the critical period of reproductive growth. This may also help in optimising the balancing of the crop growth time between vegetative and reproductive growth so that the optimum economic yield is obtained (14). This adaptive requirement will be of special interest to breeders, particularly when they are breeding for wider adaptability, as the breeding material being developed under the conditions where the vernalization requirement can naturally be met, may latently carry this requirement. Since response to vernalization is commonly controlled by a single gene, or relatively few genes, it can be modified by selection (50). This would need special screening of genetic material for vernalization. Vernalization may also be needed as an aid in the breeding programme when genotypes from divergent ecological conditions have to be crossed at a station where natural vernalization does not occur. Similar considerations may also be relevant for the genetic material being used from off-season generation advance.

*Frost susceptibility and cold tolerance:* Between 30 and 40°N, the winter planting of faba beans exposes the crop to a risk of frost damage which may arise at any stage of growth, depending upon the altitude and other geographical features. Severe damage from frost was observed at the ICARDA farm during 1979-80 (43) on the October and November-planted crop of diverse genotypes of Mediterranean and European spring types. Scarascia Mugnozza and Marzi (44) have also reported severe damage to faba beans by repeated late

Table 2.

Effect of duration of vernalization (0, 16, 32 and 48 days shown by $V_0$, $V_1$, $V_2$ and $V_3$ respectively in the table) at 5°C on the node of first flower and days to flowering in different genotypes of faba beans at ICARDA Station in Northern Syria during 1980-81 in a heated plastic house.

| Genotypes | Node of first flower | | | | Days to first flower | | | |
|---|---|---|---|---|---|---|---|---|
| | $V_0$ | $V_1$ | $V_2$ | $V_3$ | $V_0$ | $V_1$ | $V_2$ | $V_3$ |
| **European types:** | | | | | | | | |
| Wierboon CB | 12.5 ± 0.5 | 14.0 ± 1.4 | 12.3 ± 2.4 | 8.0 ± 1.0 | 61.0 ± 0.0 | 62.3 ± 3.0 | 52.5 ± 7.4 | 46.0 ± 10.0 |
| Minica | 7.8 ± 0.5 | 7.0 ± 0.4 | 7.0 ± 0.0 | 7.5 ± 0.5 | 42.8 ± 4.5 | 38.8 ± 0.8 | 35.0 ± 1.0 | 38.5 ± 2.5 |
| Ipro | 11.0 ± 2.0 | 9.0 ± 1.4 | 7.0 ± 0.0 | 9.5 ± 1.5 | 58.0 ± 1.0 | 53.8 ± 7.5 | 34.5 ± 0.5 | 53.5 ± 3.5 |
| Maris Bead | X | 11.0 ± 0.0 | 14.0 ± 1.4 | X | X | 70.0 ± 0.0 | 66.8 ± 3.3 | X |
| Maris Beagle | X | 7.5 ± 1.5 | 12.0 ± 1.3 | 12.0 ± 0.0 | X | 63.5 ± 6.5 | 54.5 ± 4.0 | 70.0 ± 0.0 |
| Throws MS | X | X | X | X | X | X | X | X |
| **Mediterranean types:** | | | | | | | | |
| Aquadulce | X | 10.5 ± 2.5 | 8.5 ± 0.3 | 7.7 ± 1.2 | X | 41.0 ± 0.0 | 35.0 ± 0.6 | 40.0 ± 4.0 |
| Syrian Local Large | X | 10.0 ± 1.5 | 8.3 ± 0.3 | 7.0 ± 1.0 | X | 53.0 ± 1.2 | 33.3 ± 0.8 | 36.0 ± 0.0 |
| Reina Blanca | 13.0 ± 0.0 | 10.3 ± 1.0 | 9.0 ± 1.4 | 8.0 ± 0.0 | 70.0 ± 0.0 | 63.0 ± 3.2 | 37.5 ± 3.5 | 70.0 ± 0.0 |
| New Mammoth | 12.0 ± 3.0 | 9.0 ± 2.0 | 8.3 ± 0.3 | 8.0 ± 0.0 | 62.5 ± 1.5 | 45.7 ± 2.3 | 33.8 ± 1.0 | 34.0 ± 0.0 |
| Giza 3 | 12.0 ± 0.4 | 7.8 ± 0.8 | 7.5 ± 0.5 | 6.8 ± 0.5 | 55.0 ± 0.0 | 39.0 ± 3.0 | 33.8 ± 1.7 | 45.0 ± 4.9 |
| Hudeiba 72 | 9.5 ± 0.3 | 6.8 ± 0.5 | 6.8 ± 0.3 | 5.7 ± 0.3 | 44.3 ± 2.6 | 35.0 ± 0.6 | 32.5 ± 2.9 | 31.3 ± 1.5 |

X No. flowering occurred up to 80 days when the experiment had to be terminated.

winter frosts in southern Italy when the bean plants are in full bloom. Sometimes there is a sudden fall in temperature during the spring as well, which damages lower pods. Late frost damage to the developing pods has also been widely seen in the highlands of Ethiopia.

Lack of adequate winter hardiness has been one of the major constraints to the expansion of winter faba bean production in France (36) (4) and Germany (27) (20). Even in U.K. (6) risk of frost and chocolate spot disease has restricted the winter faba bean area to only 25%, although the winter crop yields about 30% more than the spring one.

The level of frost resistance that is needed for more southernly latitudes experiencing sub-tropical and Mediterranean climates, is generally not as high as that for the winter planting in Europe, where the crop may experience temperatures less than $-15^{\circ}C$ without protective snow cover (27) (36) (4). Several local landraces from Syria and Lebanon, as also the other Mediterranean types from the region, showed only small frost damage when they repeatedly experienced sub-zero temperatures, at one time dropping to - $11^{\circ}C$, at ICARDA's farm (43) (Table 1). Highest suceptibility was shown by Hudeiba 72, a cv. from Sudan. Cultivars from Egypt and spring types from Europe were moderately susceptible, particularly in the earliest dates of planting. Unfortunately there were no European winter types in this test. According to Berthelm (4), most of the spring varieties in France are resistant to $-6^{\circ}C$, which is usually enough to overcome the late frost in Europe. Of the various winter types available in Europe, the French cultivar (population) Cote d'Or seems to have the best genes for frost hardiness, but even that could not stand temperatures below - $15^{\circ}C$ on bare ground (4) and its winter hardiness seems to be limited to about $-12^{\circ}C$ (36). The winter hardiness of the contemporary French varieties (e.g. Avrissot, Survoy, Soravi) is lower than that of Cote d'Or, but their recovering ability, inherited from more vigorous seedlings of English populations, is better. The English winter types, Maris Beagle, Maris Beaver and Throws MS, have been found to have relatively less frost hardiness than Cote d'Or in the 'official tests' at Hohenheim in Germany (27). According to Picard (36), natural variation in frost hardiness is limited, and mutation breeding may have to be resorted-to.

As a nodulated legume, the performance of the faba bean plant is affected not only by the direct influence of the weather conditions on the host plant itself, but also by the effect on the *Rhizobium* with which it enters in symbiotic association, and on the interaction between the host and the *Rhizobium* (50). Temperature effects on symbiotic N-fixation in faba beans have not been very intensively investigated although host-*Rhizobium* genotype interactions have been studied (33) (23). Korovin and Vorob'ev (28) showed that nodulation was greatly delayed as the soil temperature was reduced from $18\text{-}22^{\circ}C$ to $4\text{-}5^{\circ}C$. The symbiotic N fixation at flowering was less than half at the lowest temperature. Vorob' ev (53) also observed the negative effect of very low temperatures on the growth and productivity of faba beans which could be reduced by application of large rates of fertilizer nitrogen, apparently because symbiosis was not developed. Dart and Day (10) have assayed the acetylene reduction activity of faba bean root nodules incubated at various temperatures, and recorded maximum activity in a temperature range of 20 to $30^{\circ}C$. There is a need for identification of *Rhizobium leguminosarum* strains that are able to establish the symbiotic association with frost-tolerant genotypes at low temperatures.

Herzog (20) (21) (22) has given some good insight into the differences in the cold hardiness of spring and winter type faba beans and also into the mechanism of induction of cold hardiness. Cold hardiness is apparently not a permanently present character of a genotype and its appearance depends on the photoperiodic and temperature conditions in which the plant is grown. However, genotypes do differ in their ability to acquire winter hardiness. The response of three spring types (cvs. Diana, Herz Freya and Kleine Thuringer) and three winter type (cvs. Maris Beagle, Maris Bulldog and Daffa) faba beans to freezing for 20 hrs at -5.3, -5.5 and -5.8$^{\circ}$C temperature was evaluated, by Herzog (21), by monitoring the $CO_2$ - assimilation, which on an average was reduced by 17% , 50% and 77% respectively. Assimilation rates were reduced by 30%in winter types and by about 60%in spring types, suggesting the greater susceptibility of spring types. This was also confirmed by the tissue conductivity test, as the leakage of electrolytes, after the freezing test, was higher in spring than in winter types.

The $T_{50}$ value (temperature causing 50% damage) by extrapolation was about -5.4$^{\circ}$C for winter types, when all these genotypes were previously kept for 14 days in a condition which was noninductive of cold tolerance (long days of 16 h day light and 23$^{\circ}$C temperature). When the plants were kept for 25 days under inductive conditions (8h light period and 5$^{\circ}$C temperature) this induction resulted in decreased carbon assimilation, magnitude of reduction being much higher in winter than in spring types. Evaluation of frost tolerance of these hardened plants in freezing tests gave a $T_{50}$ value of less than -7.3$^{\circ}$C.

The above and subsequent studies (22) lead to the conclusion that the frost susceptibility and induction of frost tolerance are related to the rate of growth and metabolic activity and the state of tissue differentiation. Spring types seem to have a faster rate of growth and higher rate of assimilation than the winter types, and thus they show more frost susceptibility. Inductive conditions reduce the rate of growth and metabolic activity, and thus increase the tolerance to frost. The induction of frost tolerance in the growth chamber is similar to that induced in the field (21).

Field observations on frost damage as mentioned by Scarascia-Mugnozza and Marzi (44) in South Italy and as recorded at the ICARDA farm (43) are quite consistent with Herzog's observations. The crop planted at the end of October can get naturally hardened against frost as the temperature quickly falls down to above freezing point immediately after planting in many faba bean-growing areas in the Mediterranean region.

Herzog (22) does suggest that induction of frost tolerance is related to dry matter content of the tissue and the rate of ontogenic differentiation. The possibility of using these as parameters in selecting genotypes with better potentiality for frost tolerance induction, deserves further investigation.

*Response to Photoperiod*

A review of the early work on photoperiodic response by Evans (15) revealed that the responses were very diverse. Some of the studies suggested that the flower initiation could occur even in complete darkness in some stocks (8), whereas others concluded that faba beans behaved as short day (11), or day neutral plants (37). However much of the subsequent work has suggested that flower initiation shows a quantitative long day response (15) (25) (31) (51).

Evans (15) when studying the response of four *V. faba* stocks (ranging from very early Dutch Weir spring beans to the late flowering Bullocks winter beans) to increasing photoperiods from 8 to 24 h, (by supplementing natural day light for 8 h with low intensity artificial light), observed that unvernalized plants of all the four stocks showed a quantitative long-day response, at a diurnal temperature regime of 23$^{\circ}$/17$^{\circ}$C. Rise in the mean temperature of growth reduced the acceleration of flowering induced by long days. In the latest

154

flowering stock, Garton's giant winter bean showed no flower initiation even at 16 h photoperiod at 23°C-17°C temperature regime unless the seeds had previously been vernalized. This highlights the importance of both temperature and photoperiod in flower initiation particularly in the late maturing winter types.

A study on the photoperiodic responses of Mediterranean types was initiated at ICARDA with six European cultivars (4 spring and 2 winter types) and six diverse Mediterranean types during the winter season of 1980-81. Plants from unvernalized seeds were raised in a heated plastic house (Fig. 2) and were exposed to normal (10.5 to 9.6 h) and extended (16 h) photoperiod. The two winter types (Maris Beagle and Throws MS) did not flower up to 90 days after planting, when the experiment was terminated. All the remaining genotypes, flowered except a Syrian Local Large landrace (ILB 1814). Extending the photoperiod did not cause a large change in the days to first flower, and plant to plant variability was very high. Unfortunately the temperature conditions under normal and extended day treatments were not identical and this might also have affected the results.

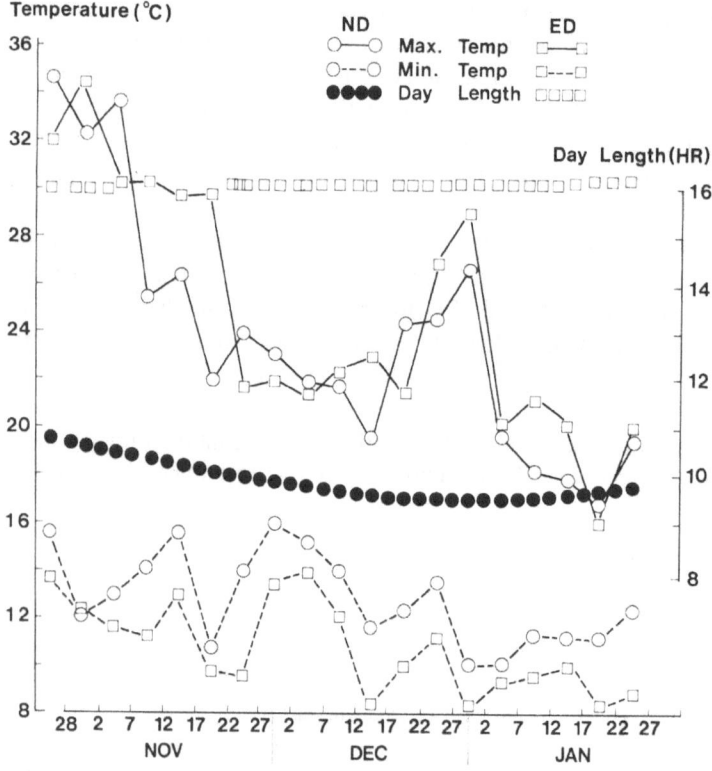

Fig. 2   Mean maximum and minimum temperatures and photoperiod in plastic house used for photoperiod x vernalization experiments.

With the same 12 genotypes, the interaction between the vernalization and photoperiodic treatments was studied in another trial in the same plastic house. (Fig.3) (Table 3). Acceleration of flowering with extended photoperiod was more consistent and conspicuous in the plants from vernalized seeds. These studies confirm that the flowering responses to day length can be conspicuously affected by the temperatures.

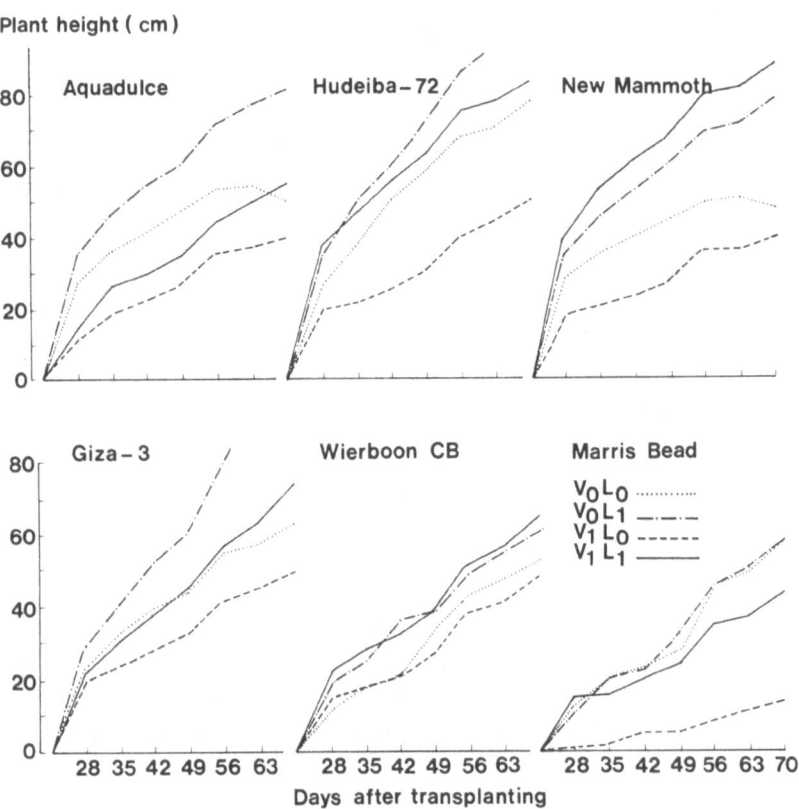

Fig. 3. Effect of vernalization and photoperiod on plant height.

The significance of day length is not only in the initiation of flowering but also, and perhaps more significantly, in the development of the flower. This is particularly evident in the relatively late maturing types as has been shown by Evans (15) with English types and by Tamaki et al. (51) with cultivar Sanuki-nagasaya. There is a need for more intensive investigations into the response of the Mediterranean types along with European types under controlled environmental conditions using factorial combinations of temperature and photoperiodic treatments to get better understanding of the physiology of adaptation of faba beans as has been emphasised by Summerfield and Wein (50). However this does not belittle the significance of well controlled field experiments at various locations with a similar set of diverse genotypes to investigate further the ecophysiology of faba beans. The start of such a multilocational study on the 'Growth and development of Vicia faba L. in relation to specific environmental components' under the auspices of the European Economic Community is a good development. This and other similar studies might help in the breeding of faba beans for better adaptation.

Table 3.

Effect of seed vernalization for 48 days at $5^{\circ}C$ and extended photoperiod (16 h) on the days to first flower in different genotypes of faba beans at ICARDA Station in Northern Syria during 1980-81 season in a heated plastic house.

| Genotypes | − Vernalization | | + Vernalization | |
|---|---|---|---|---|
| | Normal day | Extended day | Normal day | Extended day |
| European types: | | | | |
| Wier boon CB. | $59 \pm 7.7$ | $76 \pm 0.0$ | $37 \pm 0.0$ | $61 \pm 0.0$ |
| Minica | $71 \pm 0.0$ | $45 \pm 2.4$ | $36 \pm 0.7$ | $36 \pm 0.5$ |
| Ipro | $57 \pm 5.1$ | $71 \pm 1.9$ | X | $72 \pm 3.5$ |
| Maris Bead | X | X | X | X |
| Maris Beagle | X | X | X | X |
| | | | | |
| Mediterranean types: | | | | |
| Aquadulce | X | $76 \pm 0.0$ | X | $48 \pm 0.0$ |
| Syrian Local Large | X | X | $41 \pm 2.7$ | $38 \pm 0.3$ |
| Reina blanca | $76 \pm 0.0$ | $67 \pm 3.8$ | $70 \pm 0.0$ | $63 \pm 0.0$ |
| New Mammoth | $68 \pm 0.0$ | $69 \pm 2.6$ | $51 \pm 0.0$ | $38 \pm 3.1$ |
| Giza 3 | $46 \pm 1.1$ | $46 \pm 1.9$ | $35 \pm 0.0$ | $43 \pm 3.5$ |
| | | | | |
| Hudeiba 72 | $46 \pm 0.9$ | $52 \pm 3.7$ | $42 \pm 3.1$ | $35 \pm 0.0$ |

X No flowering occurred up to 90 days after planting when experiment was terminated.

## REFERENCES

1. Abdalla, M.M.F., Fischbeck, G. 1987. Growth and fertility of five stocks of field beans grown under three temperature regimes and the effect of natural water stress on seed index of a collection of *Vicia faba* L. Zeitschrift. f. Acker u. Pflanzenbau. 147: 2 81-91.

2. Ballatore, G.P. 1950. Primi risultati di alcune prove di orientamento sulla jarovizzazione della fava da granella. Annali Fac. Sci. Agri. Univ. Palermo, 1, 181-198.

3. Ballatore, G.P., Mirto, G. 1956. Ricerche sperimentali sulla jarovizzazione della fava da granella (*Vicia faba major*). Ann. sper. agr. 10, 1457-86.

4. Berthelm, P. 1980. The major agronomic problems of the cultivation of faba bean (*Vicia faba* L.) in France. FABIS 2, 9-12.

5. Bianco, V.V. 1979. Agronomic aspects of broad bean production in southern Italy. In: Some current research on *Vicia faba* in Western Europe. D.A. Bond, G.T. Scarascia-Mugnozza and M.K. Poulsen. Pub. EEC, EUR 6244 EN Luxembourg, pp. 125-143.

6. Bond, D.A. 1979. Breeding work on *Vicia faba* in the U.K. FABIS 1, 5-6.

7. Bull, T.A. 1968. Expansion of leaf per plants in field bean (*Vicia faba* L.) as related to daily maximum temperature, J. appl. Ecol. 1968, 5: 1, 61-8.

8. Chakravarty, A.K., Drayner, J.M., Fyfe, J.L. 1956. Plant breeding studies in leguminous forage crops 111. Preliminary studies on the developmental physiology of English stocks of beans (*Vicia faba*, L.). J. Agric. Sci., 48, 104-114.

9. Cubero, J. I. 1973. Evolutionary trends in *Vicia faba* L. Theoretical and Applied Genetics, 45, 47-51.

10. Dart, P.J., Day, J.M. 1971. Effect of incubation temperature and oxygen tension on nitrogenase activity of legume root nodules. Plant and Soil, 1971. Special vol, 167-184.

11. David, R., 1946. Facteurs de developpement et printinusation des vegetaux cultives. Act. Sci. et Ind. 177 pp. Hermann, Paris.

12. Dennett, M.D. 1975. Some comments on leaf growth. Proc. Physiol. Programme Formulation Workshop (IITA), Ibadan, Nigeria. 1975, 85-81.

13. Dennett, M.D., Elston, J. Milford, 1979. The effect of temperature on the growth of individual leaves of *Vicia faba* L. in the field.Annals of Botany, 1979, 43: 2, 197-208.

14. Evans, L.T., King, R.W. 1975. Factors affecting flowering and reproduction in the grain legumes. In: Report of the TAC working group on the biology of yield of grain legumes. Pub. No. DDDR: IAR/75/2, FAO, Rome, pp. 1-18.

15. Evans, L.T. 1959. Environmental control of flowering in *Vicia faba* L. Annals of Botany, N.S. Vol. 23, No. 92, 521-546.

16. Fuciman, L. 1975. Some peculiarities of the biology of the broadbeans (*Faba vulgaris* Moench). 3. The course of the organogenesis of the growing point in broad bean under various temperatures. Agronomicka. A. 1975, No. 2, 131-145.

17. Georgiev, Z. 1968. The effect of meteorological conditions on the morphology of 3 *Vicia faba* cultivars. Rast. Nauki, 1968, 5: 7, 55-63.

18. Graman, J. 1969a. Development effect of low temperatures influencing the germination and taking up of beans. Field Crop Abs. Vol. 25. Abst. No. 3723.

19. Graman, J. 1969b. Vernalization of bean seeds (*Faba vulgaris* Moench) on the mother plant. Rostl. Vjroba 1969, 15 (42) : 1, 117-24. Cf. Field Crops Abst. Vol. 23, Abst. No. 1337.

20. Herzog, H. 1978a. Wachstumsverhalten und Kaltetoleranz bei Ackerbohnen (*Vicia faba* L.) unter verschiedenen Test bedingungen. I. Mogliche indirekte Kriterien bezuglich der Frostharte wöhrend der Sprossenentwicklung. Z. Acker-u Pflanzenbau 146, 303-314.

21. Herzog, H. 1978b. Wachstumsverhalten und Kaltetoleranz bei Ackerbohnen (*Vicia faba* L.) unter verschiedenen Test bedingungen. 2. Assimilations leistung und ihre Veranderung nach Gefriertests. Zeitschrift f. Acker-u Pflanzenbau, 1978, 147: 111-120.

22. Herzog, H. 1980. Wachstumsverhalten und Kaltetoleranz bei Ackerbohnen (*Vicia faba* L.) unter verschiedenen Test Bedingungen IV. Hartungsverlauf in Abhangigkeit von der Organdifferenzierung von Sorten und verschiedenen Klimabedingunger. Z. Acker. u. Pflanzenbau 149, 271-286.

23. Islam, R. 1979. Improved nitrogen fixation in faba beans (*Vicia faba*). FABIS 1, 21-22.

24. Jaquiery, R., Gehriger, W., Keller, E.R. 1977. Physiological investigations on yield and yield stability of *Vicia faba* L. EEC Working Group Meeting in Stuttgart, Hohenheim, June 30-July 3, 1977, 81-102.

25. Kagawa, A. 1963. Effects of seed vernalization, photoperiods and plant growth regulator on flower initiation of broad bean, *Vicia faba* L. Research Bull. Fac. Agr. Gifu Uni., 18, 20-29.

26. Kay, D.E. 1979. Broadbeans. In: 'Food Legumes' TPI, Crop and Product Digest No.3, 26-47.

27. Kittlitz, E. von 1977. Projects and breeding aims in *Vicia faba* at Hohenheim. EEC Working Group Meeting in Stuttgart, Hohenheim, June 30-July 3, 1977, 110-117.

28. Korovin, A.I., Vorob'ev, V.A. 1967. Soil temperature and assimilation of atmospheric nitrogen by legumes. Fiziologiya Rast. 1967, 14: 1, 117-122.

29. Leach, J.E. 1978. Photosynthesis : laboratory studies. Rothamsted Experimental Station, U.K. Report for 1977, part 1, 202-203.

30. Moreno, M.T., Martinez, A., 1980. The divided world of *V. faba*, FABIS 2, 18-19.

31. Moursi, M.A., Abd El Gawad, A.A. 1963. Legume photoperiodism 2. Formative and photoperiodic reaction to light duration. Ann. agric. Sci. Univ. Ain. Shams 1963, 8: (1), 297-304.

32. Muratova, V.S., 1931. Common Beans (*Vicia faba* L.) Suppl. 50, Bull. appl. Bot. Genet. and Plant Breeding, Leningrad.

33. Mytton, L.R., El-Sherbeeny, M.H., Lawes, D.A. 1977. Symbiotic variability in *Vicia faba*, 3. Genetic effects of host plants, Rhizobium strain and of host x strain interaction. Euphytica 26, 785-791.

34. Pascenko, V.N. 1964. Variation in the photoperiodic response of different varieties of *Vicia faba* induced by temperature. Bot. Zurnal, 49, 1199-1202. cf. Hort. Abstr., 35, 116 (1965).

35. Pavlov, P. 1972. Growth, development and yields from broadbeans grown at different light. Comptes Rendus de l'Academie Agricole Georgi Dimitrov, 1972, 5(2), 87-92.

36. Picard, J. 1979. Some reflections on problems and prospects in *Vicia faba* breeding. In: Some current research on *Vicia faba* in Western Europe. D.A. Bond, G.T. Scaracia-Mugnozza and M.H. Poulsen, Luxembourg (eds.), pp. 23-34.

37. Ramaley, F., 1934. Influence of supplemental light on blooming. Bot. Gaz., 96, 165-175.

38. Robitzsch, J. 1938. Die Entwicklung der Ackerbohne in Abhangigkeit von Tageslange, Keimtemperatur und Aussaatzeit. J. Landw. 86, 127-162.

39. Said, H., Hegazy, T., Imam, R.M. 1967a. Growth, productivity and compositional contents of bean (*Vicia faba*) seeds as affected by different night temperatures. FLORA, Jena, (A), 1967, 158:6, 569-76.

40. Said, H., Hegazi, T., Imam, R.M. 1967b. The effect of night temperature on germination, carbohydrate and nitrogen metabolism of faba seedlings. Acta. Biol. Hung. 18(4), 421-436.

41. Saxena, M.C., Wassimi, N. 1979. A preliminary study of the response of some faba bean (*Vicia faba*) genotypes to vernalization. FABIS 1, 20-21.

42. Saxena, M.C., 1979. Some agronomic and physiological aspects of the important food legume crops in West Asia. In: 'Food Legume Improvement and Development' (G.C. Hawtin & G.J. Chancellor, eds.) IDRC, Ottawa, Canada. 155-165.

43. Saxena, M.C., Hawtin, G.C., El-Ibrahim, H. 1980. Aspects of faba bean ideotypes for drier conditions, EEC Seminar on Seed Legumes, Wageningen, June 24-26, 1980.

44. Scarascia-Mugnozza, G.T., Marzi, V. 1979. Retrospective and prospective views for the *Vicia faba* crop in southern Italy. In:Some current research on *Vicia faba* in western Europe. D.A. Bond, G.T. Scarascia-Mugnozza and M.H. Poulsen. (eds.) Pub. EEC, EUR 6244 EN Luxembourg, pp. 7-22.

45. Shinohara, S. 1959. Genecological studies on the phasic development of flowering centering on the cruciferous crops, especially on the role of vernalization on ripening seeds. Shizuoka Prefecture Agr. Expt. Sta. Tech. Bull., 6:1 (cf. Devlin, R.M. 1969 Plant Physiology, Litton Educational Publig. Inc. page 391.).

46. Shibles, R. 1980. Adaptation of soybeans to different seasonal durations. In: "Advances in Legume Science" (Summerfield, R.J. and A.H. Bunting (eds): 279-285. Kew, England.

47. Sinha, S.K. 1977. 'Food Legume Distribution, Adaptability and Biology of Yield' FAO Rome, pp. 124.

48. Souza da Camarra, A. de 1934. Subsidios para o estudo da vernalisacao. Leu. Agron. Lisboa, 22, 5-20.

49. Sutcliffe, J. 1977. Plants and temperature. Sussex Univ., Brighton. Edward Arnold (Publ) Ltd. pp. 57.

50. Summerfield, R.J., Wien, H.C. 1980. Effects of photoperiod and air temperature on growth and yield of economic legumes. In: 'Advances in Legume Science'. pp. 17-36.

51. Tamaki, K., Naka, J., Asanuma, K. 1974. Physiological studies of the growing process of broad bean plants. VIII. Effects of the length of light duration on the growth and the chemical components. Tech. Bull. Fac. Agric., Kagawa Univ. 25 (2), pp. 157-170.

52. Tanaka, Y. 1936. The effect of the preliminary low temperature treatment in accelerating the flowering of the broad bean. J. Hort. Ass. Japan 7, pp. 365-371.

53. Vorob'ev, V.A. 1967. The resistance of leguminous crops to low soil temperatures and frost and the dependence of their N-fixation on nutrient level. In: "Physiology of cold tolerance and field emergence in Siberia". Moscow, 'Nauka', 1967: 93-105. Herb. Abs. Vol. 38, Abst. No. 1040.

54. Zohary, D. 1977. Comments on the origin of cultivated broad beans, *Vicia faba* L., Israel J. Bot. 26, pp. 39-40.

15. UTILISING ALTERNATIVE MODELS IN *VICIA FABA* L.

G.P. CHAPMAN



*Wye College, University of London*

Field bean in its traditional form provides an unwieldy system, having indeterminate growth and a normally out-pollinated breeding system. The indeterminate growth makes for variable performance, and the out-pollinated breeding system, a high degree of hetero-zygosity. In seeking to improve this crop it is important to identify the most feasible strategy.

Any permanent departure toward cleistogamy and sustained inbreeding would seem to involve profound changes in the breeding system to be sufficient to incorporate and survive other changes by the breeder. By comparison, changes in growth habit appear to make more modest demands on both plant and breeder and, at the same time, leave intricacies of the current partly out-pollinated breeding system intact. However, the changes in growth habit that are now possible, provide the basis for a re-appraisal of the crop's agronomy, and old questions can be looked at in a new way.

In recent years, much interest has been given to a determinate mutant produced and described by Sjodin (4). It is important to recognise that this particular mutant is now one of several which are similar, though not identical, and that it has provided not a final solution, but the basis of a revised approach to breeding *Vicia faba*. This chapter shows in outline how a determinate mutant might be used, and what potential it has. As the various shortcomings are subsequently identified, so the requirements breeders should seek among other determinate mutants will become more clearly apparent.

*Utilisation of the Svalof mutant*

As experience accumulated, objectives which became feasible were :
i    Simplification of growth habit
ii   Adaptation to a cereal-type agronomy
iii  Accelerated genetic turnover

*G. Hawtin & C. Webb (Eds.): Faba Bean Improvement*
©1982 ICARDA. ISBN -13:978-94-009-7501-9

In sum the aim has been to produce a plant type which is not only more easily managed agronomically but is also more malleable and responsive in the hands of plant breeders (2). With the incorporation of terminal flowering and precocious flowering the above points followed in that order. What does not seem to be appreciated widely is the extent of genetic variation available in *Vicia faba* and that, with accelerated genetic turnover, this variation can be rapidly incorporated into a breeding programme. A review of genetic variation in *Vicia faba* is was recently published (1).

*i    Simplification of growth habit* : The traditional indeterminate plant produces at first bifoliate and later tri- and multi-foliate leaves. By combining determinate growth and precocious flowering, it was found possible to produce plants that completed their life cycle with few-foliate leaves and relatively short stems. The process was described in detail by Chapman and Peat (2) and the resulting plants are technically, neotenous.

*ii    Adaptation to a cereal-type agronomy* : Superficially, neotenous plants resemble determinate flowering cereals and their final size permits high density planting and development. In small experimental plots, plantings of up to 250 plants per square metre were made. This particular aspect of the work will be extended when a wider range of plant types becomes available in sufficient quantities to test.

*iii    Accelerated genetic turnover* : A feature of neotenous plants bred at Wye is the conspicuously shortened generation time; for example reducing the life cycle from about 160 to 120 days. This has an advantage in permitting the breeder more generations a year, and a disadvantage in diminishing the time available for photosynthesis in any one generation. The disadvantage is probably more apparent than real since sufficient genetic diversity is present readily to lengthen the life cycle eventually. Five points deserve mention :

*a    Three generations per year* : By growing an $F_2$ outside during the summer, the selected elites were used as a parental generation to make new $F_1$s in the first three months thereafter under glass. Seed raised from this was selfed to yield a new $F_2$ for outdoor sowing the following summer.

*b    Genetic diversification* : By utilising genetic backgrounds from *Vicia faba paucijuga* and *major*, the determinate neotenous growth habit was expressed against a wide range of genetic backgrounds, thus broadening the base of selection.

An example can be cited. A drawback to the Svalof mutant is the relatively few flowering nodes it usually produces. (The exact number is not readily decided because the terminal inflorescence is a complex structure interspersed with stipules and capable of more than one interpretation). By varying the genetic background with material from *paucijuga*, multiple branching was induced. This then permitted selection for types with the normal quota of inflorescences in the upper part of the stem supplemented by short determinate flowering branches in the lower part.

*c    Selection for agronomic characters* : In the latter stages of the work described here, selections are being made for straw stiffness, *Botrytis* resistance and *Uromyces* resistance together with subsidiary selections that vary the distribution of the crop on the plant.

*d   Physiological studies* : With the availability of contrasted phenotypes (both in determinate and a range of determinates) a study is in progress using radio-carbon to discover how assimilate is distributed and utilised, and to what extent and under what circumstances the pod is photosynthetic. With regard to contrasted phenotypes and their response to growth substances a study has recently been published (5) and utilises some of the variation referred to in this chapter. By screening a wider range of genetic variants it might be possible to show different degrees of response to the chemicals used in this study. Incorporation of such responsiveness into experimental breeding lines could be useful under some circumstances.

*e   Aphid-growth habit-virus interaction* : There is preliminary evidence that determinate types can set their crop and become aphid-unattractive before the bulk of *Aphis fabae* migration under British conditions.

Work at Wye is concerned in this respect with seasonal fecundity of aphid on contrasted growth habit in outside-raised plants. Linked with this is a study directed at the consequences for yield distribution for determinate plants of early and late virus infection.

Given, therefore, accelerated genetic turnover, it has become more feasible to apply more elaborate techniques to the crop's improvement. There is every indication that gains in multiple resistance to *Botrytis* , for example could now be more readily integrated with improved standing ability and desirable seed characteristics. More importantly for the future breeding of this crop, it is now necessary to extend these ideas to the newer determinate mutants that have recently been obtained.

### Alternative models at the chromosome level

During the last four or five years, there have been remarkable advances in 'DNA technology', and although the bulk of this work has concerned prokaryotes, there is increasing evidence that this technology can be applied (though as yet with difficulty) to eukaryotes. Van Montagu (personal communication) has shown that DNA from a modified *Agrobacterium* plasmid can survive meiosis of its host while chromosomally incorporated, and prompt a Mendelian segregation for the imparted character (octopine synthesis). This represents virtually, directed mutation and the implications for plant breeding are considerable.

1   At Wye College, we began in October 1980 a project to study aspects of this new technology. Put simply, if exogenous DNA is to be added to a crop plant's chromosome, how does it enter, where does it go and does it stay within the same chromosome region indefinitely? To approach such questions requires an understanding of chromosome ultrastructure; ultimately at the level of the chromosomal DNA. Therefore we have developed a technique allowing isolation of single whole chromosomes for electron microscope examination.

An interesting aspect of the work is that, just as with conventional phenotypic variation, there is at the chromosome level, substantial variation, and cytologists have accumulated for *Vicia faba* a range of aberrations. With the normal karyotype only the satellite chromosome can be individually recognised without differential staining as would have to be the case in material for electron microscope examination. However, Michaelis and Rieger (3) by utilising reciprocal translocations obtained an 'EF' karyotype where each chromosome became individually recognisable. At Wye, we have shown that this individuality extends to the level of electron microscopy.

Our most recent work, which is still in progress, has been to treat isolated chromosomes with various reagents that reveal details of internal arrangement.

*Future Developments*

There is no reason why *Vicia faba* breeding should abandon the proven techniques of improvement, and exploring and utilising the available variation provides the principal means of progress. At the same time, the peculiarities of the *Vicia faba* karyotype with its few large chromosomes and its accumulated aberrations make it very suitable to explore aspects of DNA technology when applied to eukaryotes, but as a long term prospect.

REFERENCES

1. Chapman, G.P. 1981. Genetic Variation within *Vicia faba*. FABIS Newsletter No. 3 Suppl. pp. 1-12.
2. Chapman, G.P., Peat, S.E. 1978. Procurement of yield in field and broad beans. Outlook on Agriculture, 9, 267-272.
3. Michaelis, A., Rieger, R. 1971. New Karyotypes of *Vicia faba*. Chromosoma (Berl) 35, 1-8.
4. Sjodin, J. 1971. Induced morphological variation in *Vicia faba*. Hereditas, 67, 155-180.
5. Chapman, G.P. and Sadjadi, A.S. 1981. Exogenous growth substances and internal competition in *Vicia faba* L. Z. Pflanzenphysiol. Bd. 104, 265-273.

# 16. THE EFFECTS OF WATER STRESS ON THE GROWTH, DEVELOPMENT AND YIELD OF *VICIA FABA* L.

PAUL HEBBLETHWAITE

*University of Nottingham, Sutton Bonington, Loughborough. United Kingdom.*

Many workers have recorded large reductions in yield due to water stress, and substantial responses to irrigation in faba bean (2) (30) (15) (27) (28) (40) (32) (7) (19) (34) (21).

Given an adequate supply of water and nutrients, the growth of a crop is largely determined by the amount of solar radiation it intercepts (25). To maximise interception, the faba bean crop must establish complete ground cover as soon as possible and maintain it for as long as possible. However, water requirements of the crop when the surface of the soil has dried out will also increase rapidly with increasing ground cover.

Once the crop is completely covering the ground, the rate of evapotranspiration is largely dependent on meteorological conditions. Good crop growth will therefore necessitate large amounts of water being available. If evaporative demand exceeds water supply for a prolonged period, the crop is stressed, leading to irreversible effects such as reduced photosynthesis and dry matter accumulation, shedding of leaves, flowers and pods, ovary abortion and consequently lower seed yields.

## The effect of water stress on leaf production

Water stress decreases the rate of leaf expansion (8) (5) and may decrease the period of expansion (8). Stress also causes premature senescence (9) (18). The results of these responses are lower potential and actual leaf area indices. Thus light interception can be decreased substantially, the amount depending on the optimum leaf area for the crop and environmental conditions in question. Light interception will also decrease when leaves wilt (4). Work at the University of Nottingham has shown that leaf area duration can be closely correlated with total dry matter and seed yield in dry years where maximum leaf areas are limited to about 3 (Fig. 1).

Water stress also decreases stem extension (37) and consequently crop height (Table 1). Irrigation increases plant length primarily because plants develop more nodes and longer internodes. As a result, irrigated crops have greater stem areas to intercept light but they are more prone to lodging which can depress yields (23).

*G. Hawtin & C. Webb (Eds.): Faba Bean Improvement*
©1982 ICARDA. ISBN -13:978-94-009-7501-9

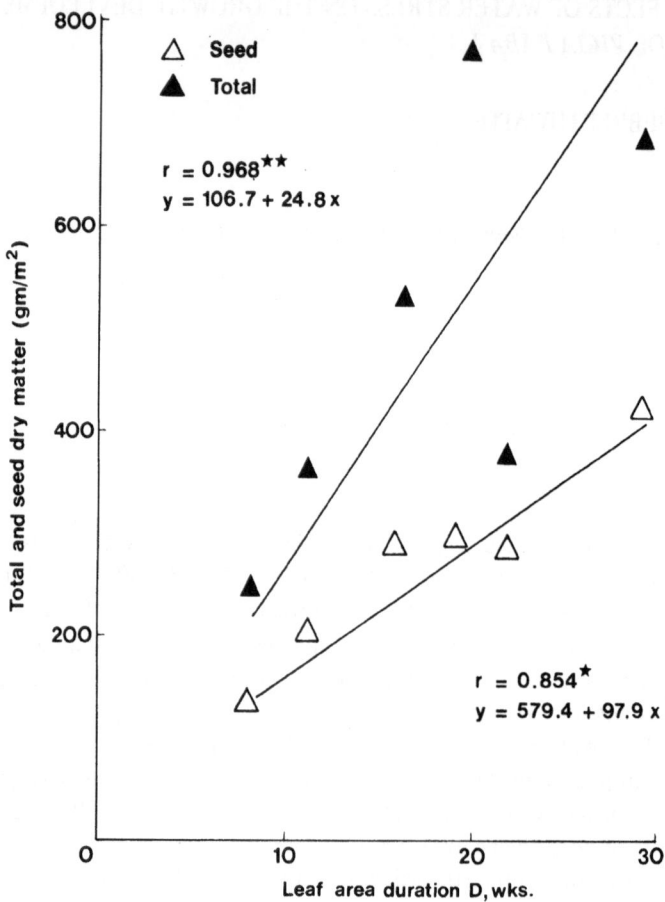

Fig. 1   The relationship between total and seed dry matter yield and leaf area duration in *Vicia faba* L.

Table 1.

The effect of irrigation on plant length, number of nodes per plant and internode length, 1969 and 1970.

| | Plant length (cm) | Number of nodes /plant | Internodes length (cm) | Max. summer potential soil moisture deficits (mm) |
|---|---|---|---|---|
| 1969 | | | | |
| Irrigated | 171* | 25* | 6.5 | 30 |
| Non-irrigated | 144 | 21 | 7.0 | 125 |
| S.E. | 9.1 | 1.1 | 0.43 | - |
| 1970 | | | | |
| Irrigated | 116*** | 23* | 5.1** | 30 |
| Non-irrigated | 64 | 16 | 3.9 | 225 |
| S.E. | 4.6 | 0.7 | 0.30 | - |

Significance of difference: 
* $0.01 < P < 0.05$; 
** $0.001 < P < 0.01$; 
*** $P < 0.001$

*Flowering and pod production*

Water stress does not affect the date at which the first flower buds appear (22) (24) (37) (20) or first open (20) but it does decrease the duration of flowering and pod set. Water stress decreases flowers, immature and mature pods, primarily because it decreases the number of flowering nodes and nodes bearing immature and mature pods rather than because of a decrease in the producitivity at individual nodes (Table 2). Water stress decreases the proportion of flowers forming mature pods. This results from a decrease in flowers setting immature pods more than from a decrease in the proportion of surviving pods (Fig. 2).

Water stress does not decrease the rate of pod shed (Fig. 2) which indicates the other internal plant factors such as high levels of ABA associated with water stress may be involved (6).

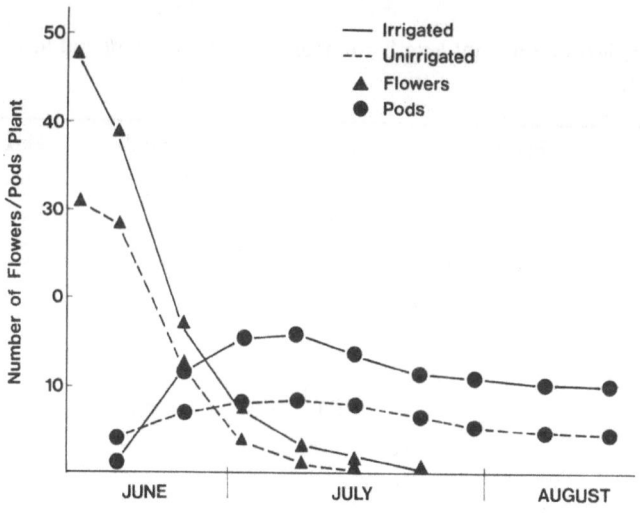

Fig. 2   Changes in time in number of flowers and pods in irrigated (max. soil water deficit 30mm) and non irrigated (max. soil water deficit 225mm) faba bean plants, 1970.

Table 2.

The effect of irrigation on flower and pod production, 1970.

| | Irrigated | Non-Irrigated | S. E. |
|---|---|---|---|
| Maximum flowers/plant | 46.3*** | 30.4 | 2.15 |
| Flowering nodes/plant | 9.2*** | 6.8 | 0.29 |
| Flowers/flowering node | 5.0** | 4.4 | 0.13 |
| Immature pods/plant | 18.4*** | 8.5 | 0.62 |
| Immature podding nodes/plant | 7.8*** | 4.0 | 0.21 |
| Immature pods/podding node | 2.4 | 2.1 | 0.08 |
| Mature pods/plant | 13.4*** | 5.1 | 0.51 |
| Podding nodes/plant | 7.0*** | 2.7 | 0.18 |
| Mature pods/podding node | 1.9 | 1.9 | 0.07 |
| Percent flowers forming immature pods | 40.2*** | 28.0 | 1.10 |
| Percent flowers forming mature pods | 29.3*** | 16.9 | 0.94 |
| Percent immature pods forming mature pods | 72.0*** | 60.4 | 1.83 |

Significance of difference:   *    $0.01 < P < 0.05$;
                             **   $0.001 < P < 0.01$;
                             ***  $P < 0.001$.

*Yield and yield components*

Water stress substantially decreases yield in faba beans and linear relationships exist between yield of total and seed dry matter and water use (Fig. 3, 4). Hence, these data give some indication of the yields that might be expected from a given amount of water use at a range of sites. Day and Legg (4), concluded that some of the variation about the line for grain yield against water use was probably caused by variations in harvest index, i.e. the proportion of total dry matter in the seed. Harvest index in faba beans is indeed variable; published data being in the range 0.24 to 0.51 (34) (21) (38) (16).

Few results of experiments with contrasting water regimes during the growth of the crop have been reported. Salter and Goode (31) concluded that there was not complete agreement on the effects of water stress in the pre-flowering period but, the data which do exist all show that soil moisture stress during flowering and pod development lead to lower yield. More recent work by Sprent *et al* (34) agrees with this conclusion. However, Stock (36) showed that water deficits at any stage of growth resulted in lower yields. Penman (29) and French and Legg (12), found no evidence of any particularly sensitive stage of growth.

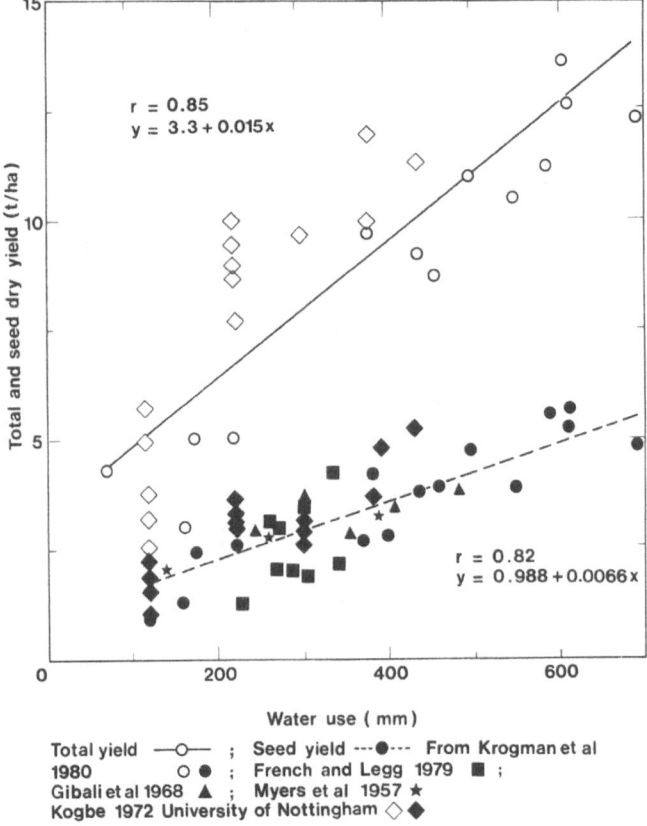

Fig. 3 The relationship between water use and total and seed dry matter yield in *Vicia faba*

Table 3.

Percentage increase in yield components as a result of irrigation at the University of Nottingham, Sutton Bonington, 1969 and 1970.

| Year | Max. summer potential soil moisture deficit for non irrigated treatments mm[1] | Flowers /plant | % flowers forming pods | Podding nodes/ plant | Pods/ podding node | Pods/ plant | Pods/ m² | Seeds /pod | wt/ seed | Seeds /m² | Seed yield |
|---|---|---|---|---|---|---|---|---|---|---|---|
| 1969 | 125 | 43*** | 21** | 45*** | 5 | 51** | 30*** | 0 | 9*** | 31** | 43*** |
| 1970 | 225 | 20* | 79*** | 128*** | -2 | 118** | 121*** | 9*** | 26*** | 139*** | 190* |
| 1971 | 75 | - | - | - | - | 17* | 17*** | 6* | -13* | 25* | 7 |

[1]For all irrigated treatments not greater than 30 mm.

Significance of difference:

     \*    $0.01 < P < 0.05$ ;

     \*\*   $0.001 < P < 0.01$ ;

     \*\*\* $P < 0.001$

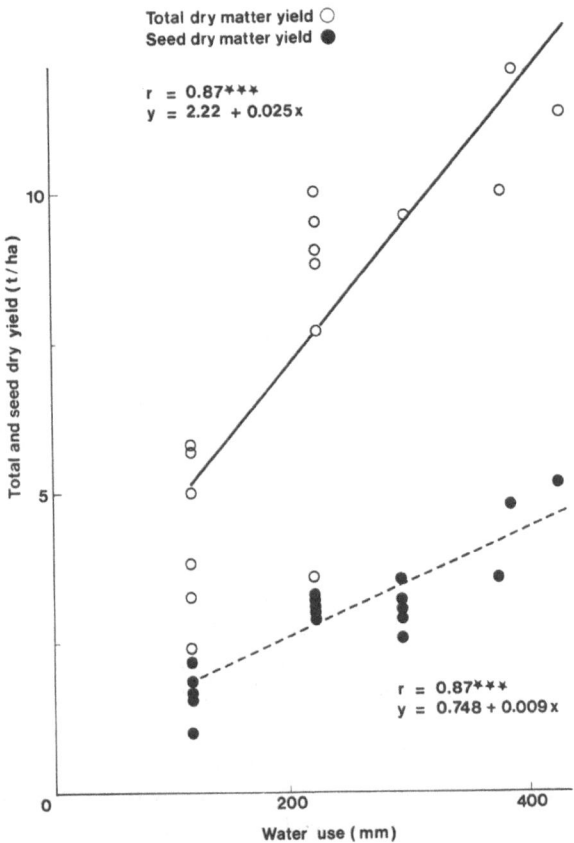

Fig. 4    The relationship between water use and total and seed dry matter yield in *Vicia faba* L., at the University of Nottingham.

Faba beans like many other crops, can remove some water from the soil before transpiration or growth rate decreases. This amount of water, termed the "limiting deficit", depends very much on soil type (12). French and Legg (12) have estimated the limiting deficits on a silty clay loam over flinty clay at Rothamsted to be 80mm, and a loamy sand over sand at Woburn to be 30mm, by relating the ratio of yields of non-irrigated to fully irrigated crops, $Y_0/Y_1$ with the maximum potential soil water deficit ( (DM) Fig. 5). A similar analysis of data was attempted at the University of Nottingham on a soil of the Astley Hall series which consists of a coarse loamy fluvial drift, often stony, passing abruptly at about 70 cm to reddish brown clay, (Fig. 5). The limiting deficit there was about 65 cm. Other workers utilize the percentage of available water used in the rooting zone as a measure of the start of yield decrease, and figures range from 75 to 60 % (21) to 62% (Woburn) and 55% (Rothamsted) (4). Consequently Day and Legg (4) have concluded that if maximum yield is required, faba beans should be irrigated when the amount of water available in the root zone falls to 70% of its value at field capacity.

172

The literature (17) (26) (32) (7) (21) (23) indicates that any or all seed yield components can be affected by water stress, and the time at which stress occurs and the degree of that stress determines which components are affected. Data from the University of Nottingham confirm this (Table 3). This strongly supports the hypothesis that the crop is sensitive at most stages of growth and does not have any specific stress stage.

*Depth of soil water extraction*

As yield is closely related to total water use, the size of the root system is extremely important because, if there is a shortage of water, the crop depends on the amount of stored water that it can extract from the soil. Neutron probe work over many years at Rothamsted Experimental Station (11), has shown that the maxium depth from which beans can take up water is about 0.9m which is much shallower than for cereals, sugar beet and grassland which can extract water from below 1m in dry years. Our results showed that no roots can be found below 80 cm, and that most of the roots are in the top 30 cm (Fig. 6). Thus it is not surprising that the crop can suffer growth checks when soil profiles have dried by an average of 30 to 80 mm compared to 100 mm or more for the deeper rooting crops.

*The effect of water stress on nitrogen fixation*

Water stress has been found to be a major factor affecting nitrogen fixation in faba beans (33) (35) (10) (1) (13) (39),  and this is at a maximum when soils are around field capacity (33). Troughs in the seasonal patterns of nodule activity often coincide with very low soil moisture (35). Irrigation has been found to increase the amount of nodular tissue, but it unexpectedly reduces the amount of nitrogen fixed per unit of nodule tissue (1), possibly because tissue has a higher moisture content. Despite this, the total amount of nitrogen fixed in the season was more than doubled by irrigation.

Nodules can rapidly adapt to new moisture requirements even if their levels were previously low because of moisture stress (33). Indeterminate nodules are better able to recover from stress than determinate ones because their meristems can resume growth, even after cessation of activity (39).

The effect of stress and its relationship with rhizobium activity is probably more acute in Mediterranean and Middle East regions, and experiments are urgently needed in these countries.

*Acknowledgement*

I wish to acknowledge some of the data collected on spring-sown faba beans at the University of Nottingham, U.K. under the supervision of Dr. R.K. Scott, now Head of Brooms Barn Experimental Station, Higham, Bury St. Edmunds, Suffolk, U.K.

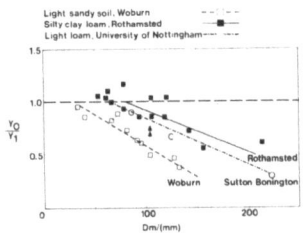

Figure 5.   Relative grain yield of non irrigated beans compared with irrigated beans versus Dm, the maximum potential soil water deficit.

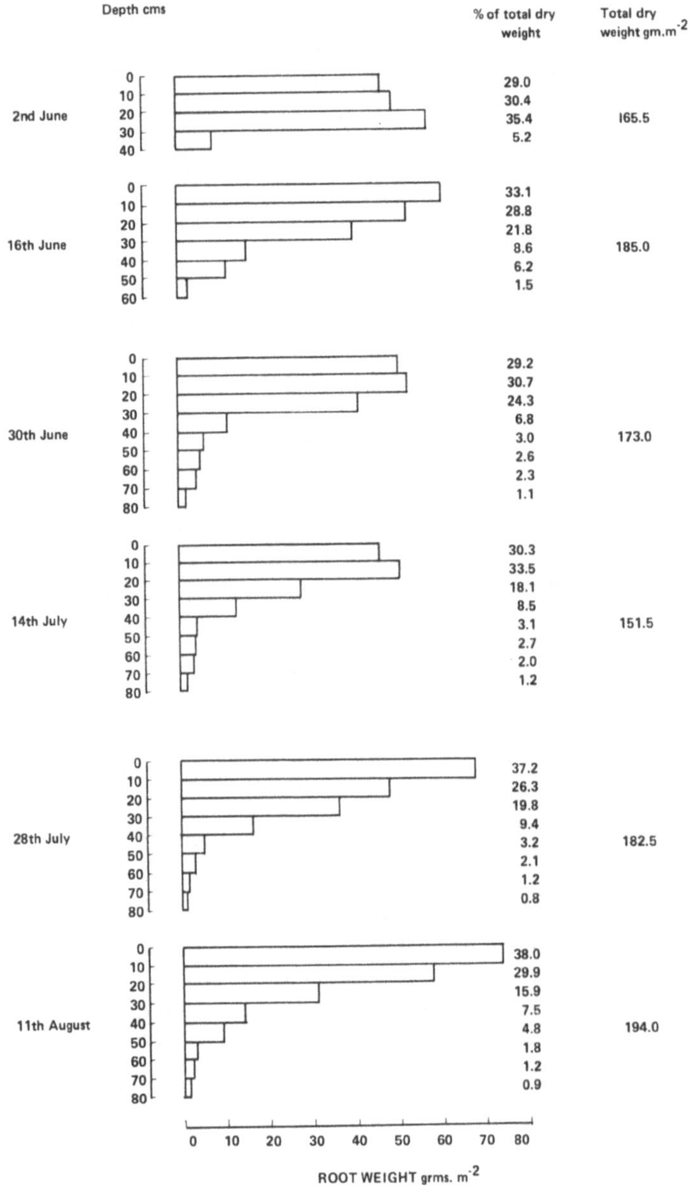

| Depth cms | | % of total dry weight | Total dry weight gm.m$^{-2}$ |
|---|---|---|---|

**2nd June**

| Depth | % of total dry weight | Total dry weight |
|---|---|---|
| 0 | 29.0 | |
| 10 | 30.4 | |
| 20 | 35.4 | 165.5 |
| 30 | 5.2 | |
| 40 | | |

**16th June**

| Depth | % of total dry weight | Total dry weight |
|---|---|---|
| 0 | 33.1 | |
| 10 | 28.8 | |
| 20 | 21.8 | |
| 30 | 8.6 | 185.0 |
| 40 | 6.2 | |
| 50 | 1.5 | |
| 60 | | |

**30th June**

| Depth | % of total dry weight | Total dry weight |
|---|---|---|
| 0 | 29.2 | |
| 10 | 30.7 | |
| 20 | 24.3 | |
| 30 | 6.8 | |
| 40 | 3.0 | 173.0 |
| 50 | 2.6 | |
| 60 | 2.3 | |
| 70 | 1.1 | |
| 80 | | |

**14th July**

| Depth | % of total dry weight | Total dry weight |
|---|---|---|
| 0 | 30.3 | |
| 10 | 33.5 | |
| 20 | 18.1 | |
| 30 | 8.5 | |
| 40 | 3.1 | 151.5 |
| 50 | 2.7 | |
| 60 | 2.0 | |
| 70 | 1.2 | |
| 80 | | |

**28th July**

| Depth | % of total dry weight | Total dry weight |
|---|---|---|
| 0 | 37.2 | |
| 10 | 26.3 | |
| 20 | 19.8 | |
| 30 | 9.4 | |
| 40 | 3.2 | 182.5 |
| 50 | 2.1 | |
| 60 | 1.2 | |
| 70 | 0.8 | |
| 80 | | |

**11th August**

| Depth | % of total dry weight | Total dry weight |
|---|---|---|
| 0 | 38.0 | |
| 10 | 29.9 | |
| 20 | 15.9 | |
| 30 | 7.5 | |
| 40 | 4.8 | 194.0 |
| 50 | 1.8 | |
| 60 | 1.2 | |
| 70 | 0.9 | |
| 80 | | |

Figure 6

ROOT WEIGHT grms. m$^{-2}$

Figure 6.   The development of root dry weight in *Vicia faba* L. at the University of Nottingham, 1970.

174

## REFERENCES

1. Bainbridge, A., Bardner, R., Cockbain, A.J. Day, J.M., Fletcher, K.E., Hooper, D.J., Legg, B.J., McEwen, J., Salt, G.A., Webb, R.M., Wilding, N. 1977. Recent work at Rothamsted on factors limiting yields of field beans (*Vicia faba* L.). Proc. Symp. on the Production, Processing and Utilization of the Field Bean (*Vicia faba* L.). Bulletin No.15. SHRI, Ed. R. Thompson.
2. Brouwer, W. 1949. Steigerung der Ertrage der Hulsenfruchte durch Beregnung sowie Fragen der Bodenuntersuchung und Dungung. Z Acker-u-pflBau, 91, 319-46.
3. Brouwer, W. 1959. Die Feldberegnung. DLG-Verlag, Frankfurt/Main, 4th Edition.
4. Day, W., Legg, B.J. 1981. Water relations and irrigation response. In the faba bean (*Vicia faba* L.) (ed. P.D. Hebblethwaite), Butterworths, London, (In preparation).
5. Dennett, M.D., Elston, J.F., Milford, J.R. 1979. The effect of temperature on the growth of individual leaves of *Vicia faba* L. in the field. Annals of Botany, 43, 197-208.
6. El Betagy, A.S., Hall, M.A. 1975. Studies on endogenous levels of ethylene and auxin in *Vicia faba* during growth and development. New Phytology, 75, 215-224.
7. El Nadi, A.H. 1970. Water Relations of Beans. II. Effects of differential irrigation on yield and seed size of broad beans. Expt. Agric. 6 (No.2), 107-111.
8. Elston, J. Karamanos, A.J., Kassam, A.H., Wadsworth, R.M. 1976. The water relations of the field bean crop. Philosophical Transactions of the Royal Society of London, B 273, 581-591.
9. Finch-Savage, W.E., Elston, J. 1976. The death of leaves in crops of field beans. Annals of Applied Biology, 85, 463-465.
10. Foulds, W. 1978. Response to soil moisture supply in three leguminous species. II. Rate of $N_2(C_2H_2)$ - fixation. New Phytologist 80, 547-555.
11. French, B.K., Long, I.F., Penman, H.L. 1973. Water use by farm crops. Test of the neutron meter on barley, beans and sugar beet, 1970. II. Spring wheat, barley, potatoes, 1969; potatoes, beans, kale 1968. III. Bare soil, short turf and crops in rotation, 1962 to 1967, 1971. Report, Rothamsted Experimental Station for 1972, part 2. pp. 5-42, 43-61, 62-85.
12. French, B.K., Legg, B.J., 1979. Rothamsted irrigation 1964-76. Journal of Agricultural Science, Cambridge, 92, 15-37.
13. Gallacher, A.E., Sprent, J.I. 1978. The effect of different water regimes on growth and nodule development of greenhouse-grown *Vicia faba*. Journal of Experimental Botany, 29, 413-423.
14. Gibali, A.A., Shenovda, N., Badawi, A.Y., Mansoor, S.F. 1968. Irrigation requirements, frequency and its effect on yield and quality of horse bean grains in Middle Egypt. Agricultural Research Review, 46, 91-98.
15. Greenwood, H.N. 1955. Experiments with field beans at the Nat. Inst. of Agric. Bot. - Interim Report J. Roy. Agric. Soc. 116, 29.
16. Hebblethwaite, P.D., Scott, R.K., Kogbe, J.O. 1981. The effect of irrigation and bees on the yield and yield components of *Vicia faba* L. In preparation for J. Agr. Sci.,Camb.
17. Jones, L.H. 1963 The effect of soil moisture gradients on the growth and development of broad beans (*Vicia faba* L.). Horticultural Research, 3, 13-26.
18. Karamanos, A.J. 1978. Water stress and leaf growth of field beans (*Vicia faba* L.) in the field : leaf number and total leaf area. Annals of Botany, 42, 1393-1402.
19. Keatinge, J.D.H., Shaykewich, C.F. 1977. The effects of the physical environment on the growth and yield of field beans (*Vicia faba, minor*) in the Canadian prairie. J. Agric. Sci. Camb., 89, 349-353.
20. Kogbe, J.O.S. 1972. Factors influencing yield variation of field beans (*Vicia faba* L.). Ph.D. thesis, University of Nottingham.

21. Krogman, K.K., McKenzie, R.G., Hobbs, E.H. 1980. Responses of faba bean yield, protein production, and water use to irrigation. Canadian Journal of Plant Science, 60, 91-95).

22. Listowski, A., Jackowska, I., Wirowski, Z. 1966. Influence of soil humidity and temperature fluctuations on development of horsebean (*Vicia faba minor*) and setting of pods. Pamietmik Pulawski, 23, 1-21.

23. McEwen, J., Bardner, R., Briggs, G.G., Bromilow, R.H., Cockbain, A.J., Day, J.M., Fletcher, K.E., Legg, B.J., Roughley, R.J., Salt, G.A., Simpson, H.R., Webb, R.M., Witty, J.F., Yeoman, D.P. 1981.. The effects of irrigation, nitrogen fertilizer and the control of pests and pathogens on spring sown field beans (*Vicia faba* L.) and residual effects on two following winter wheat crops. Journal of Agricultural Science, Camb. 96, 129-150.

24. Meriaux, S. 1972. Influence de la secheresse sur la croissance, le rendement et la composition de la feverole. Annales agronomique, 23, 533-546.

25. Monteith, J.L. 1972. J. Appl. Ecol. 9, 747-66.

26. Myers, V.I., Corey, G.L., Lebaron, M., McMaster, G. 1957. Irrigation of field beans in Idaho. Idaho Agricultural Research Bulletin No. 37, 1-16.

27. Penman, H.L. 1962. Woburn irrigation, 1951-1959. I. Purpose, design and weather II. Results for grass III. Results for rotation crops. Journal of Agricultural Science, Cambridge, 58, 343-348; 349-364; 365-379.

28. Penman, H.L. 1970. Woburn irrigation, 1960-1968. IV. Design and interpretation. V. Results for leys. VI. Results for rotation crops. Journal of Agricultural Science, Cambridge, 75, 69-73; 75-88; 89-102.

29. Penman, H.L. 1971. Irrigation at Woburn VII. Rothamsted Experimental Station Report for 1970, Part 2, pp. 147-170..

30. Reisch, W. 1952. Variabiblatsstudien an *Vicia faba* L. Z Acker-u-PflBau, 94, 281-306.

31. Salter, P.J., Goode, J.E. 1967. Crop responses to water at different stages of growth. Res. Rev. No. 2 of the Comm. Bur. of Hort. and Pltn. Crops, Kent - C.A.B.

32. Sebok, P.M., Nagy, Z. 1969. An irrigating regime in *Vicia faba* cultures. Lucr. Stunt. Inst. agron. Chij. (Ser. Agric.) (1967-1968, 1969) 23-24, 67 - 76. In Fld. Crop Abstr. (1971) 24, (No. 4), 4999.

33. Sprent, J.I. 1972. The effects of water stress on nitrogen-fixing root nodules. IV. Effects on whole plants of *Vicia faba* and *Glycine max*. New Phytology, 71, 603-611.

34. Sprent, J.I., Bradford, A.M., Norton, C. 1977. Seasonal growth patterns in field beans (*Vicia faba*) as affected by population density, shading and its relationship with soil moisture. J. Agric. Sci. Camb. 88, 293-301.

35. Sprent, J.I., Bradford, A.M. 1977. Nitrogen fixation in field beans (*Vicia faba*) as affected by population density, shading and its relationship with soil moisture. J. Agric. Sci., Camb., 88, 303-310.

36. Stock, H.G. 1977. Preliminary results obtained from studies into temporally differentiated water supply to field bean and its effects in terms of yield formation. Tag. Ber., Akad. Landwirtsch. Wiss. DDR, Berlin, 158, S, 229-241.

37. Tamaki, K., Naka, J. 1971. Physiological studies of the growing process of broad bean plants III. Effects of soil moisture on the growth and the variations of chemical components in the various organs. Technical Bulletin of the Faculty of Agriculture, Kagawa University, 22, 73-82.

38. Thompson, R. 1979. Crop growth and partition of assimilates in field beans (*Vicia faba*): responses to elimination of some major constraints. In: Some current research on *Vicia faba* in western Europe. (ed. D.A. Bond, G.T. Scarasia Mugnozza and M.H. Poulsen) pp. 407-420. European Communities Commission.

39. Wahab, A.M., Zahran, H.H. 1979. The effects of water stress on $N_2(C_2H_2)$ – fixation and growth of four legumes. Agricultura (Heverlee), 28, 383-400.

40. Winter, E.J. 1965. Irrigation of broad beans. Report on Irrigation Ann. Rep. Nat. Veg. Res. Sta. 1964, p. 51.

# 17. FLOWER AND POD DROP

MOHAMED M. EL-FOULY

*Botany Laboratory, National Research Centre, Cairo, Dokki, Egypt*

Yields of field beans are very unreliable, ranging sixfold under given climatic conditions (23).

One of the major factors which controls yield is the number of flowers which develop to pods. Only a small percentage of the flowers produced develop into fully mature pods. Under unfavourable conditions, this can be as small as 10%, or even less. The growth of faba beans is indeterminate, the flowering period is long; and while pods are set in the lower nodes, flowering continues in the upper ones.

Pollination is considered to be one of the reasons of yield unreliability. However, it is not the only reason (2).

Drop occurs at different stages, as buds, flowers or immature pods. A percentage of the dropped buds, flowers and pods might be considered to be a result of natural phenomena. Meanwhile, different environmental factors affect the percentage of dropping. There are also varietal differences in flowering pattern and drop (12, 13).

Jacquiery and Keller (16) analysed the drop at different stages during summer and winter, in pot experiments in a vegetation hall and in the field. They concluded that cultural conditions in the pot and field, influenced the drop greatly. Using the same cultivar in field and pot experiments in a vegetation hall, they obtained setting percentage of 63% and 48% respectively. Keller (19) reported on different percentages of pod setting for the same cultivar in different seasons 1971 and 1972; 31% and 40% respectively.

Keller (19) analysed the flowering and pod setting of different cultivars of beans. He was unable to get a correlation between the number of flowers produced per plant of different cultivars, pod setting percentage and yield. Comparing results of two successive years, Keller (19) concluded that metabolic processes, which are influenced by the environment, are probably important in the drop process.

## Possible factors affecting drop

Apart from the pollination aspects, there is evidence that one or more of the following factors are contributing to the drop phenomenon :

Cultivar; climatic factors such as temperature and light; agronomic practices, such as plant density; unfavourable conditions including drought, salinity, excessive soil moisture; an inadequate supply of nutrients; distribution of assimilates; and hormonal balance.

*G. Hawtin & C. Webb (Eds.): Faba Bean Improvement*
©1982 ICARDA. ISBN -13:978-94-009-7501-9

The two major metabolic factors that might affect flower drop and pod setting in one way or another are distribution of assimilates and hormonal balance.

Both can be affected, directly and indirectly, by all of the other factors.

## Distribution of assimilates

During vegetative growth, leaves above $^{14}$C-treated leaves are importing more radioactivity than those below them (27). When the lower nodes begin to produce flowers and set pods, the plant is still growing upwards and produces vegetative growth which competes for assimilates with flowers and young pods on the lower parts. The growth of the plant is also negatively influenced by the development of the lower pods (23); thus confirming the importance of competition between the vegetative and reproductive growth for the assimilates.

Jacquiery and Keller (16) (17) analysed the distribution of assimilates, using $^{14}CO_2$, at the middle of flowering, just before the end of flowering and at the end of flowering. In each stage, they fed a different leaf in each experiment. These experiments showed that the apex attracts assimilates until the end of the flowering period. By the end of flowering period, pods which are in lower influorescences act as strong sinks. The authors concluded that there was a strong competition for assimilates between reproductive and vegetative growth during this period.

In another study, Jacquiery and Keller (18) studied the qualitative distribution of assimilates after treatments during and just after the end of flowering. Pods were gaining assimilates at the cost of leaves during both periods, and at the cost of stems too. It was interesting to note that during the first phase of flowering, very low amounts of assimilates were transported to young pods. This indicates the strong competition between young pods and vegetative growth, as was also indicated by Chapman and Peat (2). Kogure et al. (22) reported on similar experiments in which the change in distribution of assimilates was analysed at different stages of growth after treating plants with $^{14}CO_2$ once during the flowering period. Also here, until the stage of green pod maturity, a strong competition between vegetative growth and the young pods was apparent in later stages (end of flowering). Pods and seeds begin to form a stronger sink, and at the seed maturity stage, seeds are a main sink. The authors concluded that the material stored in vegetative parts is then translocated to pods.

From available results, the competition for assimilates is :

a. During the entire flowering period, pods and young grains are competing with vegetative parts.

b. From the middle of the flowering period, pods are competing with each other within the same inflorescence.

c. During maturity, there is competition between pods according to their position on the plant.

Jacquiery and Keller (17) concluded that competition for assimilates is a factor which controls flower and pod drop.

*Effect of removal of plant organs*

Tamaki *et al.* (29) (22) studied the effect of different zones of vegetative growth by removing leaves on the pod-bearing zone. All treatments exhibited a negative effect on flowering, pod setting and yield.

Leaf removal was done :

i. *At the start of flowering* when it causes retardation of the development of growth as well as flower production and seed dry weight/plant.

ii. *At the end of flowering* when leaf removal led to less pod setting as well as lowering the accumulation of dry matter in seeds.

iii. *At the grain maturity stage* when generally, seed development was adversely affected.

Chapman *et al.* (1) reported that defoliation led to a decrease in the percentage of flowers and pods set and, generally, reduced production. Since the competition between vegetative growth and pod setting is well established, the effect of top removal on these two processes was examined (1) (2) (14) (15).

In general, decapitation led to a higher pod setting percentage and more pods. However, yields were not significantly affected. Gehringer *et al.* (14) concluded that this was because the total biomass was limiting grain yield. In the meantime, Chapman *et al.* (1) stated that the available photosynthetic sources were not in excess of pod requirements.

These results and conclusions reveal that, under given environmental conditions, the *Vicia faba* plants naturally adjust their yield load according to availability of assimilates.

This raises two questions with respect to increasing yields :

i. Would it be possible to increase photosynthetic efficiency and/or to enhance translocation of assimilates and thus increase pod filling in the last phase of the flowering period without affecting the existing competition in any way?

ii. Could the availability of assimilates be increased through increasing nutrition at the critical periods of normally grown plants or of those after top removal ?

*Hormonal balance*

Endogenous hormones control the different physiological processes, directly or indirectly, through various metabolic pathways. Apart from the availability of photosynthates, drop can take place through hormonal imbalance.

El-Antably (3) (4) studied bud and flower drop as well as early pod drop in *Vicia faba* and its relationship to the plant contents of endogenous hormones. Auxin content showed an opposite trend to that of abscisic acid (ABA). Auxin content was highest before flowering, decreased at flowering and again at shedding. Meanwhile the highest ABA content was found at the shedding stage (3). In this connection, it was found that plants suffering from osmotic stress, which exhibited increased drop, showed increases in ABA content (28).

As other endogenous hormones might be responsible for the shedding phenomenon, their content was also determined (4). Gibberellic acid (GA) and cytokinins (Cyt.) showed remarkably high contents at the flowering stage in comparison with the contents before flowering or at shedding. El-Beltagy and Hall (7) found that increases in endogenous ethylene contents could be correlated with flower and pod abscission. They studied also the relationship between ethylene and auxin content during abscission. No relationship could be demonstrated.

Also, differences in this relationship due to cultivar, led to the conclusion that probably, in this respect, there was no general relationship between auxin and ethylene content. However, the relationship between ethylene and drop is stressed by previous findings of El-Beltagy and Hall (6), who reported that drop was increased by either water stress or waterlogging and this was correlated with an increase in ethylene content.

El-Beltagy et al. (8) studied the effect of ethylene on drop and hormone content. Their results indicated an increase in flower, immature pods and leaf abscission due to the ethaphon treatment used. Endogenous auxin and inhibitors did not show any correlation with the effect of ethaphon on abscission. There is also no agreement between these results and those reported above by El-Antably (3) (4). However, it should be stated that El-Antably determined the hormone contents under normal development of the plants and that El-Beltagy et al. (8) treated plants with exogenous regulators which resulted in altering plant development. From this point of view, the results of El-Beltagy et al. (8) are not necessarily in contradiction with those of El-Antably, and probably both are relevant for the given conditions. In this context, it should be stressed that not only the level of one or more hormones is regulating a physiological phenomenon, but also the ratio between different hormones may play an important role.

*Effect of growth regulators*

Very little work with growth regulators has been done on faba beans (20). Growth substances can help to improve pod set. Substances used have included gibberellic acid (GA$_3$) (15a, 15b, 21); cycocel (CCC) (24) (9) (3) (4) (5) and El-Fouly et al. (unpubl.); 2,4-D (El-Fouly and El-Masry, unpublished), and other regulators (25) (5).

The idea behind using 2,4-D, a derivative of auxin, was that increasing auxin concentration may prevent flower drop to a certain extent. Field trials were conducted in 1971 using a concentration of 50ppm mixed with ferrous sulphate, (2.4 - 4.8 kg/ha) and sprayed during flowering. Drop was reduced and yield was increased by 14%. However, in a second experiment on another site, no effect could be observed (El-Fouly and El-Masry, unpubl.).

Due to the effect of 2,4-D at higher concentrations on leaf abscission, the use of 2,4-D should be studied carefully. However, the results of the first experiment described above are in agreement with the analysis of El-Antably (3) who found that during shedding, plants are characterised by a very low auxin content. Thus 2,4-D treatment might lead to increases in auxin content. This might be an explanation for the observed decrease in drop and increase in yield found after treatment with 2,4-D.

Studies by Ibrahim (15a, 15b) indicated that bean plants treated with GA exhibited earlier flowering, by about 4 days, had up to 50% more pods/plant and higher pod weight/ plant. These observations were coincident to higher polysaccharide contents in the seeds.

Recently, Keller and Bellucci (21) confirmed that the number of pods, plant as well as the average weight/seed was increased due to GA treatment. They reasoned that these effects were due to greater total dry matter production. Treatment in the field with CCC, a growth retardant, did not show any positive effect on flowering, pod setting or yield in one study (9), but caused decreases in yield in another (24).

Later reports have provided evidence that CCC treatment can lead to a decrease in the drop of buds, flowers and immature pods (3). El-Beltagy et al. (5) found that CCC induced more flower set than in the control. It is well known that CCC and probably other retardants show different effects on growth according to concentration and time of application as well as method of application. Under some conditions CCC can cause increased stem elongation as was found by El-Beltagy *et al.* (5) and this might be due to increased GA content. It had already been found that CCC increased GA content (26) (4). There may thus be no contradiction between the role of GA in flowering and the effect of CCC in increasing flowering. It has also been shown that auxin content increased, while ABA content showed a considerable decrease in CCC-treated plants at shedding (3). Aside from its effect on the content of hormones and its relation to stem elongation, CCC had shown effects on metabolism. Plants treated with CCC showed increased flow of assimilates from source towards the sink (10) (11). This was found to be partially due to increasing the activity of carbohydrates hydrolysing enzymes in leaves, and thus the mobility of assimilates.

*Other factors*

There is little need to stress or to discuss the different effects of the environmental factors. These mostly exert their effects through one or more of the metabolic process, leading to physiological phenomenon of drop or its prevention.

In general, there is an optimum level for each factor at which the plant exhibits the best growth and development, and thus its most reliable yield, within the genetic limits of the cultivar. As has been shown, these limits are expressed through the availability of assimilates to fill any increased number of pods.

*Outlook*

From the practical point of view of decreasing drop to increase the yield and its reliability, any positive treatments should be able to lead to increases in the number of pods as well as adding to their filling with assimilates. Fig. 1. illustrates a working hypothesis, which is currently being examined in pots and in the field. The idea is to use one or more factors, which are known :

i. To effect flowering and pod setting positively and/or prevent dropping.
ii. To strengthen the sink capacity of the pods, especially during the flowering period, without affecting the vegetative growth.
iii. To enhance the photosynthetic activity to be able to meet the requirements of the strengthened sink capacity.

One of the substances which can be used is CCC. It has been shown that at the proper concentration and at the proper growth stage, this chemical can increase pod setting through its effects on the hormonal balance. In the meantime, it fulfils the second requirement as it enhances assimilate translocation to the sink.

To increase the source capacity, environmental factors should be optimum, and the supply of nutrients should be higher than under conditions where no regulator is used.

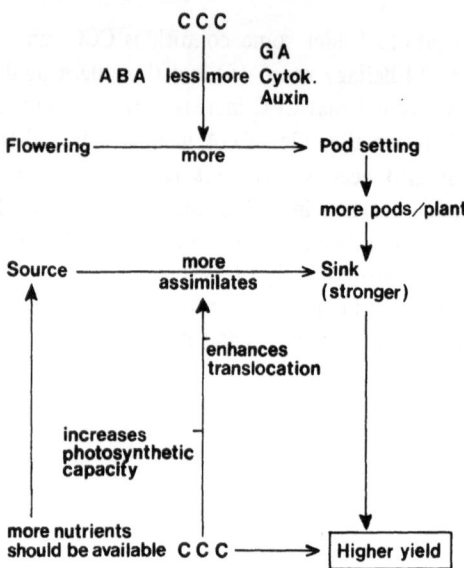

Fig. 1. A possible means of increasing yield of *Vicia faba* through increasing pod setting and sink capacity using growth regulator CCC and a supply of more nutrients.

Acknowledgements

   The author wishes to acknowledge with thanks the help of Prof. Dr. E. Keller, Zurich, Prof. Dr. K. Caesar and Dr. J. Carls, Berlin and Dr. K. Rohrmoser, Braunschweig in collecting available references.

REFERENCES

1. Chapman, G.P., Guest, H.L., Peat, W.E. , 1978. Top-removal in single stem plants of *Vicia faba* L. Z. Pflanzenphysiol., 89, 119-127.

2. Chapman, G.P., Peat, W.E. 1978. Procurement of yield in field and broad beans. Outlook in Agric., 2, 267-272.

3. El-Antably, H.M.M. 1976 a. Studies on the physiology of shedding of buds, flowers and fruits of *Vicia faba*. I. Effect of cycocel (CCC) and the role of endogenous auxin and abscisic acid (ABA). Z. Pflanzenphysiol., 80, 21-28.

4. El-Antably, H.M.M. 1976 b. Studies on the physiology of shedding of buds, flowers and fruits of *Vicia faba*. II. Effect of cycocel (CCC) and the role of endogenous gibberellins and cytokinins. Z. Pflanzenphysiol., 80, 29-35.

5. El-Beltagy, A.S., El-Beltagy, M.S., Hall, M.A. 1979. Modification of vegetative development,flowering and fruiting behaviour in *Vicia faba* L. by treatment with growth regulators. Egypt J. Hort., 6, 237-260.

6. El-Beltagy, A.S., Hall, M.A. 1974. Effect of water stress upon endogenous ethylene levels in *Vicia faba* L. New Phytol., 73, 47-60.

7. El-Beltagy, A.S., Hall, M.A. 1975. Studies on endogenous levels of ethylene and auxin in *Vicia faba* during growth and development. New Phytol., 75, 215-224.

8. El-Beltagy, A.S., Hewett, E.W., Hall, M.A. 1976. Effect of ethephon (2 -chloro ethyl phosphonic acid) on endogenous levels of auxins, inhibitors and cytokinins in relation to senescence and abscission in *Vicia faba* L. J. Hort. Sci., 51, 451-461.

9. El-Fouly, M.M., Abou-El-Lel, G., El-Hindy, M. 1975. Effect of cycocel on growth, flowering and yield of *Vicia faba* L. Agrochim., 19, 374-379.

10. El-Fouly, M.M., Garas, N.A. 1974. Amylase and invertase activities in relation to the concept of a physiological sink in potato plant grown in different seasons, and the influence of chlormequate upon these. Potato Res., 17, 249-260.

11. El-Fouly, M.M., Mohamed, B.R., Fawzi, A.F.A. 1977. Chlormequate (CCC) induced enhancement of flowering in carnation in relation tochanges in carbohydrate metabolism. Scien. Hort., 6, 241-249.

12. El-Tobgy, H.A., Ibrahim, A.A. 1968 a. A comparative study of some local and imported varieties of bean *(Vicia faba* L.). I. Flowering. Agric. Res. Rev. Cairo, 46, 1-11.

13. El-Tobgy, H.A., Ibrahim, A.A. 1968 b. A comparative study of some local and imported varieties of beans *(Vicia faba* L.). II. Fruiting and shedding. Agric. Res. Rev. Cairo, 46, 12-28.

14. Gehringer, W., Bellucci, S., Keller, E.R. 1978. Influence of decapitation and growth regulators on yield components and yield of *Vicia faba* L. Commission of European Communities, Bari, 27-29. 4.

15. Gehringer, W. and Keller, E.R. 1979. Influence of lecimage sur le development de la feverol *(Vicia faba* L.). Revue Suisse Agric., 11, 215-219.

15a.Ibrahim, Sohair, K. (1965). Effects of some hormonal treatments on yield of *Vicia faba* with special reference to its metabolism. M.Sc. Thesis, Fac. of Science, Cairo Univ.

15b.Ibrahim, Sohair, K. (1970). Studies on the effects of some growth regulators on native hormones and metabolism of *Vicia faba*, Ph.D. Thesis, Fac. of Science, Cairo Univ.

16. Jacquiery, R., Keller, E.R. 1978 a. Beeinflussung des Fruchtansatzes bei der Ackerbohne *(Vicia faba* L.) durch die Verteilung der Assimilate (Teil I). Angew Botanik, 52, 261-276.

17. Jacquiery, R., Keller, E.R. 1978 b. La chute des fruits chez la feverol *(Vicia faba* L.) en relation avec la disponibilite en assimilates marques au $^{14}$C. Revue Suisse Agric., 10, 123-127.

18. Jacquiery, R., Keller, E.R. 1980. Beeinflussun des fruchanatzes bei der ackerbohne *(Vicia faba* L.) durch dieVerteilung der Assimilate Teil II). Angew Botanik, 54, 29-32.

19. Keller, E.R. 1974. Die Ackerbohne *(Vicia faba* L.) eine vergessene Kulturpflanze mit Zukunftsaussichten Schweizer. Landwirt.. Forsch., 13, 287-300.

184

20. Keller, E.R., Bellucci, S. 1979. .Anwendung von Phytohormonon und Wachstums-regulatoren bei Kornerleguminosen Referat. Univ. Stuttgart-Hohenheim.
21. Keller, E.R., Bellucci, S. 1980. Influence of growth regulators on yield and yield structure of *Vicia faba* L. in Bond, D.A. (Ed.) *Vicia faba*. Feeding value, Processing and viruses, 385-404.
22. Kogure (Tamaki), K., Naka, J., Asannma, K. 1978. Behaviour of [14]C photosynthetic products during the reproductive growth in broad bean plant. Tech. Bull., Fac. Agric., Kagawa Univ., 30, 1-8.
23. Lawes, D.A. 1974. Field beans : improving yield and reliability. Span, 17, 21-23.
24. McEwen, J. 1970. Fertilizers and growth regulators for field beans *(Vicia faba)* II. The effect of large dressings of fertilizer nitrogen, single and split application and growth regulators. J. Agric. Sci. Camb., 74, 67-72.
25. McEwen, J. 1973. The effects of growth regulators, seed rates and row spacing on field beans *(Vicia faba* L.). J. Agric. Sci. Camb., 80, 37-42.
26. Reid, D.M., Grozier, A. 1970. CCC-induced increase of gibberellin levels in pea seedlings. Planta (Berl.), 94, 95-100.
27. Sharabash, M.T.M. 1981. Effect of $CO_2$ concentration on retention and distribution pattern of radiocarbon in *Vicia faba* L. plants. Egypt. J. Physiol. Sci. (In press).
28. Sivakumaran, S., Hall, A. 1978. Effect of osmotic stress upon endogenous hormone levels in *Euphorbia lathyrus* L. and *Vicia faba* L. Ann. Bot., 42, 1403-1411.
29. Tamaki. K., Naka, J. and Asanama, K. (1972). Physiological studies of the growing proces of broad bean plants. VI. Effects of partial leaf removal in the flowering and maturing stages on the growth and the variations of chemical components. Tech. Bull. Fac. Agric., Kagawa Univ., 24, 1-9.

## 18. TOLERANCE TO SALINITY

K. Caesar  and G. Rusitzka

*Institute of Crop Science, Technical University of Berlin, Germany*

Salinity adversely affects the growth and yield of most agricultural crops in arid, semi-arid and salt-affected regions. Highly soluble salts in the root zone cause physiological scarcity of water to plants by raising the osmotic pressure in the soil solution. The availability of water may then become so critically low that yields are reduced. Besides the reduced availability of soil water, there might be an altered availability of nutrients, due to ion-antagonism in the root zone, as well as altered physical properties of soil, restricting water movement or reducing root penetration. There may also be direct toxicity (6).

Saline soils contain soluble salts in quantities that affect plant growth at various stages and create differences between crops and differences in the ion composition of crops at maturity (8). Plants differ markedly in their ability to grow and produce economically under the saline conditions of arid and semi-arid climates. Consequently, successful crop production depends on the way that soil, water and plants are managed.

Agronomy and breeding are two possible ways of solving the salinity problem. The agronomical point of view involves different management practices in relation to salinity control e.g. leaching, irrigation and planting methods. Before saline soils can be cropped, they must usually be reclaimed by applying water to leach out excess salts. For leaching to be effective, soils must have adequate internal drainage, since leaching salts into a shallow water table may only result in a rapid resalination of the root zone (19).

When crops are irrigated at conventional frequencies, soil water contents may decrease during an irrigation cycle and both decreased matric and osmotic potentials may reduce yields (20). If water deficits are replaced frequently, roots will have a steady supply of water at the minimum salinity of the irrigation water. This can be achieved by small daily applications of water e.g. by trickle irrigation (2). If salinity cannot be eliminated, e.g. when the irrigation waters available are saline, only those crops which can tolerate salinity can be grown. Hence salt tolerance data are useful in selecting suitable crops for such conditions. The importance of screening species and varieties which give economically useful yields on salt-affected soils becomes evident. The necessary first step is to find out morphological and physiological criteria which enable a plant to tolerate certain amounts of salt. The next step would be to identify sources of salt tolerance which could be used by the plant breeder in developing salt tolerant cultivars.

*G. Hawtin & C. Webb (Eds.): Faba Bean Improvement*
*©1982 ICARDA. ISBN -13:978-94-009-7501-9*

The increasing demand for faba bean production can result in an advance of cropping into saline soils. This stresses the importance of screening for cultivars which are able to tolerate high amounts of soluble salts and to produce economically. Faba beans are known for being very sensitive to salinity; much more so than, for example, barley and sugar beet. In trials, the sensitivity of faba beans to sodium increased with age and reduced the yield by half (8). Although the mature faba bean plant tolerated smaller levels of NaCl, it was NaCl-sensitive throughout its life, and growth was retarded considerably.

Experiments have shown that NaCl uptake is proportional to soil salinity levels (17), leading to different kinds of salt injuries in this crop. Obviously there are no avoidance mechanisms in *Vicia faba* as, for example, in the *Atriplex* species, limiting the absorption of salt at high concentrations. Field experiments in the Sudan with faba beans (4) on the effect of salinity on growth and yield, also showed a 50% reduction at 9 mmho/cm. There was a negative correlation between dry matter production and yield of faba beans in the Sudan (4).

The report of Ayers, Brown and Wadleigh in 1952 that salinity depressed the yield of vegetative parts more than that of grain, seems to be in accordance with the first results of pot experiments at the Institute of Crop Science in Berlin in which the vegetative growth was depressed markedly with increasing salt (NaCl and $Na_2SO_4$) concentrations in all six cultivars tested. However increasing the salt content did not influence the harvest index, and had almost no influence on the dry weight per seed up to 50 meq of both salts. The number of seeds and the number of pods, seemed to be negatively affected by the salt concentration. Differences between the cultivars could be observed. The European cultivars seemed to be more sensitive, and NaCl showed a greater toxicity than $Na_2SO_4$.

Morphological changes in leaf anatomy, have been observed in experiments with red kidney beans (16) in response to soil salinity. There was an increase in cell number but a reduction in cell size. High salinity levels limited the growth of the bean leaves *(Phaseolus vulgaris)*. It has been found that larger Cl-concentrations increased the thickness of the leaf cuticle. Smaller but thicker leaves, due to an elongation of the palisade cells, causing increased succulence (Cl-effect), have been reported by Bernstein (3) and Meiri, A. and Poljakoff-Mayber, A. (11) (13). Similar results were found in our trial on *V. faba* in the leaf cross-sections taken during the vegetative period at high concentrations of NaCl and $Na_2SO_4$. The cultivars reacted to salt stress with a change in the thickness of the layers and elongation of the palisade cells.

As ion antagonism in saline environments is a well known phenomenon, and is usually followed by nutrient deficiency stresses, the ionic composition is very important in regard to salt tolerance. Mechanisms in plants for avoiding the uptake of excess salts (ions), are known, e.g. in soybeans, to be responsible for the tolerance of a crop to high salinity levels. Wallace *et al.* (21) reported that bush beans *(Phaseolus vulgaris)* accumulate sodium in roots, but translocate it poorly into the shoots. Sodium was translocated in young seedlings, but the effect decreased with age, exemplifying the different behaviour of plants at the different growth stages.

Other experiments with barley seedlings showed clearly, at high salt concentration levels, the depressing effect of $Ca^{2+}$ on the uptake of $K^+$; bivalent ions inhibited the uptake of monovalent ions (10). This effect could be overcome by addition of $CaCl_2$ via a stimulating effect of the readily taken-up accompanying ion $Cl^-$, giving evidence of by-effects of the counterion. Ayoub (1) studied the efficiency of calcium in enhancing the tolerance of beans *(Phaseolus vulgaris)* to sodium salinisation in both cool and hot seasons. The Na levels, both in roots and shoots, declined significantly as increased amounts of calcium were added. Ca in the range of 2.0 to 8.0 mmol/l caused competitive inhibition to Na uptake and translocation. The situation was different in the hot season. The situation with respect to *V. faba* requires further study.

The possible fluctuation of protein, due to salinity, will be studied in our experiments in Berlin. In other experiments with young barley plants, in the presence of high concentrations of NaCl (80 mm) both growth and protein synthesis were affected. There was not only an effect on protein synthesis but also on the transfer of inorganic nitrogen into amino acids. If only protein synthesis had been affected, the number of free amino acids would have increased. This was not the case. However in the presence of additional potassium, the uptake of nitrogen, the total content of nitrogen, and its transfer into protein via amino acids and protein synthesis, increased.

Experiments with faba beans and barley showed that the effect of salt on nitrogen uptake depends not only on the concentration but also on the kind of salt. Increasing the validity of the cation ($CaCl_2$ KCl $K_2SO_4$) and decreasing validity of the anion increased the uptake of nitrogen. The effect was greater in faba beans than in barley. Calcium, being of a high validity, showed the greatest influence (14). The uptake of nitrogen in *Vicia faba* is also correlated with the moisture content of the soil (18). Increased moisture supply increased not only the seed yield of faba beans but also the content of the crude protein in the seeds, showing the great importance of a sufficient supply of water for getting the optimum yield. Salinity and water supply are closely related and both affect crop yields.

Research work under semi- or entirely controlled conditions is needed to clarify the influence of particular factors, alone or in combination. More applied trials are also needed under natural conditions. To verify the results obtained so far in Berlin, experiments have been started on saline soils in Egypt, using the same cultivars.

In addition to the development of improved production methods, e.g. through irrigation or special planting methods, the development of improved cultivars having a greater tolerance to salinity would appear to be a promising approach. The experiments described above are a first step in this direction.

The possible presence of an inherited factor for salt tolerance in faba beans has already been reported by El Karouri (4). If this can be substantiated, it would open up a wide range of possibilities for breeding programs.

188

REFERENCES

1. Ayoub, A.T. 1974. Effect of calcium on sodium salinization of beans *(Phaseolus vulgaris* L.). J. Exp. Bot. 25, 245-252.
2. Bernstein, L., Francois, L.E. 1973. Comparisons of drip, furrow and sprinkler irrigation. Soil Sci. 115, 73-86.
3. Bernstein, L. 1975. Effects of salinity and sodicity of plant growth. An. Rev. Phytopath. 13, 295-312.
4. El Karouri, M.O.H. 1979. Effects of soil salinity on broad bean *(Vicia faba)* in the Sudan. Expl. Agric. 15, 59-63.
5. Fuchs, W.H. 1955. Zuchtung auf Trocken- und Salzresistenz. 14th Intern. Hortic. Congr. Report.
6. ---------------------- 1979. Management of saline soils. Outlook on Agric. 10, 13-20.
7. Greenway, H., Gume, A., Thomas, D.A. 1966. Plant response to saline substrates. VIII. Regulation of ion concentrations in salt-sensitive and halophytic species. Aust. J. Biol. Sci. 19, 741-756.
8. Hamid, A., Talibudeen, O. 1976. Effect of sodium on the growth of ion uptake by barley, sugar beet and broad beans. J. Agric. Sci. Camb., 86, 49-56.
9. Helal, H.M., Mengel, K., 1979. Der Einfluss der Versalzung durch NaCl und des Kaliums auf den Stoffwechsel des Stickstoffs von jungen Gerstenpflanzen. Kali-Briefe, 11, 1-5.
10. Marschner, H., Ossenberg-Neuhaus, H. 1970. Bedeutung des Begleitanions bei den Wechselbeziehungen zwischen K+ und Ca++ im Bereich hoher Aussenkonzentrationen. A. Pflanzenernahr., Dung. und Bodenkd, 126, 217-228.
11. Meiri, A., Poljakoff-Mayber, A., 1967. The effect of chloride salinity on growth of bean leaves in thickness and in area. Isr. J. Bot. 16, 119-123.
12. ------------------------------------------- 1969. Effect of variations in substrate salinity on the water balance and ionic composition of bean leaves. Isr. J. Bot. 18, 99-112.
13. --------------------------------------------- 1970. Effect of various salinity regimes on growth, leaf expansion and transpiration rate of bean plants. Soil Sci. 100, 26-34.
14. Metwally, A.J., El-Damaty, A. Moustafa, M., 1978. Salt influence on nitrate and phosphate uptake by broad beans and barley in sand culture. Z. Pflanzenernahr. Bodenkd. 141, 411-418.
15. Nieman, R.H., 1962. Some effects of sodium chloride on growth, photosynthesis, and respiration of twelve crop plants. Bot. Gaz. 123, 279-285.
16. -------------------- 1965. Expansion of bean leaves and its suppression by salinity *(Phaseolus vulgaris* L.). Plant Physiol. 40, 156-161.
17. Priebe, A., Jager, H.J., 1978. Einfluss von NaCl auf Wachstum und Ionengehalt unterschiedlich salztoleranter Pflanzen. Angew. Bot. 52, 331-341.
18. Shaaban, K., Hamdy, H., Omram, M., Afifi, E., 1979. The effect of soil moisture levels and nitrogenous fertilizers on the dry matter and nitrogen contents in horse beans and barley plants. Beitr. trop. Landw. u. Vet. med. 17, 267-272.
19. United States Salinity Laboratory Staff, 1954. Diagnosis and improvement of saline and alkali soils. US Dept. Agric. Handb., GO, 160 pp.

20. Wadleigh, C., Ayers, A.D., 1945. Growth and biochemical composition of bean plants as conditioned by soil moisture tension and salt concentration. Plant Physiol. 20, 106-132.

21. Wallace, A., Hemaidan, N., Sufi, S.M., 1965. Sodium translocation in bush beans. Soil Sci. 100, 331-334.

19. WEED AND *OROBANCHE* CONTROL IN EGYPT

MOHAMED KAMAL ZAHRAN


*Weed Control Research, Plant Protection Research Institute, Agricultural Research Center, Giza, Egypt*

Among the main obstacles to the optimum production of faba beans in Egypt are weeds and *Orobanche.* The reduction in faba bean yield due to weed competition, other than parasites, has ranged from 24 - 30% (9).

Early in 1908, Strasburger (16) described *Orobanche* as a dreadful pest. It inflicts serious damage upon some cultivated crops and is difficult to exterminate. He demonstrated that after the *Orobanche* seed germinates, only its haustoria penetrate the host roots and its light yellow, reddish-brown, or amethyst-coloured flower-shoot appears above the ground.

The morphological variation in populations of *O. crenata* is very high and it may be possible that there are at least two sub-species. Variation involves height, colour and size of flower, stems and stigmas, flower distribution on the spike and other characteristics (13).

Shabatay (15) listed five *Orobanche* species in Egypt, including *O. crenata* which atttacks faba bean.

*Orobanche* seeds live for more that 18 years (18). A single plant is estimated to produce 0.5 to 1 million seeds which only germinate when in contact with roots of the host plant (3) (16) (17) (2). For germination *O. crenata* seeds require a definate exudate secreted by the roots of *Vicia faba* just before and during flowering (10). The parasite spikes constantly appear during the growing period of faba bean. All of these factors contribute to the difficulty of the *Orobanche* problem.

Controlling Weeds in Faba Bean — A Study

ICARDA (6) showed in 1980 that faba beans did not compete well with weeds and that higher rates of sowing failed to increase crop competition.

As the scarcity in hand labor is impeding the optimum crop production, there is a need for acceptable means for controlling weeds in faba bean fields.

The predominant weeds of three experimental sites in Egypt (Sakha, Bahteem and Giza) where trials were conducted were *Beta vulgaris, Rumex dentatus, Chenopodium* spp. *Medicago hispida, Melilotus indica* and *Ammi majus.* The amount of weed infestation at the sites was estimated to be 11.15, 14.70 and 12.30 t/ha respectively. The mean reduction in faba bean seed yield due to annual weeds was estimated to be 34.21% (0.91 t/ha). The presence of weeds during   4, 6, 8, 10, 12 and 14 weeks after sowing accounted for

*G. Hawtin & C. Webb (Eds.): Faba Bean Improvement*
*©1982 ICARDA. ISBN -13:978-94-009-7501-9*

reducing the crop seed yield by 3.01, 1.13, 16.52, 25.56, 28.58 and 27.44% respectively as compared with the treatment in which the weeds were left only for two weeks after sowing.

In this respect, Lawson and Wiseman (9) showed that weed competion reduced faba bean yield by 24 and 30% /plant in two experiments in 1977. Table 1 suggests that the faba bean crop is not seriously affected by weeds that emerge during the first six weeks after sowing if plots are kept weedfree thereafter. Niets *et al.* (12) showed that weeds during the first 12 days from the emergence of peas depressed production by no more than 3%

Table 1. Effect of weed competion on the yield of faba beans (t/ha) 1979-1980.

| Treatment | Sakha* | Site Bahtim | Giza | Mean |
|---|---|---|---|---|
| Weedy for | | | | |
| - 2 weeks | 0.460 | 4.417 | 3.112 | 2.663 |
| - 4 weeks | 0.430 | 4.250 | 3.044 | 2.575 |
| - 6 weeks | 0.460 | 4.333 | 3.093 | 2.629 |
| - 8 weeks | 0.323 | 3.833 | 2.510 | 2.219 |
| - 10 weeks | 0.270 | 3.167 | 2.545 | 1.987 |
| - 12 weeks | 0.235 | 2.750 | 2.705 | 1.897 |
| - 14 weeks | 0.200 | 2.917 | 2.663 | 1.927 |
| Weedfree for | | | | |
| - 2 weeks | 0.275 | 2.500 | 3.138 | 1.971 |
| - 4 weeks | 0.263 | 2.668 | 2.668 | 1.866 |
| - 6 weeks | 0.238 | 3.417 | 3.263 | 2.306 |
| - 8 weeks | 0.230 | 2.883 | 2.293 | 1.802 |
| - 10 weeks | 0.295 | 3.250 | 2.638 | 2.061 |
| - 12 weeks | 0.338 | 3.833 | 2.856 | 2.342 |
| - 14 weeks | 0.373 | 4.333 | 2.125 | 2.277 |
| Weedy | 0.089 | 2.500 | 2.675 | 1.754 |
| L.S.D. (5%) | 0.002 | 1.246 | n.s. | 0.418 |

* The yield from the experimental field was exceptionally low because of the unsuitable weather.

Most likely, the critical period for weed-crop competition in faba bean begins six weeks from sowing. Post-emergence herbicides which are applied longer than six weeks after sowing, appear to have considerable promise.

The results of screening various herbicides at the Sakha Research Station are shown in Table 2. All herbicide treatments accounted for a significant reduction of weed population of the predominent annual weeds.

Although the use of Ronstar/Linuron combination was the most effective against weeds and gave the reduction of 96.14%, it also affected the crop seriously.

Taking into account the effect on seed yield, it seems that the separate use of Treflan, Cobex and Amex as PPI, as well as Linuron as post-sowing at the used rates appears to be capable of controlling annual weeds in faba bean.

Table 2. Effect of herbicidal treatments on weeds and the yield of faba beans (Sakha, 1979-1980).

| Herbicides | Rate/ha | Weeds (t/ha) | Seed yield (t/ha) |
|---|---|---|---|
| PPI | | | |
| Amex | 6.25 l | 0.238 | 0.220 |
| Cobex | 2.50 l | 0.223 | 0.238 |
| Treflan | 2.50 l | 0.225 | 0.296 |
| Stomp | 6.25 l | 0.250 | 0.148 |
| Ronstar | 6.25 l | 0.313 | 0.101 |
| Post-sowing | | | |
| Linuron | 2.00 kg | 0.275 | 0.197 |
| Amex + Linuron | 6.25 l / 2.00 kg | 0.175 | 0.185 |
| Cobex + Linuron | 2.50 l / 2.00 kg | 0.175 | 0.185 |
| Treflan + Linuron | 2.50 l / 2.00 kg | 0.150 | 0.149 |
| Stomp + Linuron | 6.25 l / 2.00 kg | 0.150 | 0.143 |
| Ronstar + Linuron | 6.25 l / 2.00 kg | 0.085 | 0.042 |
| Hand-weeding | --- | 2.200 | 0.078 |
| L.S.D. (5% ) | | 0.378 | 0.060 |

Klein (8) pointed out that cultivation followed by application of pre-emergence soil-acting and post-emergence herbicides were used in the German Democratic Republic.

Kemp (7) showed that Linuron at 1 kg (a.i.)/ha was effective against annual broad-leaved weeds and some annual grasses. Sedov and Losev (14) noted that pre-emergence application of Linuron decreased weed population by 64% in faba beans.

El-Sharkawy and Sogaier (1) showed that Trifluralin at 0.5 kg/ha decreased weed population by 40% and increased straw and seed yield by about 30% whereas the rate of 1 kg/ha reduced weeds by 65% but, at the same time, reduced straw and seed yields by 41 and 28% respectively. He added that Trifluron also reduced nodulation. Lawson and Wiseman (9) showed that trifluron at 1.12 kg (a.i.)/ha incorporated to a depth of 5 cm delayed emergence and caused some distortion and stunting of young faba bean plants.

A herbicide named Fusilade (PP009), fluazifop-butyl, was tried out in a field of faba beans as post-emergence, four weeks after sowing, in a preliminary test. Visual observations revealed that the herbicide was satisfactory against all kinds of annual grasses (mostly *Lolium termulentum* and *Avena fatua)* where it was sprayed as a 1% solution, (475 1/ha) and against perennial grasses (mainly *Cynodon dactylon)* when sprayed as a 2% solution. The crop plants proved to be completely tolerant to this new herbicide. This finding suggests

Table 3. Effect of Kerb, post-emergence on *Orobanche* and three cultivars of faba bean (three sites 1980, field trials).

| Sites | Sharkiah | | | Ismailiah | | | Minia | | | Mean | | |
|---|---|---|---|---|---|---|---|---|---|---|---|---|
| | T* | C** | Mean | T | C | Mean | T | C | Mean | T | C | Mean |
| *Orobanche* Spikes (kg/ha) | | | | | | | | | | | | |
| Giza 2 | 195 | 468 | 332 | 16 | 89 | 53 | 80 | 173 | 127 | 97 | 243 | 170 |
| Giza 4 | 289 | 1185 | 737 | 30 | 228 | 129 | 139 | 231 | 183 | 153 | 548 | 351 |
| F 402 | 188 | 305 | 247 | 22 | 42 | 32 | 7 | 46 | 37 | 72 | 131 | 102 |
| Mean | 224 | 653 | | 23 | 120 | | 75 | 150 | | 107 | 307 | |
| L.S.D. 5 % | | | | | | | | | | | | |
| Faba bean seeds (t/ha) | | | | | | | | | | | | |
| Giza 2 | 1.70 | 1.68 | 1.69 | 3.13 | 2.53 | 2.83 | 3.37 | 3.53 | 3.45 | 2.73 | 2.58 | 2.66 |
| Giza 4 | 1.77 | 1.23 | 1.50 | 2.86 | 2.75 | 2.81 | 3.69 | 2.89 | 3.29 | 2.77 | 2.29 | 2.53 |
| F 402 | 2.18 | 1.19 | 1.69 | 2.95 | 2.97 | 2.96 | 4.02 | 3.23 | 4.13 | 2.32 | 2.80 | 2.56 |
| Mean | 1.88 | 1.37 | | 2.98 | 2.75 | | 3.69 | 3.55 | | 2.61 | 2.56 | |
| Faba bean straw (t/ha) | | | | | | | | | | | | |
| Giza 2 | 5.15 | 4.76 | 4.95 | 4.93 | 4.69 | 4.81 | 3.99 | 4.03 | 4.01 | 4.69 | 4.49 | 4.59 |
| Giza 4 | 5.20 | 3.59 | 4.40 | 5.05 | 4.94 | 5.00 | 4.84 | 4.67 | 4.76 | 5.03 | 4.40 | 4.72 |
| F 402 | 6.56 | 4.24 | 5.40 | 4.91 | 5.43 | 5.17 | 5.37 | 4.57 | 4.97 | 5.61 | 4.57 | 5.18 |
| Mean | 5.64 | 4.20 | | 4.96 | 5.02 | | 4.73 | 4.42 | | 5.11 | 4.55 | |
| L.S.D. 5 % | | | | | | | | | | | | |

*Treated with 4.76 kg (a.i.)/ha in 2,500 l water/ha 4 weeks after planting.
**Control, untreated.

the potential of Fusilade/Basagran, combination against all grassy and broadleaved weeds in faba bean.

## Further Approach Towards *Orobanche* Control

Various investigations into the chemical control of *Orobanche* and on tolerant cultivars have been undertaken. Research in the past has revealed that most herbicides appeared to be useless while some others have shown a remarkable effect. The noteworthy herbicides are Lancer (glyphosate) and Kerb (pronamide : propyzamide) (19).

Recently trials have been conducted on farmers fields and research stations to investigate further the use of these herbicides and to evaluate on larger plots the cultivar F. 402 which has been reported to be tolerant to *Orobanche* (6) (1) (4).

In field trials conducted in three provinces (Sharkiah, Ismailiah and Minia) in 1980, Kerb was applied four weeks after sowing at a rate of 4.76 kg (a.i.)/ha in 2,500 l/ha water. Three cultivars were tested (Giza 2, Giza 4 and F 402) with and without herbicide application. The results (Table 3) indicated that Family 402 has some resistance to *Orobanche*, whereas Giza 4 was highly susceptible.

Yields of treated F 402 in the three sites ranged from 2.18 to 4.02 t/ha; variation which cannot be accounted for by *Orobanche*. At all sites the application of Kerb led to an increase in mean yield of the three cultivars, and a large reduction in the weight of *Orobanche* spikes.

In a trial at Samalout, Giza 4 and F 402 were tested with and without a single application of Lancer (0.357 l/ha). A significant reduction in the weight of *Orobanche* spikes was found (Table 4) but there was no significant increase in faba bean seed or straw yield. Again F 402 was superior to Giza 4.

Table 4. Effect of lancer, post-emergence on *Orobanche* in two cultivars of faba bean (Samalout, 1980, field trial).

| Cultivar | Yield of *Orobanche* spikes (kg/ha) | |
|---|---|---|
| | Treated | Control |
| Giza 4 | 14.8 | 269.4 |
| F 402 | 4.9 | 200.0 |
| Mean | 9.9 | 234.7 |
| L.S.D. % 5 | 193.6 | |

Trials were conducted at Sakha and Giza on different rates and methods of application of Kerb and Lancer. The results (Tables 5 and 6) indicated that the following treatments are very promising :

. Three sequential sprays with Lancer, each at 0.086 kg (a.i.)/ha at 3-week intervals, starting at the beginning of flowering.
. Kerb at the rate of 4.76 kg (a.i.)/ha, post-emergence at four weeks from sowing, sprayed in 2,500 l/ha of water.

Table 5. Effect of different herbicidal treatments on *Orobanche* and faba beans (Sakha, 1980, field trial).

| Herbicide | Rate per application kg (a.i.) /ha | Application | Spikes (kg/ha) | Seeds (kg/ha) | Straw (kg/ha) |
|---|---|---|---|---|---|
| Lancer* (3 applications) | 0.086 | Post-em. | 0.83 | 583 | 5.66 |
| Lancer* | 0.129 | Post-em. | 7.33 | 395 | 5.00 |
| Kerb** | 4.760 | Post-em. | 11.90 | 345 | 5.43 |
| Kerb | 4.760 | PPI | 0.60 | 571 | 5.33 |
| Kerb then Lancer | 3.570 0.086 | Post-em. | 1.67 | 357 | 3.88 |
| Kerb then Lancer | 3.570 0.086 | PPI Post-em. | 1.07 | 481 | 1.95 |
| Kerb then Lancer | 2.975 0.086 | PPI | | | |
| Kerb then Lancer | 2.975 0.129 | PPI Post-em. | 44.28 | 357 | 3.71 |
| Control | --- | --- | 1122.7 | 42 | 4.21 |
| L.S.D. 5% | | | 1.82 | 112 | 0.98 |

*The first application was made at the beginning of flowering of faba beans.
**The application took place four weeks after sowing, incorporation to 5 cms.
N.B. The yields were exceptionally low because of the unsuitable weather.

Table 6. Effect of different herbicidal treatments on *Orobanche* and faba beans (Giza, 1980, field trial).

| Herbicide | Rate kg(a.i.)/ha | Application | Spikes (kg/ha) | Crop plants (1000s/ha) | Seeds (kg/ha) |
|---|---|---|---|---|---|
| Lancer (3 applications) | 0.086 | Post-em. | 20.23 | 94.3 | 915 |
| Kerb | 4.760 | Post-em. | 209.44 | 78.1 | 680 |
| Kerb | 4.760 | PPI | 119.00 | 27.1 | 278 |
| Kerb | 0.129 | Post-em. | 195.76 | 76.8 | 753 |
| Kerb | 0.129 | PPI | 233.24 | 36.0 | 304 |
| Kerb | 2.975 | Post-em. | 199.92 | 78.6 | 681 |
| Kerb | 2.975 | PPI | 195.64 | 31.9 | 207 |
| Check | --- | --- | 696.15 | 79.1 | 408 |
| L.S.D. 5% | | | 153.96 | 16.1 | 285 |

The use of Kerb at 4.760 kg (a.i.)/ha as PPI might be beneficial, as long as it is deeply incorporated into the soil, to about 25 cm. Meanwhile the use of such herbicides at all tested rates as PPI with shallow incorporation (5 cm) has been responsible for the significant reduction in the stand of the faba beans, and hence the significant decrease in seed yield.

Results from large-scale demonstration fields in Sohag showed the evident benefit of the appropriate use of kerb as post-emergence four weeks after sowing. This treatment gave a reduction of parasitism of 81% and 58% in number and weight of the spikes respectively (Table 7). As a result of the effective control of parasitism, the faba bean seed yield was significantly increased by 61%.

Table 7. Effect of Kerb as a post-emergence herbicide on *Orobanche*. Faba bean demonstration fields Sohag, (1980).

| Site | Spikes | | | | | | Seeds (t/ha) | | |
|---|---|---|---|---|---|---|---|---|---|
| | 1000 s/ha | | | kg/ha | | | | | |
| | Treat* | Cont.** | LSD 5% | Treat. | Cont. | LSD 5% | Treat | Cont. | LSD 5 % |
| Sacolta | 4.1 | 21.4 | 1.0 | 75.4 | 184.5 | 20.9 | 3.76 | 2.18 | 0.21 |
| Tima | 4.0 | 20.1 | 1.3 | 70.0 | 142.5 | 46.7 | 4.25 | 3.00 | 0.47 |
| Shandaweel | 5.3 | 29.7 | 1.3 | 73.8 | 152.5 | 46.7 | 3.48 | 1.95 | 0.47 |
| Mean | 4.5 | 23.7 | 0.9 | 73.0 | 159.8 | 12.7 | 3.83 | 2.38 | 0.18 |

*Treatment 4.75 kg (ai)/ha, incorporated 5 cms. 4 weeks after sowing.
** Cont. - untreated.

REFERENCES

1. Basler, F., Haddad, A. 1980. Selection of *Orobanche* resistant cultivars of broadbeans and lentil. In Proceedings, Second International Symposium on Parasitic Weeds, North Carolina, 1979 (1979), 254-259.
2. Chabrolin, C.H. 1934. *Orobanche* of beans. Its life and methods of controlling. Abst. Egyptian Agric. Review. 7, 582-584. (1937) in Arabic.
3. Guidice, E. 1935. Investigations on *Orobanche* on broad beans. 1. On the seed germination. Benito Mussolini Sicilia. Publ. 7, 5-27 (Biol. Abst. 12855, 1936).
4  El-Sharkawy, M.A., Sogaier, K. 1976. Studies on chemical control of weeds in broad bean *(Vicia faba* L.) . Lybian Journal of Agricultural 5, 59-64.
5. Ibrahim, A.A., Zahran, M.K., Nassib, A.M., Farag, H.F., Farrag, H.M. 1979. The resistance of broad bean *(Vicia faba)* varieties to *Orobanche creanta* Forsk. and their response to chemic control. FABIS 1, 28.
6. International Center for Agricultural Research in the Dry Areas 1980. Weed control research in Northern Syria 1978-79. ICARDA Project Report No. 5, 43 pp.
7. Kemp, P.J. (Translator) 1968. Herbicide use in U.S.S.R. PANS (c) 14 (2), 119-131.
8. Klein, W. 1979. Problems and results of weed control in field beans. Mittel und Verfahren bei kulturpflanzen. Wissenschaftliche Beitrage der Martin-Luther-Universital Halle-Wittenberg 7 ( 5.16), 208-212.
9. Lawson, H.M., Wiseman, J.S. 1978. New herbicides for field bean. In Proceedings 20th British Crop Protection Conference, 769-776.

10. Mahmoud and Mohamed 1953. Studies on germination of seed of *Orobanche crenata*. Alex. Journ. Agric. Res. 1, 1-5.

11. Nassib, A.M., Ibrahim, A.A., Saber, H.A. 1978. Breeding for broad bean *(Vicia faba)* resistance to broomrape *(Orobanche crenata)* in Egypt. A paper presented at the Food Legume Improvement and Development Workshop held at ICARDA, Aleppo, Syria, May 2 - 7, 1978.

12. Neits, J.H. *et al.* 1968. Critical periods of the crop growth cycle for competition from weeds. PANS (c) 14 (2), 159-166.

13. Ponce De Leon, J.L., Royo, A., Cubero, J.I. 1974. El jopo de las habas *(Orobanche crenata* Forsk) Estructuray bidogia de las poblaciones del parasits. Anales Inst. Nac. Inv. Agrar. (Serie Prod. Veg.) 4, 212-237.

14. Sedov, A.I., Losev, S.I. 1970. Chemical control of weeds in legumes with prometryne and linuron. Nawchnye Trudy. Vsesoyuznyl Nauchno-issledovatel'skii, Institute Zernobobovykhi krupyanykk Kul'tur, 5, 238-297. c.f. Weed Abstr. 28 (9), 284.

15. Shabatay, Y. 1939. Agricultural Crops and Economic Plants in Egypt. Plant Breeding Sec. Giza Bull. 10.

16. Strasburger, E. 1908. Text book of Botany pp. 746. Macmillan and Co. Ltd. London.

17. Zaghloul, M.A. 1929. Halouk. Monthly Report of the Laboratory Research Committee Cotton Res. Bd. Ministry of Agriculture Egypt 1 (16) 178.

18. ------------------1945. Halouk. Monthly Report of the Laboratory Research Committee. Cotton Res. Bd. Ministry of Agriculture Egypt 78.

19. Zahran, M.K. *et al.* 1980. Chemical control of *Orobanche crenata* in *Vicia faba*. FABIS 2, 47.

# 20. BREEDING FOR RESISTANCE TO *OROBANCHE*

ABDULLAH M. NASSIB, ALI A. IBRAHIM, SHAABAN A. KHALIL

*Head, Food Legume Research Section; Director, Field Crops Institute and Researcher respectively, Agricultural Research Center, Giza, Egypt.*

Faba bean is one of the major hosts of *Orobanche crenata* Forsk. in Egypt. Infestation varies from a few scattered spikes to a large population almost attacking every bean plant in the field and thus limiting yield considerably. Infested bean fields occur in Egypt throughout the Nile Valley and Delta regions, Fayoom, and the oases in the western desert. Severe attacks of the parasite occur mostly in loamy and silt soils. Recent investigations on control measures include Glyphosate and Pronamide herbicides (30) and some agronomic practices (29).

*Sources of resistance to Orobanche in host crops*

A resistant line of faba beans, F 402, was produced in Egypt through a 3-year cycle of single plant selection starting in an $F_7$ line of a cross involving two indigenous parents (21). However since selection in the earlier segregating generations of this cross had been done throughout under open-pollination conditions, it is likely that F 402 is the outcome of a natural cross between that population and another parent probably of the *major* type (Fig 1).

Elia (12) reported on an Italian faba bean cultivar which was comparatively resistant to *Orobanche*, 'Favino palombino nero'. Kasasian (19) in UK stated that large seeded faba bean 'Express' was the most resistant among a collection of 53 cultivars. Boorsma (6), in Morocco, found that the black seeded F 331 was partially resistant. In Syria, Basler and Haddad (4), recorded 36 bean lines resistant to *Orobanche*. Cubero and Martinez (9), in Spain, used two *Paucijuga* resistant lines in their genetic study.

Sources of *Orobanche* resistance have also been found in Melon in USSR (17) (18), in hemp in France (2) Italy and USSR (28) and in eggplant, tomato, tobacco, rape and mustard in India (10, 11). Sunflower cultivars in USSR reacted variably to races of *O. cumana* having different virulence levels (5) (24). Resistant sources of *Orobanche* were also found in different species of *Nicotiana* (25) and *Helianthus* (24).

*Stability of Resistance and Effect of Natural and Artificial Selection*

The possibility of physiological races of *Orobanche crenata* and the genotype x environment interactions (8) (21) call for testing resistant material at different locations. In this

*G. Hawtin & C. Webb (Eds.): Faba Bean Improvement*
©1982 ICARDA. ISBN -13:978-94-009-7501-9

200

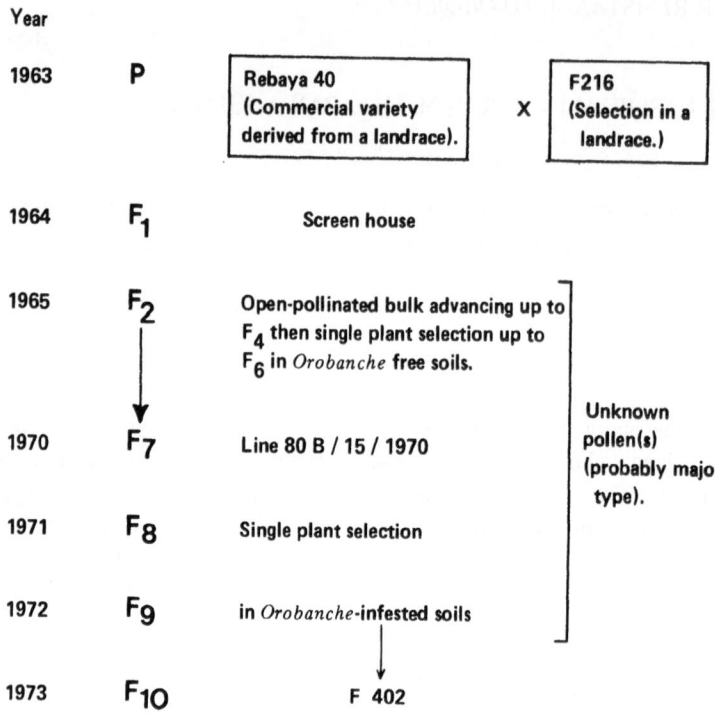

Year

1963    P

| Rebaya 40<br>(Commercial variety<br>derived from a landrace). | X | F216<br>(Selection in a<br>landrace.) |

1964    $F_1$        Screen house

1965    $F_2$       Open-pollinated bulk advancing up to
$F_4$ then single plant selection up to
$F_6$ in *Orobanche* free soils.

Unknown
pollen(s)
1970    $F_7$       Line 80 B / 15 / 1970      (probably major
type).

1971    $F_8$       Single plant selection

1972    $F_9$       in *Orobanche*-infested soils

1973    $F_{10}$          F 402

Fig. 1. Diagram of origin of *Orobanche*-resistant cv. F 402.

regard the following investigations were conducted :

Trial 1.

In a trial across seven locations, the percentage of *Orobanche*-infected bean plants and the number and total dry weight of *Orobanche* spikes/plot above the soil were lower in F 402 as compared with three other cultivars (Table 1). The gap narrowed between F 402 and Giza 4 where the soil had the highest rate of *Orobanche* infestation. However, the seed yield and Harvest Index (Table 2) of the resistant cultivar were 5.8 and 2.2 times those of the check Giza 4 respectively. This was probably due to much higher *Orobanche* population attacking the susceptible variety under the surface.

Giza 2 and Rebaya 40 were rather more tolerant to *Orobanche* than the most susceptible cultivar Giza 4. Both cultivars had been selected from Egyptian landraces, and probably inherited resistance from the same source as F 402. These landraces had probably been grown under selection pressure favouring *Orobanche*-resistance for a long time. On the other hand, Giza 4 which was developed from a cross between landraces and an introduction from Holland, had never been subjected to selection pressure for *Orobanche* resistance.

Trial 2.

In 1980, 22 pure lines, originating at ICARDA, Syria (Basler and Haddad, 1978) were screened in *Orobanche* infested soil at Giza. Most of the top lines originated in Egypt (Table 3). These lines can be considered to have stable resistance against both Syrian and Egyptian *Orobanche* cultures. Rebaya 40 stood midway between the resistant and susceptible lines whereas Giza 3, a sister line to Giza 4, was susceptible.

These trials indicate that the Egyptian landraces may have developed resistance mechanisms through natural selection. Artificial selection has been effective in identifying resistant lines from these land races. Cubero (8) stressed the effect of natural selection in accumulating genes for resistance in faba beans. The availability of *Orobanche*-resistant sunflower cultivars in USSR was the result of mass and recurrent selection (14) (20) (24).

*Mode of Inheritance and Nature of Resistance Mechanisms*

Recent studies on faba beans (9) (26) have indicated that susceptibility to broomrape is partially dominant.

In sunflower, resistance to individual races of *Orobanche* may be controlled by a single dominant gene (7) (14) (23) or may be more complex with two or more genes influencing resistance (14).

Evidence that F 402 might have a low production of a germination stimulant and/or some mechanical barriers, was reported by Nassib *et al.* (21). The resistance appears to be associated with slower tap root growth, less production of lateral roots and an altogether more compact root mass than in more susceptible cultivars.

In sunflower, resistant plants deposit layers of lignin in cells damaged by broomrape haustoria to arrest its development and later causing its death. Some haustoria die in the parenchyma while some reach as far as the vascular bundle (3).

*Screening Techniques for Resistance*

Smaller plots, more replicates, and planting a repeating check line within each replicate have been used at Giza to overcome the problem of fluctuating parasite population and

Table 1. Percentage of *Orobanche*-infected faba bean plants and number and total dry weight of *Orobanche* spikes/plot of F 402 and commercial cultivars in 1979 and 1980*.

| Location and Season cv. | *Orobanche*-infected bean plants (%) | | | | *Orobanche* spikes/plot Number | | | | Wt.** relative to Giza 4 or Giza 2 | | | |
|---|---|---|---|---|---|---|---|---|---|---|---|---|
| | F 402 | R 40 | G 2 | G 4 | F 402 | R 40 | G 2 | G 4 | F 402 | R 40 | G 2 | G 4 |
| Giza 1979 | 12.1 | 31.1 | — | 43.7 | 21.3 | 55.1 | — | 99.6 | 14.9 | 32.4 | — | 100 |
| Sharkiia 1980 | — | — | — | — | — | — | — | — | 25.7 | — | 39.5 | 100 |
| Ismailia 1980 | — | — | — | — | 10.0 | — | 32.6 | 99.8 | 18.4 | — | 39.0 | 100 |
| Samalloot 1980 | 22.4 | — | — | 33.4 | 122.8 | — | 198.2 | 198.2 | 74.2 | — | — | 100 |
| Minia 1980 | — | — | — | — | 37.7 | — | 149.5 | 172.5 | 19.9 | — | 74.9 | 100 |
| Abo Korkas "A" 1980 | 2.3 | — | 25.1 | — | 25.0 | — | 209.8 | — | 8.6 | — | 100 | — |
| Abo Korkas "B" 1980 | 78.5 | — | — | 98.9 | 768.3 | — | — | 1000.7 | 85.2 | — | — | 100 |

*Average of 3 to 8 replications.   **Air dry wt.

Table 2. Faba bean seed yield and harvest index of F 402 and commercial cultivars at different locations in 1979 and 1980.

| Location and Season | Seed Yield t/ha | | | | Seed Yield relative to Giza 4 or Giza 2 | | | | Harvest Index relative to Giza 4 or Giza 2 | | | |
|---|---|---|---|---|---|---|---|---|---|---|---|---|
| cv. | F 402 | R 40 | G 2 | G 4 | F 402 | R 40 | G 2 | G 4 | F 402 | R 40 | G 2 | G 4 |
| Giza 1979 | 4.597 | 3.148 | -- | 2.231 | 206.1 | 141.1 | -- | 100 | 158 | 131 | -- | 100 |
| Sharkiia 1980 | -- | -- | -- | -- | 96.6 | -- | 136.8 | 100 | 86 | -- | 103 | 100 |
| Ismailia 1980 | 2.971 | -- | 2.531 | 2.745 | 108.2 | -- | 92.2 | 100 | 99 | -- | 98 | 100 |
| Samalloot 1980 | 2.989 | -- | -- | 1.937 | 154.3 | -- | -- | 100 | 120 | -- | -- | 100 |
| Minia 1980 | 4.232 | -- | 3.530 | 2.893 | 146.3 | -- | 122.0 | 100 | 95 | -- | 76 | 100 |
| Abo Korkas "A" 1980 | 4.236 | -- | 2.467 | -- | 171.7 | -- | 100 | -- | 140 | -- | 100 | -- |
| Abo Korkas "B" 1980 | 1.420 | -- | -- | 0.247 | 574.9 | -- | -- | 100 | 216 | -- | -- | 100 |

other soil factors. To enable better estimates to be made of host resistance a technique has been developed at Giza involving the inoculation of the root mass of a 25-day old plant in a soil core of 5 cm in diameter with sufficient *Orobanche* seed and then growing the plant in a 25 cm pot (Nassib *et al.* 1978). *Orobanche* seeds placed in muslin bags close to the roots of faba beans grown in pots were used by (Abou-Raya *et al.* (1)). This technique could be valuable in testing genotypic differences in stimulant production.

Table 3. Screening of lines found resistant in Syria (ICARDA) to *Orobanche crenata* in Giza, Egypt, 1980.

| Rank | ICARDA BPL no. | Country of origin | *Mean *Orob.* Spikes/pl. relative to F 402 | Mean Visual rate |
|------|----------------|-------------------|---------------------------------------------|------------------|
| 1 | 582 | Egypt | - 0.21 | 1.00 |
| 2 | F 402 (1278) | " | - 0.16 | 1.25 |
| 3 | 472 | Lebanon | - 0.16 | 1.75 |
| 4 | 561 | Egypt | - 0.01 | 1.25 |
| 5 | 1021 | " | +0.18 | 2.50 |
| 6 | 562 | " | +0.33 | 1.75 |
| 7 | (Rebaya 40) | " | +0.47 | 1.75 |
| 8 | 4 | Ethiopia | +0.65 | 3.00 |
| 9 | 579 | Egypt | +0.81 | 1.75 |
| 10 | 853 | Canada | +0.91 | 2.25 |
| 11 | 730 | Sudan | +1.31 | 2.25 |
| 12 | 587 | Egypt | +1.44 | 2.00 |
| 13 | 50 | Iraq | +1.49 | 2.25 |
| 14 | 490 | Spain | +1.60 | 2.00 |
| 15 | 733 | Afghanistan | +1.83 | 3.00 |
| 16 | 375 | Turkey | +2.02 | 3.50 |
| 17 | 811 | Ethiopia | +2.08 | 2.75 |
| 18 | 911 | Canada | +2.09 | 3.25 |
| 19 | 54 | Iraq | +2.14 | 2.00 |
| 20 | (Giza 3) | Egypt | +2.41 | 3.25 |
| 21 | 485 | Spain | +2.58 | 3.50 |
| 22 | 530 | Japan | +2.82 | 2.00 |
| 23 | 734 | Afghanistan | +3.28 | 3.66 |
| 24 | 553 | Egypt | +3.54 | 3.00 |
| 25 | 108 | UK | +5.65 | 3.75 |
| LSD 0.05 | | | 2.93 | |

*Each entry was tested against F 402 in two halves of a row (5 - 7 plants each half) x 4 reps.

## Looking Ahead in Breeding for Orobanche Resistance

Now it is evident that Egyptian landraces may constitute an important source of resistance genes, efforts will be made to collect and screen more landraces from remote and isolated areas. To strengthen resistance *per se* , a recurrent selection program has been started in F 402. Inter-crossing different resistant local and introduced lines will be done. Transferring resistance to other commercial cultivars and promising breeding material is under way. Studies on the nature of resistance in available sources are required to help the breeding program.

## The Integrated Approach to Orobanche Control

Several field trials are being conducted in 1980-81 in cooperation with the Weed Research Section, ARC, to test the combination of cultivars, herbicides (Glyphosate and Pronamide) and some agronomic practices e.g. planting in untilled soils, or delaying sowing dates.

## REFERENCES

1. Abou Raya, M.A., Radi, A.F., Heikal, M.M. 1973. Host-parasite relationship of *Orobanche* species. Proc. Eur. Weed Res. Coun. Parasitic weeds, 167-176.
2. Antonets, N.P. 1970. Phytomyza against broomrape. Zashch. Rast., 15 (7) : 13-14.
3. Antonova, T.S. 1978. The development of the haustoria of *Orobanche cumana* Wallr. in the roots of immune and susceptible forms of sunflower. Botanicheskii Zhurnal 63 (7) 1025-1030 (Ru) Weed Abst. 1979, 27 (1) : 406.
4. Basler, F., Haddad, A. 1978. Selection of *Orobanche* resistant cultivars of broad bean and lentil, Striga Workshop, Khartoum (unpublished).
5. Beilin, I.G. 1968. Flowering hemi-parasites and parasites. Publishing house "Nauka", Moscow.
6. Boorsma, P.A. (n.d.) Preliminary work on *Vicia faba* breeding. IPHR, Morocco (unpublished).
7. Burlov, V.V., Kostiuk, S.V. 1976. Inheritance of the resistance to the local race of broomrape *(Orobanche cumana* Wallr.) in sunflower (Ru). Genetica 12 (2) : 44-51.
8. Cubero, J.I. 1973. Resistance to *Orobanche crenato* Forsk. in *Vicia faba* L. Proc. Eur. Weed Res. Coun. Parasitic Weeds, 205-217.
9. Cubero, J.I., Martinez, A. 1980. The genetics of the resistance to faba beans to *Orobanche crenata* . FABIS 2 : 19-20.
10. Dalela, G.G., Mathur, R.L. 1971. Resistance of varieties of eggplant, tomato and tobacco to broomrape *(Orobanche cernua* Loefl.) PANS, 17 (4) 482-483.
11. ----------------------------- 1971a. Resistance of varieties of rape and mustard to broomrape *(Orobanche aegyptiaca).* Indian Phytopathology, 24 (2) 417-418.
12. Elia, M. 1964. Indagini preliminari sulla resistenza varietale della Fava *(Vicia faba)* all' Orobanche, Phytopathol. mediter : 3 : 31-32. In Kasasian, L. Orobanche spp. PANS 17 (1) 35-41.
13. Food Legume Research 1979. Food Legume Improvement Project in Egypt. Progress Report to IDRC, 1978-79, ARC, Giza. (unpublished).
14. Gundaev, A.L. 1971. Basic principles of sunflower selection, 417-465. In Genetic principles of plant selection. Nauka, Moscow.
15. ICRISAT 1978. Sorghum Breeding. Report of work 1977-78, pp. 157, ICRISAT, Hyderabad, India.
16. ICRISAT 1978a. "Sorghum". In Annual Report 1977-78, p. 27-61. ICRISAT, Hyderabad, India.
17. Kabulov, D. 1966. Broomrape-dangerous weed. Kartofel Ovoshichi, 14 (12) 38-40.
18. Kabulov, D., Mukumov, Kh. 1967. The broomrape resistance of melon crops in Uzbecistan. Kartofel' Ovoshichi, 12 (8) : 39-40.

19. Kasasian, L. 1973. The chemical control of *Orobanche crenata* in *V. faba* and the susceptibility of 53 cultivars of *V. faba* to *O. crenata*. Proc. Eur. Weed Res. Coun. Parasitic Weeds 224-230.

20. Morozov, V.K. 1947. Sunflower breeding in the USSR, Moscow. Piscepromlzdat. p. 272 (Extended summary prepared by IBPG, Cambridge, England).

21. Nassib, A.M. Ibrahim, A.A., Saber, H.A. 1978. Broomrape *(Orobanche crenata)* resistance in broad beans. Proc. "Seed Legumes" 133-135. ICARDA Workshop, Aleppo, Syria.

22. Panchenko, A. Ya. 1975. Early diagnosis of broomrape resistance in the selection and breeding of sunflower. Vestnik, Selskokhozyaistvennoi Nauki, Moscow, USSR., 20 (2) 107-115 (Ru) Weed Abst. 1977, 26 (1) 77.

23. Pogorletskii, B.K., Geshele, E'.E'. 1975. On the immunity of sunflower to broomrape. Genetika 11 (7) 18-26 (Ru). Weed Abstr. 1977, 26 (4) 1061.

24. Pustovoit, G.V. 1975. Sunflower breeding for groups immunity by the method of interspecific hybridization. In "Sunflower", V.S. Pustovoit (ed.) Kolos, Moscow, pp. 590 (Ru).

25. Racoviia, A.E. 1973. Experiments on the susceptiblity of various species of *Nicotiana* and of different tobacco cultivars to *Orobanche ramosa*. Proc. Eur. Weed Res. Coun. Parasitic Weeds 194-204.

26. Suso, M.J. 1980. Studies on quantitative inheritance in *Vicia faba major*. FABIS 2, 30.

27. Tackholm, Viva, Drar, M. Fadeel, A.A. 1956. Student's Flora of Egypt. Anglo-Egyptian Bookshop, Cairo, pp 649.

28. Walker, H.C. (n.d.). Review of literature on *Orobanche* with suggestions for its control. Publ. FAO 61/B/1288. pp. 6.

29. Zahran, M.K. 1973. Control of broomrape in field and vegetable crops in Egypt. Final Report on PL 480 Project pp. 40 ARC, Giza, (unpublished).

30. Zahran, M.K., Ibrahim, T.S., Farag, F.H., Hassannien, E.E. 1980. *Orobanche* and Weed Control in faba beans. Progress Report 1979-80. ICARDA/IFAD Nile Valley Project. pp. 26 (unpublished).

## 21. CHARACTERISTICS OF A LOCAL FABA BEAN COLLECTION AND ITS REACTION TO *OROBANCHE*.

MAZHAR M.F. ABDALLA

*Faculty of Agriculture, Cairo University, Giza, Egypt*

Improvement of the faba bean crop is slow because of several factors including lack of useful variability and susceptibility to pests and environmental conditions. Hence it is important to seek for useful variability, particularly among locally adapted ecotypes.

Broomrape *(Orobanche crenata* Forsk.) is an annual plant which is an obligate parasite on different legumes and causes heavy losses to faba beans. It is confined mostly to the Mediterranean area and its seeds may live in the soil for many years. It is reported that seed germination is caused by a stimulant excreted from the host roots, but nodule bacteria, rhizospheric microflora, extracts of roots of some other species, e.g. flax and certain chemicals may also stimulate germination (14) (13) (5) (10) (4) (2).

In 1979, in cooperation with GTZ and Prof. G.Fischbeck of Weihenstephan, Germany, a program was established to collect and evaluate landraces of faba beans in Egypt. During that summer, a collecting team travelled through Egypt and collected more than 200 samples from farmers in 16 provinces who did not use improved cultivars. The origin of these samples is shown in Fig. 1.

During the 1979-80 season, the collection was sown at Giza in mid-November. One of the four replicates which were sown for agronomic evaluation was covered with an insect-proof cage to ensure selfing; another two or three were sown in the naturally infested *Orobanche* sick plot at Giza.

Data on various plant characteristics were obtained from 10 randomly chosen plants from each replicate.

Variability was observed for a range of seed characteristics (seed weight, length, width and thickness), and in general seeds from southern locations were relatively smaller. Other agronomic characters also showed considerable variability, however there was no clear trend in variability from north to south as had been observed earlier (1).

At that time it was observed that samples from southern locations were characterised by fewer tillers, earlier flowering and maturity, lower flowering nodes, less malformed plants, more pods and seeds but less yield/plant, and the seeds were smaller. It is possible that the size of the present collection is more representative, but it is also certain that more heterogeneity is present particularly in samples from southern Egypt. Possibly after the High Dam had been built and irrigation became available throughout the year, local seed stocks have become mixed with other materials which may be more productive under the new conditions.

Comparisons were made for yield between plants raised in the cages and in the open. Variability was observed in the relative ability to yield in cages, which some authors con-

*G. Hawtin & C. Webb (Eds.): Faba Bean Improvement*
©1982 ICARDA. ISBN -13:978-94-009-7501-9

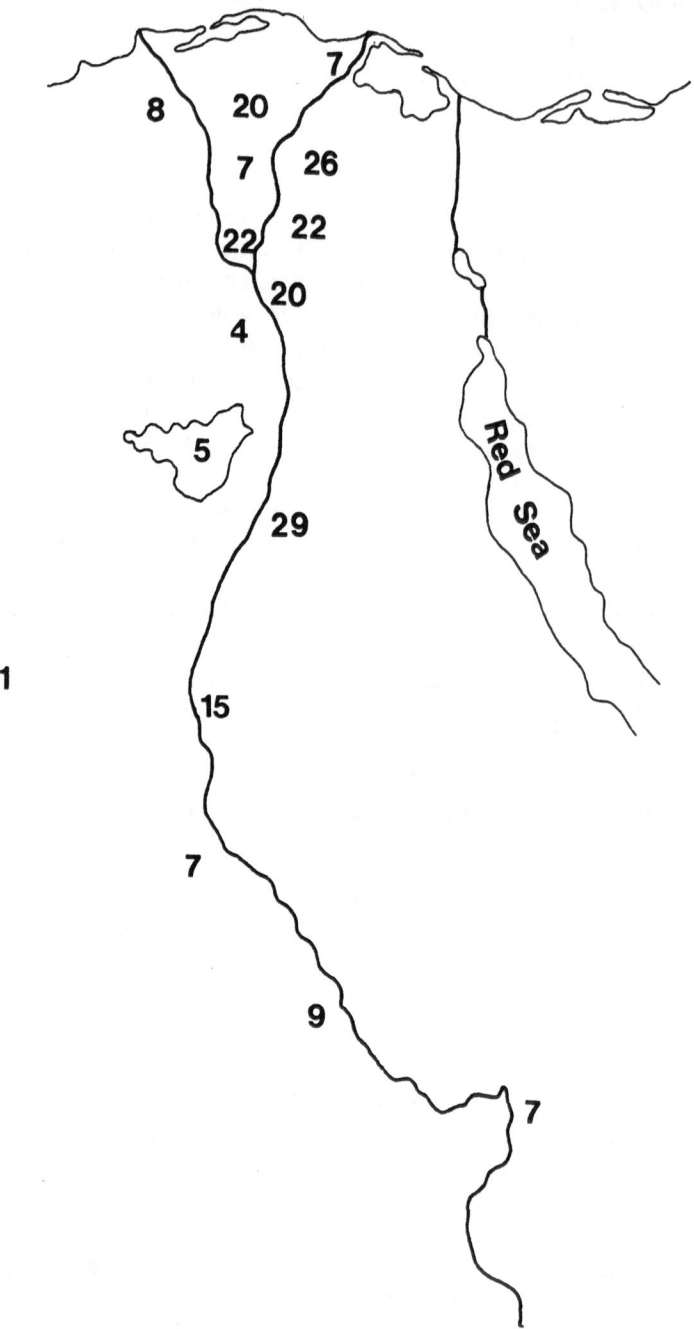

Fig. 1 Places of collecting in Egypt (number of samples).

sider to be a measure of autofertility. However, as discussed by Abdalla and Fischbeck (in press),caging effects may be due to differences in environmental factors as well as differences in pollen vectors. Giza samples were relatively the best yielders under cages and in general samples from southern Egypt showed less caging effects than those from the far north. Caging effects appeared to be independant of yield potential under open pollination.

Counts were made of *Orobanche* heads which emerged from March 1 until harvest. Samples were taken for determining the fresh and dry weight.

Table 1 summarizes the data from more than 17,000 plants grown in the *Orobanche*-infested plot. Considering the mean number of *Orobanche* plants which had emerged by March 22, it appears that samples originating from e.g. El-Faiyum had three times as many as those from certain other regions e.g. New Valley. Differences between samples originating in the same province were also observed.

The percentages of surviving plants tolerating, or escaping, *Orobanche* were generally low (mean less than 40%); however the relative yield per plant on surviving individuals was generally low. Some of the surviving plants produced no yield at all whereas, in other samples, yields were almost the same as those obtained in *Orobanche*-free plots. Samples from El-Faiyum were greatly affected by *Orobanche*.

Table 1. Number of *Orobanche* heads and percentage of surviving plants and their relative seed yield.

| Sample origin | No. of *Orobanche* spikes per ridge (22 March) | | Percent of surviving plants | | Relative yield per plant in the *Orobanche* field compared to the free one (%) | |
|---|---|---|---|---|---|---|
| | Mean | Range | Mean | Range | Mean | Range |
| New Valley | 65 | – | 18 | -- | 85 | - |
| Kena | 116 | 70 - 176 | 30 | 21 - 47 | 44 | 18 - 66 |
| Sohag | 121 | 23 - 170 | 39 | 19 - 76 | 59 | 11 - 99 |
| Assiut | 141 | 48 - 129 | 27 | 15 - 42 | 32 | 21 - 48 |
| El-Faiyum | 190 | 37 - 255 | 21 | 9 - 36 | 13 | 2 - 28 |
| El-Minya | 108 | 19 - 214 | 29 | 9 - 55 | 52 | 12 - 92 |
| Beni Suef | 137 | 10 - 316 | 26 | 8 - 45 | 44 | 0 - 102 |
| Giza | 124 | 28 - 188 | 38 | 35 - 41 | 48 | 18 - 84 |
| Kualyubia | 163 | 19 - 399 | 28 | 13 - 46 | 36 | 0 - 98 |
| Minufiya | 135 | 15 - 240 | 25 | 9 - 37 | 34 | 0 - 89 |
| Sharkia | 113 | 35 - 195 | 32 | 19 - 51 | 58 | 17 - 96 |
| Gharbia | 114 | 50 - 148 | 32 | 21 - 41 | 48 | 2 - 97 |
| Kafr El-Sheikh | 119 | 26 - 223 | 25 | 9 - 41 | 27 | 0 - 82 |
| Dakahlia | 91 | 22 - 277 | 33 | 15 - 56 | 38 | 0 - 96 |
| Domiat | 112 | 45 - 225 | 26 | 18 - 38 | 33 | 0 - 65 |
| Behera | 115 | 36 - 214 | 25 | 6 - 37 | 31 | 0 - 58 |
| Improved Cultivars | | | | | | |
| Giza 1 | 172 | -- | 14 | -- | 0 | -- |
| Giza 3 | 216 | -- | 19 | -- | 0 | -- |
| Giza 4 | 249 | -- | 34 | -- | 2 | -- |

In another trial, 46 samples from 15 provinces were sown in three replicates. Some plants were sprayed with Lancer (glyphosate) to inhibit *Orobanche*, while others were sprayed only with water as a control treatment. Table 2 shows that *Orobanche* resulted in reduced plant height, pod set and yield, but more tillers per plant. In the *Orobanche* plot it was observed that when few spikes were present they were generally taller, thicker and heavier than when many spikes were present.

Table 2. Relative effect of *Orobanche* on various characters (free plants = 100), and the number of *Orobanche* spikes per ridge at harvest.

| Samples from | No. of samples | Stem height | | Tillers/ plant | | Pods/ plant | | Yields/ ridge | | *Orobanche* spikes/ridge at harvest | |
|---|---|---|---|---|---|---|---|---|---|---|---|
| | | Mean | C.V. | Mean | C.V. | Mean | C.V. | Mean | C.V. | Mean | C.V. |
| New Valley | 1 | 100 | – | 116 | – | 76 | – | 39 | – | 159 | – |
| Kena | 5 | 99 | 12.5 | 91 | 12.1 | 60 | 22.4 | 47 | 34.4 | 173 | 22.4 |
| Sohag | 5 | 89 | 9.7 | 103 | 17.9 | 65 | 49.7 | 45 | 35.7 | 188 | 23.5 |
| Assiut | 4 | 80 | 43.5 | 109 | 15.1 | 49 | 21.3 | 32 | 32.2 | 216 | 12.1 |
| El-Faiyum | 1 | 89 | – | 138 | – | 41 | – | 17 | – | 138 | – |
| El-Minya | 3 | 88 | 6.7 | 108 | 35.0 | 42 | 45.6 | 22 | 41.4 | 178 | 40.5 |
| Beni Suef | 4 | 85 | 12.1 | 112 | 16.5 | 34 | 97.9 | 18 | 55.6 | 240 | 16.0 |
| Giza | 1 | 89 | – | 138 | – | 41 | – | 17 | – | 154 | – |
| Kualyubia | 3 | 89 | 10.6 | 98 | 27.5 | 43 | 3.6 | 42 | 38.4 | 187 | 16.7 |
| Minufiya | 4 | 87 | 10.5 | 95 | 22.0 | 36 | 37.8 | 45 | 69.9 | 226 | 27.9 |
| Sharkia | 4 | 82 | 15.1 | 108 | 29.5 | 55 | 79.6 | 36 | 81.7 | 213 | 20.5 |
| Gharbia | 3 | 98 | 7.3 | 103 | 27.5 | 88 | 16.8 | 54 | 28.3 | 173 | 34.2 |
| Kafr El-Sheikh | 4 | 94 | 4.1 | 104 | 18.1 | 65 | 21.1 | 50 | 10.8 | 195 | 9.4 |
| Dakahlia | 3 | 90 | 5.8 | 102 | 27.2 | 48 | 21.8 | 33 | 22.0 | 163 | 11.4 |
| Domiat | 1 | 93 | – | 138 | – | 79 | – | 35 | – | 219 | – |

The variability observed in reaction to *Orobanche* was encouraging. Cubero (6), Kasasian (11), Nassib *et al.* (12) and Cubero and Martinez (7) have all reported similar variability.

Nassib *et al.* (12) who selected *Orobanche* -free and less infected plants, showed that the percentage of "tolerant" lines in the progenies was 44.5% compared to 21.0% in other nonselected entries. Certain selections from the cross Rabaya 40 x F 216 (landrace) were reported to be tolerant, particularly Family 402 which was considered highly tolerant.

Ibrahim *et al.* (9) reported that in a naturally infested field, when Giza 4 was given a score of 100 for *Orobanche* attack, the cultivars Rebaya 40 and F. 402 were given relative scores of 55 and 21 respectively. Zahran *et al.* (15) reported that F. 402 "in some cases was found with potential capability of tolerance to *Orobanche* parasitism as compared with Giza 2 and Giza 4".

Ibrahim *et al.* (8) referred to results of 25 lines tested in the "international nursery". Three lines, two from Egypt and one from Lebanon, showed good visual rating for resistance and a low number of *Orobanche* spikes per faba bean plant. These lines, however, did not differ significantly from F. 402 in *Orobanche* attack and their yield was slightly lower.

In view of the variability in reaction, the complex genetic systems of resistance (6) (7), environmental influences, known systems of resistance to other pests (3) and the breeding system, it is not logical to search for immunity to *Orobanche* in faba beans. We have tested different entries and segregating populations from crosses between *paucijuga* and *eu-faba* types of different geographic origin, in the *Orobanche* field at Cairo University, Giza. "Tolerant" plants were found in different stocks. I believe that we have to confine our efforts to selecting for tolerance and to raising the level of tolerance through appropriate handling of the materials. Such an approach may be analogous to the breeding for "uniform resistance".

It appears that tolerance (or resistance) to *Orobanche* is a quantitative trait controlled by a complex genetic system. There is evidence that resistance is controlled by recessive genes (6) (7). It is known that resistance is an interaction between the host, parasite and the environment, e.g. it has been shown that several factors can affect the germination of *Orobanche* seeds and the susceptibility of faba beans (10) (2).

In my opinion, it is safer in a breeding program to avoid selecting *Orobanche*-free plants, which may be escapes, and to restrict selection to slightly infected ones. At Cairo University, selection is practised between and within samples for plants that are infected, but at the same time produce a good yield and have other desired characteristics. The relative yield of infected and free plants is considered in selection. Some of the selections made so far have produced almost double the yield of F. 402 when grown in the *Orobanche* plot.

Some progress has been made; however further research is need particularly to answer the following questions : Are there different 'races' or 'biotypes'? What is their distribution? How do they develop? What determines their virulence? What is the nature and genetic basis of resistance? What are the best conditions in which to screen for resistance? and how do various environmental conditions and cultural practices affect *Orobanche* germination, and host-parasite interaction?

## REFERENCES

1. Abdalla, M.M.F. 1964. Variation of some agronomic characters in different collections of *Vicia faba* L. M.Sc. Thesis., Fac. Agric. Cairo University, Egypt.
2. Abdalla, M.M.F. 1979. A bibliography of field beans *(Vicia faba)* research in Egypt. ICARDA.
3. Abdalla, M.M.F., Hermsen, J.G.Th. 1971. The concept of breeding for uniform and differential resistance and their integration. Euphytica 20 : 351-361.
4. Abd-El-Hafeez, S. 1975. Physiological studies on the germination of *Orobanche* seeds with relation to root exudates of the host. M.Sc. Thesis, Fac. Agric. Ain Shams Univ., Egypt.
5. Al-Menoufi, O.A. 1971. Studies on the parasitism of species of *Orobanche*. Ph.D. Thesis, Fac. Agric. Alexandria University, Egypt.
6. Cubero, J.I. 1973. Resistance to *Orobanche crenata* Forsk. in *Vicia faba* L. Proc. Eur. Weed-Res. Coun. Symp. Parasitic Weeds : 205-217.
7. Cubero, J.I., Martinez, A. 1980. The genetics of the resistance of faba beans to *Orobanche crenata* . FABIS 2 : 19-20.
8. Ibrahim, A.A., Nassib, A.M., Khalil, S.A. and El-Sherbeeni, M.H. 1980. Faba bean breeding and agronomy. ICARDA/IFAD Nile Valley Project on Faba Beans. Ann. Coord. Meeting, Cairo, Egypt, August 25 to 27.
9. Ibrahim, A.A., Zahran, A.A., Nassib, A.M., Farag, F.H., Farag, H.M. 1979. The resistance of broad bean *(Vicia faba)* varieties to *Orobanche crenata* Forsk. and their response to chemical control. FABIS, 1: 28.

10. Kasasian, L. 1973 a. Miscellaneous observations on the biology of *Orobanche crenata* and *O. aegyptiaca*. Proc. Eur. Weed Res. Coun. Symp. Parasitic Weeds : 66-75.
11. Kasasian, L. 1973 b. The chemical control of *Orobanche crenata* in *Vicia faba* and the susceptibility of 53 cultivars of *V. faba* to *O. crenata*. Proc. Eur. Weed. Res. Coun. Symp. Parasitic Weeds : 224-230.
12. Nassib, A.M., Ibrahim, A.A., Saber, H.A. 1978. Breeding for broad bean resistance to broomrape *(Orobanche crenata)* in Egypt. Food Legume Improvement and Development Workshop, ICARDA, May 2 to 7, Aleppo, Syria.
13. Salama, E.A. 1959. Studies on the parasitism of *Orobanche* Sp. on bean *(Vicia faba*, Linn.) M.Sc. Thesis, Fac. Agric. Ain Shams University, Egypt.
14. Tewfic, H.A. 1956. Developmental stages of seedlings of *Orobanche crenata* (Forsk.) in relation to its host *Vicia faba* (Linn.). M.Sc. Thesis, Fac. Agric. Ain Shams Univ., Egypt.
15. Zahran, M.K., Ibrahim, T.S.E., Farag, F.H., Hassannien, E.E., 1980. *Orobanche* and weed control in faba beans. ICARDA/IFAD Nile Valley Project on faba beans, Ann. Coord. Meeting, Cairo, Egypt, August 25 to 27.

## 22. MAJOR DISEASE PROBLEMS OF FABA BEANS IN EGYPT

HOSNI ABDEL RAHMAN MOHAMED


*Plant Pathology Institute, Agricultural Research Center, Giza, Egypt*

Faba bean is one of the most important crops in Egypt. However, several diseases cause serious damage to the plants and consequent decrease in the seed yield and quality.

Faba beans are attacked by both foliar diseases, and seedling and root diseases. Foliar diseases are the limiting factor for faba bean production in the northern parts of Egypt. During 1979-80, the loss was estimated to be more than 55% for the susceptible Rebaya 40 which was left for natural infection at Sakha (26).

*Leaf spots*

Leaf spots are the most predominant and widespread diseases of faba beans in Egypt. El Helaly (13) in 1936 reported a brown spot of faba beans caused by *Botrytis* sp. which he isolated from leaves and stems. He considered the disease to be destructive in the delta where it rains heavily. He proved further (14) the causal fungus to be *Botrytis fabae,* and showed that the optimum conditions for growth on artificial media were $21^{o}C$ and pH 4 for production of sclerotia, $20^{o}C$. Lethal temperature for spores and sclerotia was $51^{o}C$.

Leaf spot is caused in Egypt by *Botrytis fabae* Sard (14), *B. cinerea* (Pers) Er. (27), *Stemphylium botryosum* Wall (3), and *Alternaria tenuis* Nees (*Alternaria alternata* Kiesl). The disease is favoured by lower temperatures and high relative humidity.

It is very destructive on Egyptian cultivars but the reaction of imported cultivars has varied (6). Recently, several attempts were made to test locally-bred lines and cultivars and imported entries for their reaction to these diseases. During the last two years, several methods have been used in testing these entries.

*The detached leaf test :* During 1978-79, leaflets were washed, placed on moistened clean sand in plastic trays then sprayed with the spore suspension and covered with polyethylene cloth (18). This year, detached leaflets were washed with sterilized distilled water then the lower surface was placed on moistened filter paper. Only one drop of the spore suspension was placed in the center of the leaflet (26). Notes were recorded after 24, 48 and 72 hours on the percentage of the leaflet covered by the leaf spot(s).

Table 1 shows that entries subjected to the detached leaf test, differed in their reaction to chocolate leaf spot *(Botrytis fabae).* NEB 519 was the least infected, while Rebaya 40 was the most susceptible.

Table 2 shows the reaction of several entries and $F_1$ resulting from crossing them.

*G. Hawtin & C. Webb (Eds.): Faba Bean Improvement*
*©1982 ICARDA. ISBN -13:978-94-009-7501-9*

These preliminary data indicate that resistance to chocolate spot, under these conditions, tended to be dominant.

Table 3 shows that entries differed in their reaction to *Botrytis* leaf spot. After 72 hours, the infection ranged from 14.3% for NEB 938 to 62.3% for family 402. Some of the other entries were as resistant to the disease as Giza 2.

Table 1. Reaction of detached leaflets of faba bean entries to chocolate leaf spot *(Botrytis fabae)*.

| Entry | % infection | Remarks |
|-------|-------------|---------|
| NEB 519 | 14 | lesions small |
| Giza 1 | 20 | lesions small |
| Giza 2 | 28.3 | lesions small and numerous but did not coalesce |
| Giza 3 | 25 | lesions large, some coalesced |
| Rebaya 40 | 92 | leaflets black due to heavy infection |

Table 2. Reaction of detached leaflets of some entries and their crosses to chocolate leaf spot.

| Entry or cross | Reaction of leaflets in replicate | | | | |
|----------------|------|------|------|------|------|
| | I | II | III | IV | V |
| 139A/214/77 | Ms-S | Ms | R | Ms-S | Mr-Ms |
| NEB 938 x 139A/2141/77 ($F_1$) | VR | VR | R | R | Mr |
| NEB 938 | R | VR | VR | VR | R |
| NEB 938 x G3 ($F_1$) | R | R | R | Mr | Mr |
| Giza 3 | Ms | R | Ms-S | Mr | Mr |

VR = highly resistant, no lesions.
R  = resistant, very few lesions.
Mr = moderate resistant, few lesions.
Ms = moderate susceptible, lesions covering the leaflets, and larger in size (2-3 m).
S  = susceptible, lesions covering the leaflet, lesions larger, start to coalesce.
VS = very susceptible, lesions covering the leaflet, lesions larger and coalesce, part of the leaflet black in colour.

Table 3. Average percentage of infection of detached leaflets of some faba bean entries, 24, 48 and 72 h after inoculation with *Botrytis fabae*.

| Entry | Av. % infection after (hr) | | |
|---|---|---|---|
| | 24 | 48 | 72 |
| NEB 938 | 3.3 | 7.7 | 14.3 |
| Giza 1 | 6.6 | 20.2 | 28.4 |
| Giza 2 | 5.5 | 13.2 | 19.8 |
| Giza 3 | 7.7 | 17.6 | 20.9 |
| Giza 4 | 17.6 | 28.7 | 47.6 |
| Rebaya 40 | 15.1 | 31.1 | 58.9 |
| Fam. 402 | 16.5 | 36.4 | 62.3 |

Comparison after 24, 48 and 72 h (Table 3), showed that percentages of infection increased with time, however 24 hours was long enough to determine the reaction of the entry. Most of the entries which showed less than 10% infection after 24 h, would be either resistant or moderately resistant. This indicates that this short period (24 h) is enough to show the behaviour of the entry to leaf spots, whether resistant or susceptible.

Thirteen faba bean entries were tested for their reaction to isolates of *B. fabae* collected at different localities from the northern parts of the Delta, where leaf spots are a problem. Entries differed in their reaction to the five isolates (Table 4). Some entries were more resistant to one isolate than the other. Many of the entries reacted differently to the five isolates. However, Rebaya 40 was the most susceptible to the five isolates, while introduction 302 was the least susceptible.

Table 4. Average percentage infection of 13 faba bean entries detached leaflets by five isolates of *Botrytis fabae* after 24 h.

| Entry | Av. % infection by isolate | | | | |
|---|---|---|---|---|---|
| | Sakha | Nubaria | Ismailia | Gemeiza | Alexandria |
| Giza 1 | 8.8 | 24.2 | 2.2 | 15.4 | 11.0 |
| Giza 2 | 11.0 | 41.8 | 6.6 | 11.0 | 30.8 |
| Giza 3 | 6.6 | 50.6 | 13.2 | 15.4 | 33.0 |
| Giza 4 | 4.4 | 38.9 | 6.6 | 19.8 | 15.4 |
| Rebaya 40 | 13.2 | 57.2 | 26.4 | 28.6 | 44.0 |
| Fam. 402 | 8.8 | 24.2 | 15.4 | 24.2 | 15.4 |
| Equado 12 | 8.8 | 52.8 | 11.0 | 24.2 | 19.8 |
| Intr. 289 | 8.8 | 35.2 | 2.2 | 19.8 | 22.0 |
| 300 | 2.2 | 44.0 | 11.0 | 8.8 | 26.4 |
| 301 | 6.6 | 23.0 | 0 | 22.0 | 28.6 |
| 302 | 2.2 | 13.2 | 4.4 | 8.8 | 22.0 |
| 295 | 6.6 | 33.0 | 4.4 | 22.0 | 41.8 |
| NEB 938 | 0 | 35.2 | 2.2 | 33.0 | 15.4 |

Also isolates differed in their virulence on the tested faba bean entries. The most virulent was the Nubaria isolate, followed by the Alexandria isolate. The least virulent was the Sakha isolate (Table 4). This may be because the Sakha isolate had been maintained for a period on artificial media.

It may be concluded that *B. fabae* isolates differed in their virulence on faba bean entries. This indicates the possible existence of races within *B. fabae*. Also, the detached leaf technique proved to be an efficient method of testing the reaction of faba bean entries to the disease as well as virulence of isolates of the fungus.

*Seedling test :* In a test conducted on artificially inoculated ($12.5 \times 10^4$ spores/ml) seedlings, results showed that faba bean entries differed in their reaction to *Botrytis*. They were divided into five groups — highly resistant, resistant, intermediate, susceptible, and very susceptible, according to their disease reaction :

i.   Highly resistant : infection on leaves and flowers not more than 1%, e.g. NEB 938.
ii.  Resistant : infection on leaves, flowers and stems up to 10%, e.g. Giza 1, Giza 3, NEB 368, NA 67, NA 174, NA 176, Selection 23, Selection 27.
iii. Intermediate reaction : infection on leaves, flowers and stems 10 to 20%, e.g. Giza 2, S. 402, 8 B/41/70, 8B/2528/70, 136 B/2125/77, NEB 315, NEB 18, NEB 431, NEB 676, NEB 864, NEB 680, NEB 811, NEB 655, NEB 679, NA 146, NA 158, NA 161, Sel. 1, Sel. 2, Sel. 15.
iv.  Susceptible : average infection on leaves, flowers and stems ranged from 20 to 60%, e.g. Giza 4, 61/1887/69, 139A/2141/99, NEB 640, NEB 643.
v.   Very susceptible : average infection on leaves, stems and flowers exceeded 60%, e.g. Rebaya 40.

In subsequent tests, the effects of the age of the plants, and temperature obscured genetic differences between cultivars and highlighted the need to conduct tests under uniform and controlled conditions.

*Field test :* Several entries were tested under field conditions, at Sakha and Nubaria Agricultural Experiment Stations. No artificial inoculation was done and the trials depended on natural infection in both locations. Infection began depending on the amount of inoculum present in the location and environmental conditions, especially temperature.

Some of the entries which showed some resistance, or were less infected than the others, were M 272, M 276, M 277, M 279, M 280, M 283, M 284, M 285, M 292, M 294, M 295, M 296, M 297, M 305, M 308, M 299, M 300 and NEB 938, in addition to some other introductions, and some of the lines in the breeding nursery. Some of these entries have already been included in the breeding program.

*Rust*

Rust of faba beans has been destructive in many parts of the country, especially in Beheira (Nubaria), Sharkia and Ismailia governorates where the temperature is comparatively higher and relative humidity is less than the Delta. It is caused by *Uromyces fabae* (Pers) de Bary which was first reported in Egypt by Briton-Jones in 1925 (10).

Several methods were used in testing entries for their reaction to rust.

*Detached leaf test :* This method appears promising and will be used in the future to test genotypes. Rust uredia appeared after about 10 days.

*Seedling test :* Seedlings, under controlled greenhouse conditions (20°C), showed differences between entries in their reaction to rust. Entries such as NEB 439 and NA 174 were resistant to this rust isolate.

However, the age of the plants at the time of rust inoculation affected rust reaction. The average number of pustules on the leaf and the stem was higher when plants were 75 days old compared to 50 days. Also rust pustules were often larger and of the susceptible type on the older plants, indicating that plants increased in susceptibility to rust with age. However, the opposite effect was observed in certain lines and clearly requires further study.

*Field test :* Several segregating generations, introductions and cultivars were tested under field conditons at both Sakha and Nubaria Agricultural Experiment Stations. No artificial inoculation was done but natural infection in both locations reached 100%. Infection with rust usually appears later than leaf spots. However, during 1979-80, rust infection appeared earlier especially at Nubaria where it was more severe. It is planned to study the appearance of rust during this coming growing season in more detail.

Entries which were rated resistant to rust included M 288, M 311, M 299, M 300 and NEB 938.

## Downy mildew

Downy mildew, caused by *Peronospora viciae* (Berk) de Bary, was reported in 1925 (10). It was found occasionally but did not cause any severe damage to the plants (21). However, during 1978-79, it was observed in the breeding plots at Sakha where it spread fast on both the introductions and local cultivars. It was also observed in the Sharkia governorate during a survey in 1979-80 (26).

This disease is very dangerous as none of the entries seemed to be resistant. It can kill the growing point and dwarf the plant.

## Tolerance to foliar diseases

In field tests, using the fungicide Dithane M 45, at Sakha, losses caused by foliar diseases ranged from 22.82% to 55.70% (Table 5). Entries having a small yield reduction due to diseases can be regarded as tolerant, as has been reported in other crops (e.g. 9). The line 90/1966/72 when rated 82.5% rust, 88% leaf spots and 79.75% downy mildew had only a 22.82% yield reduction compared to the sprayed plants which were rated 22, 30.3 and 44% for the three diseases respectively. Such lines could be of value if used in the breeding program, at least until resistant lines are developed.

## Agricultural practices

Several agricultural practices were studied in an attempt to devise a farming program that will result in a decrease in severity of infection by both leaf spots and rust.

*Date of planting :* Bekheit (6) found that infection with chocolate spot was most severe when faba beans were planted on October 1. It decreased in late plantings. Also, Hegazy (19) indicated that infection with chocolate spot decreased with delaying planting until

December 1. Bekheit *et al.* (8) showed that infection with chocolate spot was heavy in plantings on November 1.

In 1979-80 at Sakha Agricultural Experiment Station, it was found that the percentage of infection with the three foliar diseases (leaf spots, rust and downy mildew), exposed to natural infection, was higher in November plantings compared to early and mid October, (Table 6) when rated mid-season. Similar results were obtained when ratings were made at near maturity.

Table 5. Average percentage of infection of 10 faba bean entries with rust, leaf spots, and downy mildew and average seed yield when plants were sprayed or not sprayed with Dithane M 45, Sakha, 1979-80.

| Entry | Average % infection when plants were | | | | | | Av. seed yield (T/H) when | | % Loss in seed yield |
|---|---|---|---|---|---|---|---|---|---|
| | Sprayed | | | Not sprayed | | | | | |
| | R | L | D | R | L | D | Sprayed | Not sprayed | |
| Giza 1 | 27.5 | 33 | 33 | 88 | 99 | 63.75 | 3.778 | 2.271 | 39.88 |
| Giza 2 | 22 | 33 | 35.8 | 88 | 99 | 57.75 | 3.710 | 2.110 | 43.11 |
| Giza 3 | 44 | 38.5 | 44 | 90.75 | 99 | 66 | 3.054 | 2.075 | 32.05 |
| Rebaya 40 | 44 | 33 | 44 | 96.25 | 99 | 79.75 | 2.606 | 1.154 | 55.70 |
| Intr. 420 | 38.5 | 49 | 44 | 93.50 | 99 | 99 | 3.283 | 1.733 | 47.21 |
| 90/1966/72 | 22 | 30.3 | 44 | 82.50 | 88 | 79.75 | 3.155 | 2.435 | 22.82 |
| 80B/41/80 | 22 | 27.5 | 33 | 79.75 | 77 | 77 | 3.000 | 2.069 | 31.03 |
| 112/3200/74 | 33 | 30.3 | 33 | 77 | 77 | 68.75 | 3.762 | 2.213 | 41.15 |
| 129/1807/76 | 33 | 24.8 | 27.5 | 71.50 | 68.75 | 66 | 3.762 | 2.273 | 27.95 |
| 130/1881/76 | 22 | 24.8 | 33 | 71.50 | 66 | 66 | 3.892 | 2.184 | 43.90 |

| LSD 0.05 (seed yield) between fungicidal spray | | | | | | | | 0.357 | |
| entries | | | | | | | | 0.480 | |

R = rust, L = leaf spots, D = downy mildew

Table 6. Average percentage of infection of faba bean cultivar Giza 3, on February 13, when planted at different dates.

| Date of planting | Average % of infection with | | |
|---|---|---|---|
| | leaf spots | rust | downy mildew |
| October 15 | 16.50 | 0 | 0 |
| October 30 | 22.00 | 0 | 41.25 |
| November 15 | 46.75 | 30.25 | 22.00 |

*Plant density*

Hegazy (19) showed that the percentage of infection with chocolate spot increased with decreasing distance between plants. Ashour *et al.* (5) likewise indicated that a positive correlation existed between plant population and infection with chocolate spot.

During 1979-80, it was observed at the Nubaria Agricultural Experiment Station that there were more leaf spots in plots with high plant populations while more rust was observed in plots with a lower plant population.

*Fertilizers*

Hegazy (19) reported that phosphate decreased the percentage of infection and raised yield, but potash, by itself was of no value. The addition of 476 kg superphosphate + 78.5 kg potassium sulphate/ha gave the lowest infection. Hegazi (20) further showed that phosphorus and potassium decreased the percentage of infection but nitrogen had no effect. Mansour and Kamel (24) indicated that the best treatment to provide high yield and to decrease chocolate spot infection included addition of calcium superphosphate 15% at the rate of 238 kg/ha and calcium nitrate 15.5% at the rate of 238 kg/ha.

At Sakha Agricultural Experiment Station, during 1977-78 and 1978-79, the percentage of rust decreased with increasing the amount of phosphatic fertilizer (Table 7). Rust infection was not affected by the addition of low amounts of nitrogen but increased only at the highest rate of ammonium nitrate (240 kg/ha).

Table 7. Effect of phosphate on rust severity and yield of faba beans.

| Amount of Calcium Superphosphate (kg/ha) | Av. % of rust infection | | Av. yield (T/ha) |
| --- | --- | --- | --- |
| | 1977-78 | 1978-79 | 1977-78 |
| 0 | 24.75 | 23.65 | 1.882 |
| 120 | 19.85 | 19.53 | 1.976 |
| 240 | 16.65 | 19.43 | 2.123 |
| 480 | 12.58 | 14.95 | 1.157 |

*Chemical control*

As early as 1936 biosul, sociol and Bordeaux mixture, applied at three week intervals, gave almost 100% control of rust as reported by El-Helaly (11) who in another report, (12) mentioned that sulphur decreased yield but yield increased when plants were sprayed with Bordeaux mixture, biosul, or solsul. El-Helaly (15) further showed that seed treatment with fungicides was not effective but spraying the plants with solsul, lime and sulphur was effective. El Helaly (16) (17) and Bekeit (7) indicated that Bordeaux mixture was the most effective in controlling rust and chocolate spot.

Later, Mansour *et al.* (25) showed that spraying with 0.25% Dithan M 45 plus 0.05% Triton B-1956, three to four times at two week intervals starting by mid-January, resulted in the best control of chocolate spot and rust diseases, and increased yield.

Recently Mohamed *et al.* (26), studied the chemical control of foliar diseases of faba beans under greenhouse conditions for rust and under field conditions at both Sakha and Nubaria Agricultural Experiment Stations for rust and leaf spots :
Greenhouse test : Plantvax (300 cc/100 L water) K.W.G. 0599 (75 cc/100 L water), Dithane M 45 (250 g/100 L water), Sicarol (100 g/100 L water), and Rovral (100 g/100 L water), were tested at different rates for their effect on bean rust on seedlings under greenhouse conditions.

The results (Table 8) indicate that Dithane M 45 and Rovral have a protective action as they only controlled rust effectively if sprayed before rust inoculation. On the other hand, KWG 0599, Plantvax and Sicarol had an eradicative action as they controlled rust both when sprayed after rust inoculation as well as before rust inoculation.

Field test : The five fungicides tested in the field differed in their effect on rust attack (Table 9). Differences between fungicide-sprayed plots and unsprayed plots were significant, also between fungicidal treatments. Results were similar at Sakha and Nubaria Stations except for Sicarol.

Table 8. Average number of pustules per leaflet on faba bean seedlings inoculated with isolate 1 of *Uromyces fabae*.

| Fungicide | Average No. of pustules/leaflet sprayed with concentrations | | | |
|---|---|---|---|---|
| | 0 | 1 | 2 | 3 |
| Sprayed one day before inoculation : | | | | |
| Dithane M 45 | 80.1 | 0 | 0 | 0 |
| Rovral | 80.1 | 1.62 | 0.96 | 0.12 |
| Sicarol | 80.1 | 0 | 0 | 0 |
| K.W.G. 0599 | 80.1 | 0 | 0 | 0 |
| Plantvax | 80.1 | 0 | 0 | 0 |
| Sprayed one day after inoculation : | | | | |
| Dithane M 45 | 80.1 | 28.82 | 13.41 | 6.37 |
| Rovral | 80.1 | 63.73 | 17.50 | 3.29 |
| Sicarol | 80.1 | 4.94 | 3.88 | 0 |
| K.W.G. 0599 | 80.1 | 0.25 | 0.59 | 0.57 |
| Plantvax | 80.1 | 3.01 | 1.23 | 0.49 |

1. ½ recommended dose
2. recommended dose
3. twice recommended dose

Table 9. Average percentage of rust and yield of faba bean Giza 3 cultivar sprayed with fungicides, 1979-80.

| Fungicide | Dose per 100 L | Average % infection | | Average yield (t/ha) | |
|---|---|---|---|---|---|
| | | Sakha | Nubaria | Sakha | Nubaria |
| Plantvax | 300 cc | 11.00 | 44.00 | 4.715 | 1.230 |
| K.W.G. 0599 | 75 cc | 22.00 | 55.00 | 4.984 | 1.139 |
| Dithane M 45 | 250 g | 35.75 | 66.75 | 5.065 | 1.420 |
| Sicarol | 100 g | 35.75 | 91.00 | 4.192 | 0.766 |
| Rovral | 100 g | 71.50 | 88.25 | 4.169 | 1.153 |
| Not sprayed | -- | 94.00 | 100.00 | 3.223 | 0.805 |
| LSD 0.05 | | 11.74 | 10.72 | 0.665 | 0.513 |

The highest yield at both stations was obtained from plots sprayed with Dithane M 45 followed by those sprayed with KWG 0599, Plantvax 20 and Rovral. However Plantvax 20 and KWG 0599 gave the best control of rust.

In another trial on 15 fungicides against leaf spots, significant differences were recorded between most of the fungicides and the control for both leaf spot infection and yield at both Sakha and Nubaria (Table 10). Some fungicides, e.g. Dithane M 45, gave good control at both locations, some showed a differential reaction e.g. Miltox and White zineb, while others were ineffective at both locations e.g. Cuprozan super D and Bavistan.

Overall, Dithane M 45 was the best fungicide and its use resulted in 36% and 50.8% yield increases over the control pots at Sakha and Nubaria respectively. This fungicide is being recommended to farmers for the control of both leaf spots and rust.

Table 10. Average percentage of leaf spot infection and yield when plants of Giza 3 were sprayed with different fungicides, 1979-80.

| Fungicide | dose per 100 L water | Average % infection | | Average yield (t/ha) | |
|---|---|---|---|---|---|
| | | Sakha | Nubaria | Sakha | Nubaria |
| Ronilan | 100 g | 77.25 | 100 | 3.468 | 0.789 |
| Rovral | 100 g | 82.75 | 82.50 | 3.718 | 1.093 |
| Cuprozan super D | 250 g | 85.75 | 88.25 | 3.611 | 1.694 |
| Sicarol | 100 g | 77 | –– | 2.942 | –– |
| Bayleton | 250 g | –– | 71.50 | –– | 1.649 |
| Dithane M 45 | 250 g | 41.25 | 41.25 | 5.132 | 1.355 |
| Cupronyl | 65 g | 80 | 85.50 | 3.811 | 0.954 |
| Macroprax | 350 g | 52.25 | 86 | 3.758 | 0.910 |
| Agrimycin | 600 g | 33 | 97 | 3.530 | 0.762 |
| Miltox | 500 g | 88.25 | 52.25 | 3.743 | 1.066 |
| Bemidin | 250 g | 55 | 88.25 | 4.412 | 0.701 |
| White Zineb | 250 g | 33 | 88 | 4.649 | 1.125 |
| Manesan 80 | 250 g | 66 | 68.75 | 4.429 | 1.235 |
| Calxin M | 200 g | 82.50 | 63.25 | 4.780 | 1.145 |
| Bavistin | 100 g | 77 | 88.50 | 3.928 | 0.942 |
| Mancozan | 350 g | 71 | 71.5 | 4.401 | 1.295 |
| Not sprayed | ––– | 94 | 100 | 3.283 | 0.667 |
| LSD 05 | 19.03 | 19.03 | 16.70 | 0.550 | 0.257 |

## Seedling and root diseases

Several fungi have been reported to cause root rot on faba beans in Egypt. Jones (23) reported *Rhizoctonia* spp to cause root rots, while Abdel Rehim (2) showed that *Fusarium oxysporum* f. *fabae* and *F. solani* f. *fabae* infected faba bean plants and caused different symptoms. Aly (1967) reported that *Rhizoctonia solani*, *Fusarium solani* f. *fabae*, and *F. avenaceum* were pathogenic on faba beans. Losses from *R. solani* infection were severe in the early but not in the late plantings. Abdallah (1969) also reported that *Rhizoctonia solani*, *Fusarium solani* f. *fabae*, *F. oxysporum* and *Sclerotium rolfsii* had been isolated from faba bean seedlings.

Recent isolations from different governorates in Egypt have included *Fusarium oxysporum*, *Rhizoctonia solani*, *Macrophomina* sp. and *Sclerotium rolfsii*.

*Cultivar evaluation* : The technique suggested by Roberts and Kraft (28) was used in testing entries of faba beans. Roots of faba bean seedlings were submerged in spore suspension or Hogland solution for one week before notes were taken.

Table 11 shows that the percentage of infection of plants submerged in Fusarium wilt spore suspension ranged between 40 and 95% . Giza 3 and Fam. 402 were the least infected (40 to 50% ), while Rebaya 40, Fam 63/1051 and Romi 83 were the most susceptible (90 to 95%). On the other hand, seedlings submerged in Hogland solution showed almost no symptoms.

Table 11. Percentage of Fusarium wilt on faba bean entries, using the rapid technique of Roberts and Kraft, (28).

| Entry | Spore suspension of *F. oxysporum* * | | | Hogland solution | | |
|---|---|---|---|---|---|---|
| | % infection | Av. root length (cm) | Av. stem length (cm) | % infection | Av. root length (cm) | Av. stem length (cm) |
| Giza 1 | 80 | 12 | 7 | 0 | 18 | 16 |
| Giza 2 | 75 | 10 | 9 | 1 | 22 | 13 |
| Giza 3 | 40 | 14 | 11 | 0 | 18 | 13 |
| Giza 4 | 60 | 7 | 4 | 0 | 17 | 11 |
| Rebaya 40 | 90 | 15 | 7 | 0 | 12 | 20 |
| Fam. 402 | 50 | 16 | 12 | 0 | 16 | 18 |
| Fam. 63/1051 | 95 | 15 | 12 | 1 | 18 | 16 |
| Romi 83 | 95 | 18 | 7 | 1 | 19 | 16 |

* No. of conidia in the spore suspension 1 x $10^6$ conidia/ml.
Infection rated as percentage of discoloured roots.

*Field test* : Field tests at Sids Agricultural Experiment Station, showed that Fam. 402 was the least infected with seedling diseases (19.9%); followed by line 63/1051 and Giza 1. The most susceptible was Giza2, which had 42.1% infection (Table 12).

Fam. 402 was the least infected and produced the highest yield. All other entries produced lower yields although some showed a comparatively lower incidence of disease (Table 12). Yields of all entries were low, however, as a result of a hot spell near maturity.

From Tables 11 and 12, it is clear that the two least wilt-infected entries under laboratory conditions, Fam. 402 and Giza 3, produced the highest yield in the field test. The comparatively lower yield of Giza 3 was probably due to lower seedling emergence (Table 12) which may have been due to low seed viability and not to disease infection. On the other hand, entries identified as susceptible to wilt using the rapid technique (Table 11), Romi Intr. 83, Fam. 63/1051 and Rebaya 40, produced the least seed yield under field conditions.

Table 12. Average percentage of pre- and post-emergence damping off and seed yield of eight faba beans entries-Sids, 1979-80.

| Entry | % Damping-off | | Yield (t/ha) |
|---|---|---|---|
| | Pre-emergence | Post-emergence | |
| F. 402 | 7.1 | 12.8 | 1.185 |
| Giza 3 | 26.6 | 9.9 | .782 |
| Giza 1 | 13.6 | 10.1 | .754 |
| Giza 2 | 28 | 14.1 | .749 |
| Giza 4 | 16.3 | 18.3 | .717 |
| 63/1051 | 10.8 | 11.7 | .676 |
| Rebaya 40 | 18.8 | 13.4 | .632 |
| Romi Intr. 83 | 13.3 | 15.3 | .495 |

*Seed treatment*

Studies on seed treatment to control soil borne pathogens have been limited. Abdallah (1969) reported that Ceresan, Rhizoctol combi and Orthocide 75 at the rate of 3 g/kg seed gave the lowest percentage of infection.

In 1979-80, five fungicides tested differed in their effectiveness at controlling seedling diseases. The most effective was Vitavax/captan especially at the highest dose (3 g/kg seed). Other rates were also effective, and the differences between all rates and the untreated control were significant (Table 13). The highest yield was also obtained when seeds were treated with Vitavax/captan at 3 g/kg seed.

Table 13. Average percentage of seedling infection and seed yield of Rabaya 40 when seeds were treated with three doses of five fungicides.

| Fungicide | dose g/kg seed | Percentage damping off | | Yield (t/ha) |
|---|---|---|---|---|
| | | pre-emergence | post-emergence | |
| Vitavax/captan | 1 | 15.29 | 2.36 | 1.08 |
| | 2 | 13.23 | 3.82 | 1.26 |
| | 3 | 10.76 | 1.11 | 2.01 |
| R H 50 | 1 | 19.12 | 2.99 | 1.79 |
| | 2 | 21.76 | 1.85 | 1.83 |
| | 3 | 18.97 | 2.16 | 1.59 |
| Vitavax/thiram | 1 | 16.70 | 3.84 | 1.09 |
| | 2 | 19.10 | 5.27 | 1.48 |
| | 3 | 25.44 | 4.51 | 1.26 |
| Vitavax 100 | 1 | 27.49 | 2.25 | 1.27 |
| | 2 | 16.32 | 3.96 | 1.23 |
| | 3 | 20.59 | 3.68 | 1.13 |
| Terachlor | 1 | 24.38 | 4.61 | 1.15 |
| | 2 | 23.53 | 1.35 | 1.77 |
| | 3 | 20.52 | 2.06 | 1.37 |
| Not treated | – | 24.41 | 5.42 | 1.09 |
| LSD .05 | | 4.96 | --- | 0.55 |

REFERENCES

1. Abdallah, A.M.A. 1969. Studies on root rot diseases of broad bean in Egypt. M.Sc. thesis, Fac. Agric., Ain Shams Univ.
2. Abdel Rehim, M.A. 1962. Studies on the organisms causing root rot and wilt of horse beans *(Vicia faba* var. *equina)* in U.A.R. Ph.D. thesis, Fac. Agric., Alex. Univ.
3. Abdou, Y.A., Fahim, M.M. 1969. A stemphylium leaf spot of *Vicia faba* in the United Arab Republic. Plant Dis. Reptr. 53 : 157-159.
4. Aly, M.D.E.H. 1967. Studies on root rot disease of broad bean and its control in U.A.R., Ph.D. thesis, Fac. Agric., Ain Shams Univ.
5. Ashour, W.A., Sirry, A.R., Hegazy, M.F., 1966. Studies on the fungus *Botrytis fabae* Sard. causing chocolate spot to broad bean *(Vicia faba).* Ann. Agric. Sci. 11 : 143-158.
6. Bekheit, M.R. 1950 a. Remarks on reaction of field bean cultivars to brown spot and relation of date of planting to disease severity. Report Agric. Research, Min. Agric. (January), 44-45 (In Arabic).
7. Bekheit, M.R. 1950 b. Control of chocolate spot and rust of beans. Report Agric. Research Min. Agric. (May), 174-178 (In Arabic).
8. Bekheit, M.R., Fahmy, I., Rizk, Z., El Yamany, T. 1951. Testing some fungicides to control chocolate spot and rust of faba beans and also the effect of date of planting and different agricultural practices. Report Agric. Research, Min. Agric. (October) 21 (In Arabic).
9. Borlaug, N.E., Rodriguez, J., Curiel, F. 1956. Field resistance and tolerance of certain varieties to stem rust and difficulties of adequately evaluating this very useful type of resistance. Report of the third International Wheat Rust Conference, Mexico, 64-65.
10. Briton-Jones, M.R. 1925. Mycological work in Egypt during the period 1920-1922. Tech. Bull., Min. Agric. 49, 129 pp.
11. El-Helaly, A.F. 1936 a. Bean rust. Report Agric. Research, Min. Agric. (March), 103-104 (In Arabic).
12. ──────────── 1936 b. Bean rust. Report Agric. Research, Min. Agric. (July), 190-191 (In Arabic).
13. ──────────── 1936 c. Studies on the brown spot disease of field bean *(Vicia faba)* in Egypt. Report Agric. Research, Min. Agric. (March), 105 (In Arabic).
14. ──────────── 1938. Brown spot of field beans Part 1. Studies on a fungus from genus *Botrytis* causing this disease in Egypt. Tech. Bull. 191, Min. Agric., 12 pp (In Arabic).
15. ──────────── 1939 a. Preliminary studies on the control of field bean rust. Tech. Bull. 201, Min. Agric. (In Arabic).
16. ──────────── 1939 b. Further studies on the control of bean rust with some reference to the prevention of chocolate spot of beans. Tech. Bull. Min Agric. 236.
17. ──────────── 1950. Bordeaux mixture for the prevention of rust and chocolate spot of beans. Phytopathology 40, 699-701.
18. El-Sherbeeny, M.H., Mohamed, Hosni A. 1980. Detached leaf technique for infection of faba bean plants *(Vicia faba* L.) with *Botrytis fabae.* FABIS 2, 44-45.
19. Hegazy, M.F.H. 1964. Studies on some factors affecting *Botrytis fabae* Sard. causing chocolate spot disease of *Vicia faba* L. M.Sc. thesis, Fac. Agric. Ain Shams Univ.
20. ──────────── 1968. Effect of some factors affecting infection with *Botrytis fabae* Sard on broad bean with special reference to nutrition. Ph.D. thesis, Fac. Agric., Ain Shams Univ.
21. Hussein, M.R.E. 1963. Diseases of legume and oil crops. Min. Agric. Handbook 39, 54 pp. (In Arabic).

22. Ibrahim, I.A., Abd-El-Rehim, M.A. 1965. Fusarium root rot and wilt of horse bean *(Vicia faba* var. *equina)* in U.A.R. Alex. Jour. Agric. Research 13, 415-426.

23. Jones, G.H. 1935. Egyptian plant diseases. Summary of research and control methods. Tech. Bull., Min. Agric. 146.

24. Mansour, K., Kamel, B. 1975. Studies on the effects of phosphorus and nitrogen fertilization and spraying with the fungicide Dithane M 45 on the control of chocolate spot and amount of yield of horse beans. Agric. Research Rev. 53, 123-134.

25. Mansour K Rizk, Z., Fouad, W. 1975. Studies on the control of chocolate spot and rust of horse beans. Agric. Research Rev., 53 : 89-95.

26. Mohamed, Hosni, A., El Rafei, M.E., Abou Zeid, N.A., Omar, S.A., Habib, Wadaia F., Ismail, I.A., Raof, M., Khidr, H. 1980. Plant pathology research studies. ICARDA/ IFAD Nile Valley Project on faba beans. Annual Coordination Meeting 1979-80, Cairo, Mimeograph, 49 pp.

27. Naguib, K. 1948. Physiological studies on the chocolate spot disease of broad beans *(Vicia faba)* in Egypt. M.Sc. thesis, Fac. Sci., Cairo Univ.

28. Roberts, D.D., Kraft, J.M. 1971. A rapid technique for studying Fusarium wilt of peas. Phytopathology 61, 342-343.

23. MAJOR DISEASE PROBLEMS OF FABA BEANS IN SUDAN

MUSTAFA M. HUSSEIN


*Hudeiba Research Station, P.O.Box 31, Ed-Damer, Sudan*

Some of the diseases which attack the faba bean crop in Sudan are of great economic importance; others are only potentially important.

*Wilt and Root rot Diseases*

Wilt (*Fusarium oxysporum*) and root rot (*Fusarium solani* f. sp. *fabae*) diseases are both responsible for early deaths in faba bean in the Sudan. The incidence of these widespread diseases in all faba bean production areas is closely related to sowing date (temperature). Losses, caused by direct reduction in the crop stand, can be tremendous especially when hot spells occur just after germination.

Symptoms usually appear two to three weeks after sowing. In wilt, they consist of yellowing of leaves and characteristic discoloration of xylem vessels of the roots. This is accompanied by marked reduction in plant vigour and, in severe cases, death of the growing tip. In root rot, they consist of yellowing of the tips and margins of older leaves which become brown and dry up. This is usually accompanied by rotting of the roots which may extend as black discoloration just above soil level, and the plant soon collapses (10). Mixed infections of wilt and root rot in the same plant are quite common; wilt usually coming first.

The rate of infection increases steadily in the early counts then drops quickly to negligible levels with the onset of cooler temperatures.

No effective control has been found. Since the pathogens are essentially soil-borne, repeated trials using seed-dressing fungicides such as Benlate, Topsin, Aldrex T, Fernasan, and Vitavax, had no effect on the incidence of the diseases (5).

All local and introduced materials tested were susceptable. However Freigoun (5) claimed that single plant selections during the past four years have given promising results as far as a high degree of tolerance and high yields are concerned.

So far, the sowing date is the main factor which has been found to affect the level of incidence of both diseases. The incidence is high in early sowings (October 10-20) and drops

*G. Hawtin & C. Webb (Eds.): Faba Bean Improvement*
*©1982 ICARDA. ISBN -13:978-94-009-7501-9*

sharply with later sowings (Table 1). The table shows that the effect of sowing date was very highly significant (P≤ 0.001), but that of cultivars and the interaction between sowing date and cultivars was not significant. Results of the following seasons showed a similar trend (8).

Table 1.

Incidence of wilt and root rot in relation to cultivar and sowing date at Hudeiba Research Station. (Percentage of dead plants transformed to degrees).

| Cultivar | Sowing date (1976-77) | | | Mean |
|---|---|---|---|---|
| | Oct. 8 | Oct. 22 | Nov. 4 | |
| H.72 | 39.20 | 21.56 | 12.65 | 24.47 |
| BF2/2 | 40.60 | 24.11 | 13.01 | 25.91 |
| Giza 2 | 38.10 | 22.14 | 13.63 | 24.62 |
| Selaim | 45.26 | 24.95 | 15.01 | 28.41 |
| Beladi | 37.59 | 23.53 | 13.03 | 24.72 |
| SE | | +1.58 | | SE = ± 0.91 |
| Mean (+ 0.71) | 40.59 | 23.26 | 13.46 | |

The disease continues to be a big problem to both researchers and farmers. Due to unstable climatic conditions, it is difficult to fix a suitable sowing date to minimise incidence. The recommended sowing date for faba bean in the Hudeiba region in the early 1970's used to be around mid-October (1). This no longer holds, and sowing on this date results in almost total loss of stand because of the persistence of hotter temperatures for most of October. Accordingly, the recommended date has now been moved to the end of October and beginning of November.

Wilt and root rot are also the main limiting factors in extending the faba bean production area south of Khartoum where there is more land and cheap irrigation.

*Mosaic Diseases*

Mosaic disease was first observed in faba beans in 1959 in Khartoum Province (14) and on Hudeiba Research Station in 1968 (1). Thereafter it has been commonly found in all faba bean growing areas in Sudan. It is caused mostly by pea mosaic virus (PMV) and to a lesser extent by broad bean mottle virus (BBMV). Both viruses can be present separately or together in the same plant. PMV has a wide range of cultivated as well as wild plant hosts while BBMV has a more limited host range (9).

Symptoms usually begin with mild chlorosis and vein clearing in the young leaves which soon change into chlorotic mottling in fully expanded leaves. This gets more pronounced with the onset of cooler temperatures. A slight reduction in growth is observed as well as flower and pod shedding (14). The disease can also be carried symptomless in faba beans.

Both viruses could be transmitted through seed, and both are readily mechanically transmissible. Greenhouse tests showed about 0.2 % seed transmission for the mixture of both viruses as detected by visual symptoms, but field observations indicated a much higher rate (9). Only PMV is aphid-transmitted in the non-persistent manner. *Aphis craccivora* and *Acyrthosiphon gossypii* (=*sesbaniae*) readily transmit the disease under both natural and experimental conditions (14) (1) while *Longiugius sacchari* and *Aphis gossypii* (14) are able to transmit the disease under experimental conditions.

Incidence is usually low at first but rises to a peak during late December and early January, coinciding with favourable temperatures and peak infestation by aphid vectors. Thus the disease is especially severe on late sown plants which are more attractive to the aphids than older plants from earlier sowings. All local and introduced cultivars as well as a number of Hudeiba-improved lines tested were susceptible under greenhouse and field conditions; giving symptoms varying from mild to severe mottling.

Table 2 shows that the effect of sowing date was very highly significant (P=0.001); but that of cultivars and the interaction between cultivars and sowing date were not significant (8).

Table 2.

Incidence of mosaic in relation to cultivar and sowing date at Hudeiba Research Station. (Percentage of infected plants transformed into degrees).

| | Sowing date (1976-77) | | | |
|---|---|---|---|---|
| Cultivars | Oct. 8 | Oct. 22 | Nov. 4 | Mean |
| H. 72 | 26.06 | 36.15 | 40.31 | 34.17 |
| BF2/2 | 29.41 | 33.91 | 41.70 | 35.01 |
| Giza 2 | 25.93 | 35.65 | 41.22 | 34.17 |
| Selaim | 22.38 | 36.58 | 43.13 | 34.03 |
| Beladi | 25.59 | 37.24 | 38.22 | 33.68 |
| SE | | +2.4 | | +1.4 |
| Mean (+ 1.06) | 25.87 | 35.91 | 40.92 | 34.22 |

It is generally understood that virus-induced mosaics can influence the process of photosynthesis by lowering the chlorophyll content and hence directly affecting the plant growth and development. Nour & Nour (14) reported flower and pod shedding as a result of infection which could cause severe yield reductions. Table 3 shows that artificial infection by PMV, BBMV and by a mixture of both, after two weeks from sowing under field conditions, highly significantly reduced the total number of pods and the seed weight per plant. The effect of PMV alone was most pronounced (9).

Abu Salih (2) reported that four sprayings with Folimat against the aphid vectors had no effect on the incidence of the disease. So far sowing at the optimum time appears to be the only effective method of control. This enables the plants to be in an advanced stage of development by the the time the outbreak of aphid infestation occurs. Older plants, being less attractive to aphids, also get less spread of virus than young late sown plants.

Table 3.

Effect of PMV, BBMV and the mixture of both on some yield components of faba bean at Hudeiba Research Station.

| Treatment | No. of pods | Seed wt.(gm.) |
|-----------|-------------|---------------|
| PMV | 9.0 | 4.9 |
| BBMV | 17.5 | 13.2 |
| PMV + BBMV | 17.7 | 15.7 |
| Control | 28.5 | 23.4 |
| SE | $\pm$1.38 | $\pm$1.42 |

*Powdery Mildews*

Powdery mildews are generally more common in the dry northern Sudan than in the wetter parts and thus affect a wide range of cultivated and wild plants (16). In faba beans the disease is caused mainly by *Leveillula taurica* and *Erysiphe polygoni* (13). The fact that both pathogens appear to be highly xerophytic, may explain their common occurrence in the drier desert conditions of the north (16).

*E. polygoni* was first recorded by Boughy in 1946 on faba bean and certain other legumes and brassicas in northern Sudan . By contrast *L.taurica* is more common and infects a wide range of hosts in the northern and central Sudan. Indeed Tarr (16) listed it on more than 50 plant species including cultivated plants in the *Leguminosae, Solonaceae, Liliaceae*, and other plant families, as well as wild annuals and perennials such as *Abutilon* spp., *Solanum* spp., *Calotropis procera* and even trees such as *Acacia* spp., including *A. seyal*.

Powdery mildew is prevalent in all faba bean growing areas in Sudan. *L. taurica*, which is endoparasitic, usually infects the crop at an earlier stage than *E. polygoni* which is ecto-parasitic. This sequence of infection is found to be the same in all sites of production (5). Disease development is favoured by the onset of cooler temperatures in December-January when symptoms can be seen as greyish-white areas on the aerial parts of the plant after which they spread and coalesce.

Conidia accumulate as dusty powder on the host surface, causing chlorosis of the infected part and a gradual decrease in host vigour. The conidia are easily disseminated by wind to start a secondary spread of infection. The pathogens are not known to produce perithecia under Sudanese conditions, and the disease cycle is maintained only asexually in the off-season on alternate hosts.

Most of the research work on this disease has been conducted at Shambat (Faculty of Agriculture) and, later, at Hudeiba Research Station. Almost all the local and introduced cultivars tested were found to be susceptible. However, one introduction from Germany and another from Russia were found to remain completely free from the disease.

None of the agronomic factors studied at Hudeiba for the last two decades such as watering interval, application of nitrogen, and plant population, seem to have any definite effect on the incidence of the disease. However sowing date was different. Although early sown crops became heavily infected at a later stage of their development they consistently gave better yields than late sown crops which were less heavily infected.

A lot of work has been done on the use of fungicides. Khalifa (11) found that 10 sprayings with sulfinette (25 cc/litre) at weekly intervals beginning four weeks after sowing, completely checked the disease. Freigoun (5) found that Sofril repeatedly gave better results than other fungicides, irrespective of the time of application. However, neither partial nor complete control of the disease was accompanied by significant increase in yields (15). This may partly be due to the fact that *L. taurica*, being essentially endoparasitic, is more difficult to control by dusting or spraying non-systemic fungicides (16). It may also explain the statement by Freigoun (5) that delayed sprayings after the symptoms appear, may result in less efficient control.

El-Karouri (4) has reviewed the research work on breeding and selection for powdery mildew resistance which has been conducted mostly at Hudeiba.

Mutwakil (12) tested 21 introductions, some of which showed reasonable tolerance. However, he found that all the new selections from the tolerant introductions were susceptible, indicating that the tolerance obtained in the original selections was recessive.

Ibrahim (10) suggested that the basis of tolerance may be due to physiological reactions, e.g. phytoalexin formation, rather than anatomical. Two of the apparently resistant lines he tested produced phytoalexins which could possibly, he argued, affect the powdery mildew pathogens. He argued further that susceptibility to powdery mildew increased with leaf age, possibly due to the older leaves losing their capacity to form phytoalexins. This may also be in line with an earlier statement by Heipko and Kaufmann (7) that heavier infections were always associated with earlier sowings and higher yields.

Salih (15) crossed the two sources of resistance from Germany and Russia with good yielding, but susceptible, cultivars. He found that the selected lines which were completely free from the disease did not significantly outyield other lines which were heavily infected. Further yield tests were not consistent.

*Potentially Important Diseases*

Leaf roll disease has been recently observed in faba beans in experimental plots at Wad Medani, Shendi and Hudeiba as well as in farmers' fields in almost all areas of production (3). No systematic studies have yet been made but on a recent tour, Dr. L. Bos (Consultant virologist) and I noticed that the incidence was high at Medani(more than 50 %)and decreased northwards to a minimum of 11% in the Selaim area. Most of the infections appeared to be early ones and the plants seemed to have recovered completely. No direct economic importance is evident yet, but the disease could be potentially important since it is reported in the literature that yellows diseases caused by luteoviruses are among the most economically devasting diseases. Like the root rot and wilt diseases, leaf roll could well be a limiting factor in the drive to extend the production area south of Khartoum.

Phyllody or green flower disease is another example of   potentially important diseases of faba beans. At present its incidence is low and does not exceed 2 % (under Hudeiba conditions). No secondary spread is known to occur but the disease is potentially serious since infected plants are rendered completely sterile.

## Conclusion

Powdery mildews, mosaics, wilt and root rots are widespread diseases in faba bean production areas in Sudan and can cause considerable loss in yield. The pathogens involved have wide host ranges and are able to maintain their life cycle on alternate hosts or survive independently in the soil. Thus control measures such as eradication of alternate hosts or crop rotation seem to be of limited value.

Extensive research efforts directed towards developing control by chemicals applied either directly against the pathogens or indirectly against the vectors or as seed dressings, were mostly ineffective or uneconomic.

The close relationship between the incidence of these disease and sowing date has made it possible to develop a certain degree of (passive) control. Through careful manipulation, a compromised sowing date may be chosen which will greatly minimize the disease losses. However unstable climatic conditions, make it difficult to fix or recommend a specific sowing date.

Research on breeding for resistance has not been consistent, but has indicated promising signs. In view of the inefficiency of the other methods of control, concerted and continuous breeding and selection programmes are all the more urgent.

## REFERENCES

1. Abu Salih, H.S., Ishag, H.M., Siddig, S.A. 1973. Effect of sowing date on incidence of Sudanese broad bean mosaic virus in, and yield of *Vicia faba*. Ann. appl. Biol.74, 371-378.
2. Abu Salih 1973-74. Annual Report, Hudeiba Research Station.
3. Bos, L. 1980. Virus diseases of *Vicia faba* and some other crops in the Nile Valley (Sudan and Egypt) and the involvement of ICARDA. A consultant report based on field visits to Egypt and Sudan during 1980, ICARDA, Aleppo : 17 pp.
4. El-Karouri, A.M.O. 1979. A review of literature on research carried out on broad beans (*Vicia faba*) in the Sudan. Special report prepared for ICARDA. (mimeo).
5. Freigoun, S.O. 1974-75. Annual Report, Hudeiba Research Station.
6. Freigoun, S.O. 1980. Report on fungal diseases situation, season 1979-80 (Nile Valley Project on Faba beans).
7. Heipko, G.H., Kaufmann, H. 1964-65. Annual Report, Hudeiba Res. Station.
8. Hussein, M.M. 1976-77. Annual Report, Hudeiba Res. Station.
9. Hussein, M.M., Freigoun, S.O. 1978. Diseases of broad beans (*Vicia faba*) in the Sudan. A paper presented at the food legume improvement and development workshop, May 2-7, 1978, Aleppo, Syria. Pub. IDRC-126e Ottawa, Canada.
10. Ibrahim, G. 1972-73. Annual Report, Hudeiba Res. Station.
11. Khalifa, O. 1966-67. Annual Report, Hudeiba Res. Station.
12. Mutwakil, A. 1963-64. Annual Report, Hudeiba Res. Station.
13. Nour, M.A. 1957. Control of powdery mildew diseases in the Sudan with special reference to broad beans. Emp. J. Exp. Agric. 25, (98) 119-131.
14. Nour, M.A., Nour, J.J. 1962. Broad bean mosaic caused by pea mosaic virus in the Sudan. Phytopathology 52, 398-403.
15. Salih, F.A., 1975-78. Annual Reports, Hudeiba Research Station.
16. Tarr, S.A.J. 1955. The fungi and plant diseases of the Sudan. CMI, Kew, England.

## 24. VIRUS DISEASES OF FABA BEANS

L. BOS

*Research Institute for Plant Protection, P.O. Box 42, 6700 AA Wageningen, The Netherlands*

The fact that virus diseases of Man and animals are still shrouded in mystery, often leads to misidentifications and misinterpretation; but the unpleasant and often disastrous effects of viruses are widely known. The same holds for viruses and virus diseases of plants.

This chapter aims to show that research workers working with faba beans cannot afford to ignore viruses and that these diseases will continue to need attention.

### Background

*Vicia faba* is being used widely in research as a test plant which is susceptible and highly sensitive to a wide range of viruses. Thus, as a crop it is potentially vulnerable to virus diseases. In fact, in the early 1920's with a rapidly increasing interest in "mosaic diseases", mosaic-diseased faba bean crops were among the first to attract the attention by virologists in U.S.A. (8) (14), the Netherlands (33), Germany (3), Bermuda (41), and Japan (18). Quantz (43) summarises the work in many other countries.

Boning (3) was the first to distinguish between different types of discolouration but it was not until the second half of the 1930's that understanding of the viruses behind virus diseases gradually developed.

With increasing knowledge, it was gradually understood that viruses do not only cause mosaics but also types of necrosis which have long been attributed to fungi. Stubbs (52) in Australia was the first to describe a destructive vascular wilt of faba bean caused by broad bean wilt virus. The virus is now known to be widespread in various parts of the world as a very damaging pathogen of several crops, including non-legumes and even woody ornamentals (53). Gradually other polyphagous viruses, such as alfalfa mosaic virus and cucumber mosaic virus, were also associated with necrotic diseases of faba beans. Special strains of bean yellow mosaic virus that cause such severe necrosis were detected. (6) (17)

Other continuously confusing diseases now known to be due to virus infection, are characterised by yellowing, leafrolling and plant dwarfing. These are widespread in many parts of the world and affect a wide range of wild species and crops, including faba beans and other legumes. Growers mostly attribute these disorders to premature ageing or mere mineral deficiencies, e.g. presumed magnesium deficiency in subterranean clover with red-leaf virus in Australia (30) and pathologists identify them as diseases caused by root rot fungi or other pathogens that only invade plants which have been weakened by virus. Ex-

*G. Hawtin & C. Webb (Eds.): Faba Bean Improvement*
©1982 ICARDA. ISBN -13:978-94-009-7501-9

amples include leafroll-infected faba beans in Iran by *Botrytis fabae* and root fungi (27). Other well known diseases are potato leafroll and barley yellow dwarf.

Böning (3), in Germany, was the first to clearly describe such a leafroll disease of faba bean. The disease, caused by a non-saptransmissible aphid-borne virus, has been studied in considerable detail in some Western European countries (45) (16) (56) (54) (47). Similar diseases in faba bean, pea and some other legumes have been identified later in Australia and New Zealand (subterranean clover stunt virus (19); subterranean clover red-leaf virus (2) (25) (26)(58));Japan (milk vetch dwarf virus(22));USA (a persistent aphid transmitted virus from alfalfa (55); legume yellows and beet western yellows viruses (11)), and may already be prevalent in the Middle East, as reported for Iran (27).

Table 1 summarises some information from the literature on virus diseases of faba bean in the Middle East. However very little information is available on their incidence and importance except in Iran and, to some extent, Sudan.

Table 1. Virus infections of faba beans in the Middle East

| Author and country | virus | incidence |
|---|---|---|
| **Sudan** | | |
| Nour and Nour 1962a | BYMV | +++ |
| 1962b | AMV | ++ |
| Hussein and Freigoun 1978 | BYMV | +++ |
| Hussein 1979 | BBMV | ++ |
| **Palestine** | | |
| Nitzany and Cohen 1962 | BYMV | +++ |
| 1964 | BYMV | +++ |
| | AMV | ++ |
| **Egypt** | | |
| Allain and El-Kady 1966 | BYMV | ++ |
| Nour-Eldin, El-Banna, El Attar and Eid 1966 | BYMV | + |
| El-Attar, Nour-Eldin and Ghabrial 1971 | BYMV | + |
| El-Attar, Ghabrial and Nour-Eldin 1971 | AMV | + |
| Mazyad, El-Hammady and Tolba 1975 | BBTMV | + |
| Kishtah, Russo, Tolba and Martelli 1978 | BBWV | + |
| **Iran** | | |
| Kaiser, Danesh, Okhovat and Mossahebi 1968 | BYMV | +++ |
| | AMV | ++ |
| | CMV | ++ |
| do; Kaiser 1972, 1973 | BLRV | +++ |

AMV - alfalfa mosaic virus
BBMV - broad bean mottle virus
BBTMV - broad bean true mosaic virus
BLRV - bean leafroll virus
BYMV - bean yellow mosaic virus
CMV - cucumber mosaic virus

*Viruses Reported from Field-infected Faba beans*

Adequate control of virus diseases requires knowledge of their etiology. Very little was known about these pathogens for many years.

Symptomatology soon turned out to be highly unreliable for distinguishing between virus diseases. Symptoms of virus infections in plants depend not only on virus and virus strain, but on host species, cultivar as well as the environment. Moreover, multiple infections with other viruses and completely different pathogens may complicate the ultimate syndromes. In complex infection, interactions between viruses may range from antagonism to synergism. Viruses that are normally symptomless when on their own in plants, may greatly aggravate the effect of other viruses. Finally, symptoms of virus diseases may easily be mistaken for the effect of certain fungi, which in fact may more severely attack virus-infected plants as for example, leafroll-diseased faba bean plants with *Botrytis fabae* (56) and root rot-inducing fungi (28).

Artificial virus transmission to test plants soon helped to distinguish between different incitants of virus diseases on the basis of differences in host ranges and symptoms in test plants growing under standardized conditions. Many viruses are transmissible "mechanically" in sap expressed from diseased plants. Most of these viruses are also insect-transmitted but this is also more or less "mechanical" and in the so-called non-persistent manner. Viruses that cause yellows diseases are harder to study. They cannot be transferred from plant to plant in sap because within plants they are limited to the phloem. They are only spread by insects, usually aphids, that feed on the contents of the phloem. Transfer to healthy plants is only after virus circulation through the body of the vector. Such transmission is in the persistent manner.

Knowledge of artificial transmission not only assisted manipulation and study of the viruses in laboratory and greenhouse but also provided insight into the natural spread and ecology of the diseases concerned. Long distance spread may also take place in seed from infected plants although for a long time, this has been considered to be a rare phenomenon. Quantz (42) (43) was the first to detect seed transmission of a special mosaic virus of faba bean which he called *Echtes Ackerbohnenmosaikvirus* (broad bean true mosaic virus), which could pass via up to 15% of the seed from a diseased plant. He soon detected that bean yellow mosaic virus, the incitant of the oldest known faba virus disease, could do the same, although at much lower rates (44). Seed transmission in *V. faba* is now known in six out of 18 viruses reported from naturally infected faba beans (Table 2).

With the advent of modern physio-chemical techniques, derived from molecular biology, to isolate viruses from diseased plants and purify and characterize them *in vitro* , information on plant viruses themselves is rapidly increasing. This allows classification of diseases on an etiological basis. The utilization of results of research and control in different parts of the world is also greatly enhanced by modern international communications. Table 2 lists the pertinent viruses with some of their characters and other information.

Table 1 lists viruses which have already been detected in the Middle East. Many virus infections in these countries have not yet been recognised as such. They are still being overlooked as mere nutritional disorders, or have not been properly diagnosed as to which virus is involved. Syria is a virtually virgin country as far as plant viruses are concerned but, in two neighbouring countries, faba bean crops are known to be riddled with viruses.

Table 2. Viruses reported from naturally infected faba beans*

| name of virus | main symptoms | main other natural hosts | type of virus | mode of transmission | | | geographical distribution in faba beans (1 = Egypt, 2 = Sudan, 3 = Iran, 4 = Israel) |
|---|---|---|---|---|---|---|---|
| | | | | sap | seed | vectors | |
| | leaf and stem necrosis, mosaic | numerous legumes and non-legumes (including some woody species) | bacilliform 3 components 18 x 36, 40 and 58 nm | + | (+)** | several aphids non-persistent | 1, 2, 3, 4, Worldwide |
| bean common mosaic virus | mostly symptomless | *Phaseolus bean* | potyvirus flexuous 750 nm | + (+) | | several aphids non-persistant | Syria, USA |
| bean leafroll virus (synonym pea leafroll virus) | leaf chlorosis and rolling, erect growth and plant stunting | chickpea, clovers (including alfalfa) cowpea, lentil, pea *Phaseolus* bean, soybean | luteovirus spherical 27 nm | – | – | some aphid spp. not by *Myzus persicae* persistent | 3. Eur, USA |
| closely related if not identical : | | | | | | | |
| chickpea dwarf virus | | | | | | | India |
| legume yellows virus | | | | | | | Calif. |
| milk vetch dwarf virus | | | | | | | Japan |
| **Soybean dwarf virus** | | | | | | | Japan |
| suoterranean clover red leaf virus | | | | | | | Austr, NZ |
| more distantly related : | | | | | | | |
| beet western yellows virus | do | do, and numerous non-legumes | do | | | some aphid species, esp. *Myzus persicae* | Calif, Eur |
| subterranean clover stunt virus | | | | | | | Austr. NZ |
| bean yellow mosaic virus | yellow mosaic (pea mosaic strains) green mosaic (bean mosaic strains) necrosis (pea necrosis strains) | several legumes some non-legumes (e.g. gladiolus) | potyvirus flexuous 750 nm | | | | 1, 2, 3, 4, Worldwide |
| broad bean mottle virus | mosaic | unknown | bromovirus spherical 26 nm | + | + | some beetles | 2. UK, India Portugal |
| broad bean necrosis virus | leaf and stem necrosis leaf flecking and recovery | pea, sweet pea | tobra- or tobamovirus rigid rods 2 components 25 x 150 and 250 nm | + | – | via soil | Japan |

| Virus | Symptoms | Host range | Virus group / particle | Seed transm. | ** | Vectors | Distribution |
|---|---|---|---|---|---|---|---|
| broad bean stain virus closely related if not identical: MF-virus | malforming mosaic, necrotic staining of seed testa | unknown | comovirus spherical 28 nm | + | + | Apion vorax (Sitona lineatus) | UK. Sw. DDR, Austr. Morocco, DDR Fr |
| broad bean true mosaic virus | malforming mosaic | Vicia narbonensis, V. sativa, Lathyrus tingitanus | comovirus spherical 28 nm | + | + | Apion vorax (Sitona lineatus) | 1. W.Germ.,DDR,Pol,UK,Sw USSR, Leb., |
| broad bean wilt virus possibly related: pea dwarf mosaic virus | necrosis and wilt, often plant death, sometimes partial recovery | numerous crops and wild species | ungrouped virus spherical 25 nm | + | – | several aphid spp. non-persistent | 13 Austr., DDR, Fr, It, Yug. Jap. Pol. |
| cucumber mosaic virus | leaf and stem necrosis, sometimes mottle or mosaic, premature death | numerous crops and wild species | cucumovirus spherical 30 nm | + | (+) | several aphid spp. non-persistent | 3. Bulg, It, DDR Morocco UK |
| pea early-browning virus | symptomless or mild mottle | pea, Phaseolus-bean several other cultivated and wild species | tobravirus | + | + | nematodes | Pol |
| pea enation mosaic virus | translucent leaf flecks and veins with enations pod enations | pea, sweet pea, clovers | ungrouped virus spherical 30 nm | + | – | aphid spp. persistent | Eur. USA |
| pea seed-borne mosaic virus | leaf narrowing and rolling. plant stunting | pea, lentil and some other legumes | potyvirus flexuous 750 nm | + | + | aphid spp. non-persistent | Japan, DDR |
| red clover vein mosaic virus | | pea, clovers | carlavirus rigid 650 nm | + | – | aphid spp. non-persistent | DDR USA |
| tomato spotted wilt virus | necrosis | pea, peanut, several non-leguminous cultivated and wild species | ungrouped virus membrane-bound spherical 70-90 nm | + | – | thrips spp. | Austr, China |
| Vicia cryptic virus | symptomless | | ungrouped virus spherical 30 nm | – | + | | UK |
| watermelon mosaic virus | mild vein chlorosis, spotting and curling | mainly non-legumes | potyvirus flexuous 750 nm | + | – | several aphid spp. non-persistant | Japan |
| white clover mosaic virus | | pea, clovers | potex-virus flexuous 480 nm | + | – | no vectors | DDR N2, Japan, |

*Publications used for assembling this table have not been included in list of references.

** (+) = seed transmission not yet detected in faba beans.

My recent visits to Sudan have revealed that a yellows disease indistinguishable from bean leafroll virus infection in the Netherlands is often prevalent, particularly in the tirals at Wad Medani where more than 50% of the plants were affected. From such plants found in Egypt, Dr. M.A. Tolba and co-workers (personal comm.) have already been able to transmit an infectious agent with help of aphids. In contrast to the description by Kaiser (27) (28), plants in Sudan tended to recover from disease. The variation may be due to difference in temperature or to involvement of a different virus or strain. A possible role by the more polyphagous beet western yellows virus (11) cannot be excluded. Leaves from necrotic faba bean plants sampled in 1980 at Shambat, Sudan, by Dr. A. Hashim, and at Hudeiba, Sudan, by Dr. M.M. Hussein both contained high concentrations of a carlavirus (650 nm long) which was different from any other faba bean virus reported.

A few reviews or surveys of faba bean viruses are available from other countries. Inouye (23) summarises reports on viruses detected on legumes including *V. faba* in Japan. Recent reviews of the viruses reported from faba crops in England and Sweden with some information on incidence and importance are by Cockbain (7) and Sjödin (51), respectively.

Two papers from Poland and East Germany (DDR) are of particular interest. Fiedorow (15) made observations for three years in 19 fields at 12 different locations in Poland and fond that virus diseases were common with 1 to 94% of the plants infected. Bean yellow mosaic, pea enation mosaic and broad bean true mosaic viruses were most frequently isolated, and alfalfa mosaic, pea early-browning and broad bean wilt viruses were less common. A survey by Schmidt *et al.* (48), during a series of years in 49 representative fields in all faba bean growing districts of the DDR, revealed the presence of 11 viruses. Bean yellow mosaic, pea enation mosaic, broad bean true mosaic and bean leafroll viruses were the most important. A virus resembling pea seed-borne mosaic virus and broad bean stain virus were isolated less often. Broad bean wilt and alfalfa mosaic viruses were of local importance and red clover vein mosaic, cucumber mosaic and white clover mosaic viruses were isolated frequently. Average yields losses during 1973 to 1975 were estimated at 11.3%.

Progress in studying viruses *in vitro* and their detection by electron microscopy in ultrathin sections of diseased plant tissues, has revealed that a number of diseases hitherto ascribed to viruses in actual fact are associated with and most probably caused-by very minute micro-organisms called mycoplasmas (MLO). They are systemic in plants, although phloem-limited, and are transmitted by insect vectors (usually leafhoppers) in a persistent manner. Diseases caused by such virus-like pathogens are characterized by hormonal disorders such as witches' broom growth and floral abnormalities (phyllody, virescence, sterility and proliferation) but these may be accompanied or dominated by chlorosis and reduced growth due to phloem degeneration or phloem necrosis, as in virus-incited yellows diseases.

Such a virus-like disease is common in faba beans in Sudan where it has been reported under the names "witches broom" and "phyllody" to occur usually in 2 to 8% of the plants, and sometimes to affect 16% of the plants of single fields (37). The disease which also affects some other species, is still practically omnipresent in faba bean growing areas of Sudan but with low incidence. In the North, attack is claimed never to exceed 2% of the plants (Dr. M.M. Hussein, personal communication). Infection occurs early during crop development, with little or no secondary spread. The disease continues to be of potential importance since MLO have wide host ranges, and infected plants always turn sterile.

So far, very little information exists on the incidence of viruses and yield losses in faba beans and consequently on their actual importance in Egypt. Most research has been on virus isolates from plants at research stations, but during my visit in 1980, mosaic was already causing concern in the Nile delta. Incidence of mosaic is often high in Sudan, particularly late in the season when all crops may be infected. Incidence in Khartoum province has been found to range from 20 to 75% , and the disease caused highly significant reductions in pod number and plant height (38). In Iran, where the virus is prevalent, inoculation of faba bean plants with bean yellow mosaic virus at pre-bloom and full-bloom, decreased seed yield by 44 and 42%, respectivley (28). The virus also considerably reduces root nodulation in faba beans (17).

Bean yellow mosaic may come from other crops such as clovers and even from gladiolus, but its non-persistent transmission by aphids is over short distances only. Seed infection may be the main, but very efficient, source of infection. In field trials in Iran, with less than 1% of the seed infected, subsequent spread by aphids was to 4 to 34% of the plants in 15 weeks, and to 93 to 100% after 22 weeks (28).

The occurrence of leafroll is quite alarming. The disease is already widespread in food and fodder legumes in Iran (27) (28) and infection of faba beans at the pre-bloom stage may lead to yield reductions of 100% . In Europe, yield reductions of infected plants by up to 50% have been recorded, and pod set is considerably reduced (56). Incidence of leafroll-like symptoms was high in certain parts of Sudan particularly at Wad Medani, but its effect on yield seemed less than reported elsewhere. The disease has also been observed in Turkey (28). It has been reported to be of higher incidence in North Africa, together with related diseases in chickpeas (up to 50% in Algeria), lentil and peas, Reddy *et al.* (46) and is most likely to be widespread throughout the Middle East. In Iran, lucerne *(Medicago sativa)* was a major source of infection without showing symptoms (27) (28).

The incidence of insect-transmitted viruses may vary considerably from year to year; depending greatly on vector population densities.

Reasons why viruses are of special potential importance to faba bean improvement include :

(i)  Crop diversification and breeding require transfer of germplasm with the attendant risks of introduction of seed-borne viruses as indicated in Table 2. Information on seed transmission is still limited (5) and infection percentages may be so low that the viruses easily escape attention. Certain viruses that are not normally seed-transmitted may do so in complex with other viruses as was demonstrated for broad bean mottle virus when in broad bean together with bean yellow mosaic virus (34). Whether newly introduced viruses will get established depends on the presence of vectors and of other natural hosts. Broad bean stain and true mosaic viruses have been found to be self eliminating in Scotland because of absence of the main natural vector *Apion vorax* (26).

(ii) Viruses often become established more easily and reach epidemic proportions more rapidly in breeding nurseries than in farmers' fields as the result of variable genetic material and smaller plots and more open stands. Such nurseries may then act as sources of further spread, especially in the case of seed transmission. Apart from this, viruses often are a nuisance or even a limiting factor to the breeding programme itself.

(iii) Crop improvement also means the introduction of new cultivars which may have previously unknown susceptibility or sensitivity to viruses, which were hitherto of no or little importance, but are present in the natural habitat.

During January 1981, we noticed that at Hudeiba, Sudan, incidence of necrosis was higher in introduced foreign breeding material than in local cultivars. Change of farming methods such as sowing date may also have a striking effect on the incidence of viruses.

This brings us to the ecological relationships between different crops as far as virus diseases are concerned. Most viruses are polyphagous and introduction with seed or planting material of a certain crop and the consequent epidemic build-up may have bearing on the health of other crops. Examples are bean yellow mosaic virus in clovers and gladiolus and bean leafroll virus in lucerne and other clovers, which threaten several food legumes. Infection in such hosts is often symptomless e.g. bean leafroll virus in lucerne (57), or permanentely so under conditions of high temperature (27). Hence, interference with existing agro-ecosystems is bound to lead to ever changing virus problems.

*Control*

Viruses basically differ from other crop pathogens and pests because they cannot be eradicated chemically. Control of virus diseases in crops is preventative by avoiding or removing sources of infection, by controlling spread and/or by increasing plant resistance.

Sources of infection such as lucerne fields with bean leafroll virus, or red clover or gladiolus fields with bean yellow mosaic virus, can only be avoided through spatial isolation e.g. (50). With non-persistent viruses such as bean yellow mosaic virus, a few hundred meters will suffice, but the persistent bean leafroll virus can be carried by aphids over long distances although the chances of infection rapidly decrease with distance. In New Zealand, Ashby *et al.* (2) showed that 41% of *Aulacorthum solani*, trapped in a mixed agricultural area for four years and tested singly on subterranean clover plants, carried subterranean clover red leaf virus.

With respect to the importance of virus-free seed, two risks have to be distinguished. If the introduction of new viruses has to be avoided, no seed infection can be tolerated. In case of infection by a virus which already occurs, economic damage must be prevented. In this case, a low level of seed infection may be acceptable; the degree of which depends on the chances of spread and thus on vector population densities. In Europe, infection percentages up to 0.1% of aphid-borne viruses, as of bean common mosaic in *Phaseolus* beans and of lettuce mosaic in lettuce, usually do not lead to appreciable crop loss, but in California, with high aphid densities, lettuce seed is required to be practically free of virus (5) If infected seedlings show clear symptoms, their timely removal may be advisable. In most cases a small percentage of seeds developed on fully infected mother plants get infected. Seed infection may then be self eliminating in an area where the vector is absent, as of broad bean stain and true mosaic viruses in Scotland (26).

Since the viruses that are transferred in the embryo, as with all seed-borne faba bean viruses, cannot be eliminated from seed in any way, seeds have to originate from virus-free plants grown under vector-free conditions or from virus-free crops. Germplasm for international exchange should ideally be from plants which have been individually tested for freedom from seed-borne viruses. Certification of commercial seed practically never guarantees absolute freedom from virus. Kaiser (28) detected 0.1 to 0.9% infection with bean yellow mosaic virus in seed of 12 out of 20 faba bean lines from nine countries.

Spread of viruses and epidemic development of disease can sometimes be prevented by vector control. Insecticides are known to have a limited effect on persistent viruses and little or no effect on non-persistent viruses which are acquired from diseased plants and transferred to healthy ones in short probes. However, large area aphid control leading to an overall reduction in aphid population density may also reduce infection by the non-persistent viruses. Three-year trials in large areas with insecticide application by aircraft reduced the average infection of faba bean in East Germany (DDR) by bean yellow mosaic virus, pea enation mosaic virus and bean leafroll virus by 63.2, 72.2, and 70.6% , respectively (49).

The spread of broad bean stain and true mosaic viruses within crops grown from infected seed lots can be checked by applying insecticides such as fenitrothion and malathion(7).

Sometimes peaks of aphid flights can be avoided by appropriate choice of sowing date as in Tasmania where yield losses in faba bean with leafroll due to subterranean clover red-leaf virus in 1972-73 were 21, 30, 61 and 8% in plots sown in May, July, September and November, respectively. However, the next season, yield losses were 79 and 91% for plots shown in May and September, respectively, because of earlier and more rapid colonisation of plants by *Aulacorthum solani* (25).

Mention should also be made of the tendency towards lower infection percentages with insect-transmitted viruses in dense stands as compared with more open crops (50).

Much spread of virus, of course, is by Man himself as in seed and germplasm; over long distances. This requires special attention.

Developing countries mostly lack infrastructures to provide farmers with certified virus-free seed, and farmers mostly lack knowledge about how to apply hygienic measures of virus control. Under these circumstances, breeding for resistance is the only or most efficient way of control. However, very little has been done so far, on breeding for resistance to viruses in faba beans.

A number of scientists have tried to find resistance to bean yellow mosaic virus in faba beans but all 25, 85 and 53 cultivars and lines tested by Nitzany and Cohen (36), Kaiser *et al.*, (29) and Fiedorow (15) respectively, were highly susceptible.

Likewise Kaiser (27) did not find resistance in Iran to bean leafroll virus in any of 115 faba bean lines tested. However, in the Netherlands, testing and breeding for resistance to the virus has been successful for pea (9) (10) and in faba bean germplasm ., some promising differences have been found.

For such breeding work appropriate knowledge is required (1) of the viruses and of their natural variation within the area to be covered, (ii) of which strain(s) can be representatively used for screening for resistance, and (iii) of whether screening should be done under field conditions for proper symptom expression or in the laboratory or greenhouse for optimal chance of infection.

We should also realise that, in resistant material, new strains or viruses that hitherto were of no importance may come to the fore. Even breeding may not provide a lasting answer, but requires continuing breeders' attention.

*Conclusion*

Information on viruses of faba beans is rapidly increasing internationally and several viruses have already been found to play an economically important role in faba bean production.

A number of destructive virus diseases have been reported in the area covered by ICARDA, and new problems are likely to come to the fore as a consequence of changing agriculture. Further crop diversification, and intensification through improved cultural methods and breeding will affect crop ecology and thus the incidence and spread of viruses which already occur, and will unavoidably lead to the introduction and establishment of new ones. Virus problems in different crops are often interrelated and legume fodder crops may have an important effect on virus diseases of annual legumes. Breeding for resistance may provide a major way of control, but other viruses and other diseases may become prominent in new cultivars.

Hence continuously changing modern agriculture will constantly suffer from viruses and will continuously require support by virological research. Much applied research including virus identification and study of the complicated ecology of virus diseases belongs to the domain of the scientific laboratory because of the very special techniques and expertise required to detect viruses and to work with them. Shortage or total lack of such expertise and facilities poses special problems to developing countries (4).

## REFERENCES

1. Allain, E. K. El-Kady, E.A. 1966. A virus causing mosaic disease of broad bean and its vector *Aphis craccivora* in Egypt. Ent. Exp. and App. 9 : 413-418.
2. Ashby, J.W., Teh, P.B., CLose, R.C. 1979. Symptomatology of subterranean clover red leaf virus and its incidence in some legume crops, weed hosts, and certain alate aphids in Canterbury, New Zealand. N.Z. J. Agric. Res. 22 : 361-365.
3. Böning, K. 1927. Die Mosaikkrankheit der Ackerbohne *(Vicia faba* L.). Ein Beitrag zu dem Mosaik der Papilionaceen. Forsch. Geb. PflKrankh. Imm. PflReich 4:43-111.
4. Bos, L. 1976. Research on plant virus diseases in developing countries; possible ways for improvement. FAO Plant Prot. Bull. 24 : 109-118.
5. ──────── 1977. Seed-borne viruses. In: W.B. Hewitt & L. Chiarappa (eds.). Plant health and quarantine in international transfer of genetic resources. CRC Press, Cleveland, Ohio : 39-69.
6. Bos, L., Kowalska, C., Maat, D.Z. 1974. The identification of bean mosaic, pea yellow mosaic and pea necrosis strains of bean yellow mosaic virus. Neth. J. Plant Pathol. 80 : 173-191.
7. Cockbain, A.J. 1980. Viruses of spring-sown field beans *(Vicia faba)* in Great Britain. In: D.A. Bond (ed.) *Vicia faba* , feeding value, processing and viruses. Martinus Nijhoff The Hague: 297-308.
8. Dickson, B.T. 1921. Studies on mosaic. Phytopathology 11 : 202.
9. Drijfhout, E. 1968. Testing for pea leafroll virus and inheritance of resistance in peas. Euphytica 17: 224-235.
10. Drijfhout, E., van Steenbergen, A. 1972. De vatbaarheid van een aantal erwterassen voor topvergeling. Med. Inst. Vered. Tuinbouwgew., Wageningen 348: 4 pp.
11. Duffus, J.E. 1979. Legume yellows virus, a new persistent aphid-transmitted virus of legumes in California. Phytopathology 69 : 217-221.
12. El-Attar, S., Ghabrial, S.A., Nour-Eldin, F. 1971a. A strain of alfalfa mosaic virus on broad bean in the Arab Republic of Egypt. Agric. Res. Rev. Cairo, 49: 277-284.
13. El-Attar, S., Nour-Eldin, F., Ghabrial, S.A. 1971b. A strain of bean yellow mosaic virus naturally occurring on broad bean in the Arab Republic of Egypt. Agric. Res. Rev. Cairo, 49: 285-290.
14. Elliot, J.A. 1921. A mosaic of sweet and red clovers. Phytopathology 11: 146-148.

15. Fiedorow, Z.G. 1980. Some virus diseases of horse bean. Tag.Ber., Akad. Landwirt-sch.-Wiss. DDR, Berlin, 184: 361-366.

16. Fluiter, H.J. de, Hubbeling, N. 1955. Observations on top yellows of peas (Dutch with English summ.) T. PlZiekt. 61: 165-175.

17. Frowd, J.A., Bernier, C.C. 1977. Virus diseases of faba beans in Manitoba and their effects on plant growth and yield. Can. J. Plant Sci. 57: 845-852.

18. Fukushi, T. 1930. On the mosaic disease of broad beans. J. Plant Protect. 17 (11) 707-712; 17 (12) 779-784.

19. Grylls, N.E., Butler, F.C. 1959. Subterranean clover stunt, a virus disease of pasture legumes. Aust. J. Agric. Res. 10 : 145-159.

20. Hussein, M.M. 1979. Recent research on certain broad bean (Vicia faba) diseases in the Sudan. FABIS Newsletter 1: 25.

21. Hussein, M.M., Freigoun, S.O. 1978. Diseases of broad beans (Vicia faba) in the Sudan.Leg. Imp. Dev. Workshop, Aleppo, Syria, May 1978: 3 pp.

22. Inouye, T., Inouye, N., Mitsuhata, K. 1968. Yellow dwarf of pea and broad bean caused by milk-vetch dwarf virus. Ann. Phytopath. Soc. Japan, 34: 28-35.

23. Inouye, T. 1969. The legume viruses of Japan. Rev. Plant Prot. Res. 2 : 42-51.

24. Johnstone, G.R. 1978. Diseases of broad bean (Vicia faba L.) and green pea (Pisum sativum L.) in Tasmania caused by subterranean clover red leaf virus. Austr. J. Agric. Res. 29 : 1003-1010.

25. Johnstone, G.R., Rapley, P.E.L. 1979. The effect of time of sowing on the incidence of subterranean clover red-leaf virus infection in broad bean (Vicia faba). Ann. appl. Biol. 91: 345-352.

26. Jones, A.T. 1978. Incidence, field spread, seed transmission and effects of broad bean stain virus and echtes ackerbohnenmosaik-virus in Vicia faba in Eastern Scotland. Ann. appl. Biol. 88: 137-144.

27. Kaiser, W.J. 1972. Diseases of food legumes caused by pea leaf roll virus in Iran. - FAO Plant Prot. Bull. 20: 127-133.

28. --------------- 1973. Biology of bean yellow mosaic and pea leafroll viruses affecting Vicia faba in Iran. Phytopathol. Z. 78: 253-263.

29. Kaiser, W.J., Danesh, D., Okhovat, M., Mossahebi, H. 1968. Diseases of pulse crops (edible legumes) in Iran. Plant Dis. Rept. 52: 687-691.

30. Kellock, A.W. 1971. Red-leaf virus, a newly recognised virus disease of subterranean clover (Trifolium subterraneum L.) . Austr. J. Agric. Res. 22 : 615-624.

31. Kishtah, A.A., Russo, M., Tolba, M.A., Martelli, G.P. 1978. A strain of broad bean wilt virus isolated from pea in Egypt. Phytopath. Mediterr. 17 : 157-164.

32. Mazyad, H., El-Hammady, M., Tolba, M.A. 1975. Broad bean true mosaic disease in Egypt. Ann. Agric. Sci. Moshtohor 4: 87-94.

33. Meulen, J.G. van der, 1928. Voorlopig onderzoek naar de specialisatie en de infectiebronnen der mozaiekziekten van landbouwgewassen. Tijdschr. PlZiekt. 34: 155-176.

34. Murant, A.F., Abu-Salih, H.S., Goold, R.A. 1974. Viruses from broad bean in the Sudan. Rept. Scott. Hort. Res. Inst. 1973: 67.

35. Nitzany, F.E., Cohen, S. 1962. Pea mosaic virus on broad beans in Israel. Abstr. 16th Int. Hort. Congr. Brussels.

36. ----------------------- 1964. Viruses affecting broad beans in Israel. Phytopath. Mediterr. 3 : 1-8.

37. Nour, M.A. 1962. Witches' broom and phyllody in some plants in Khartoum Province, Sudan. FAOPlantProt. Bull. 10: 49-56.

38. Nour, M.A., Nour, J.J. 1962a. Broad bean mosaic caused by pea mosaic virus in the Sudan. Phytopathology 52 : 398-403.

39. ———————————1962b. A mosaic disease of *Dolichos lablab* and diseases of other crops caused by alfalfa mosaic virus in the Sudan. Phytopathology 52 : 427-432.

40. Nour-Eldin, F., El-Banna, M.T., El-Attar, S., Eid, S.A. 1966. Pea and broad bean mosaic virus in U.A.R. J. Microbiol. UAR 1 : 237-242.

41. Ogilvie, L. 1928. Virus diseases of plants in Bermuda. Agric. Bull. Bermuda 7 : 4-7.

42. Quantz, L. 1950. Beobachtungen zur Samenübertragbarkeit eines Mosaikvirus der Ackerbohne *(Vicia faba* L.) NachrBl. dt. PflSchD. (Braunschweig) 2: 172-173.

43. ——————— 1953. Untersuchungen uber ein samenubertragbares Mosaikvirus der Ackerbohne *(Vicia faba).* Phytopath. Z. 20: 421-448.

44. ——————— 1954. Untersuchungen uber die Viruskrankheiten der Ackerbohne. Mitt. Biol. Bundesanst. Land - Forstwirtsch. Berlin Dahlem 80: 171-175.

45. Quantz, L., Volk, G. 1954. Die Blattrollkrankheit der Ackerbohne und Erbse, eine neue Viruskrankheit bei Leguminosen. NachrBl. dt. PflSchD. (Braunschweig) 6: 177-182.

46. Reddy, M.V., Gridley, H.E., Daack, H.J. 1980. Major diseases of chickpea in North Africa. Intern. Chickpea Newsletter. ICRISAT 3 : 13-14.

47. Schmidt, H.E., Karl, E., Rollwitz, W. 1975. Haufigkeit des Vorkommens des Ackerbohnenblattroll-Virus an Ackerbohne *(Vicia faba* L.) in der Deutschen Demokratischen Republik. NachrBl. dt. PflSch. (Berlin) 10: 204-208.

48. Schmidt, H.E., Schmidt, H.B., Karl, E., Rollwitz, W. 1977a. Untersuchungen uber Virosen der Ackerbohne *(Vicia faba* L.) in der Deutsche Demokratischen Republik. NachrBl. PflSch. D.D.R., 31: 23-24.

49. Schmidt, H.E., Dubnik, H., Karl, E., Schmidt, H.B., Kamann, H. 1977b. Verminderung von Virusinfektionen der Ackerbohne *(vicia faba* L.) im Rahmen der Blattlausbekampfung auf Grossflachen. NachrBl. PflSchutz DDR. 31: 247-250.

50. Schmidt, H.E., Karl, E., Rollwitz, W., Klein, W., Kastner, H.F. 1979. Moglichkeiten zur Einschrankung von Virusbefall bei der Ackerbohne *(Vicia faba* L.) durch agrotechnische Massnahmen unter Berucksichtigung des Pflanzenschutzes. Arch. Acker-PflBau Bodenk., Berlin 23: 389-396.

51. Sjodin, J. 1980. Viruses of *Vicia faba* in Sweden. In: D.A. Bond (ed) *Vicia faba,* feeding value, processing and viruses. Martinus Nijhoff, The Hague.

52. Stubbs, L.L. 1947. A destructive vascular wilt virus disease of broad bean *(Vicia faba* L.) in Victoria. J. Dept. Agric. Victoria 46 : 323-332.

53. Taylor, R.H., Stubbs, L.L. 1972. Broad bean wilt virus. CMI/AAB Descr. Plant viruses 81 : 4 pp.

54. Thottappilly, G. 1968. Untersuchungen uber die Beziehungen zwischen dem Erbsenblattrollvirus und seinen Vektoren sowie uber ein neues pilz - und blatt-lausubertragbares Virus der Erbse. Diss. Justus Lieb ig. Univ. Giessen : 145 pp.

55. Thottappilly, G., Kao, Y-C, J., Hooper, G.R., Bath, J.E. 1977. Host range, symptomatology, and electron microscopy of a persistent, aphid-transmitted virus from alfalfa in Michigan. Phytopathology 67: 1451-1459.

56. Tinsley, T.W. 1959. Pea leafroll, a new virus disease of legumes in England. Plant Pathol. 8: 17-18.

57. Want, J.P.H. van der, Bos, L. 1959. Geelnervigheid, een virusziekte van luzerne. Tijdschr. PlZiekt. 65: 73-78.

58. Wilson, J., Close, R.C. 1973. Subterranean clover red leaf virus and other legume viruses in Canterbury. N.Z. J. Agric. Res. 16: 305-310.

NB. Publications used for assembling data of Table 2 are not included in this list of references.

## 25. SCREENING FOR RESISTANCE TO CHOCOLATE SPOT CAUSED BY *BOTRYTIS FABAE*

S. B. HANOUNIK* AND G.C. HAWTIN**

\* *Tobacco Research Institute, P.O. Box 507, Lattakia, Syria*
\*\**ICARDA, P.O.Box 5466, Aleppo, Syria*

Chocolate spot is responsible for reduced yields of faba beans in many parts of the world (1), (8), (15), (16). Severe epiphytotics have been reported in several countries e.g. Syria (5), Tunisia (2) and England (12). Such epiphytotics have frequently resulted in farmers having to abandon the crop. In Egypt, Ibrahim *et al.* (6) have reported that up to 50% loss in yield can occur due to the effects of chocolate spot and rust diseases which are frequently found in association with each other. It is widely held that chocolate spot is more damaging than rust. Annual losses in Egypt were reported to vary from 5 to 20%.

Chocolate spot is caused by two species of *Botrytis; B. fabae* Sard., and *B. cinerea* Pers. Both species have been identified in Egypt, in other parts of the Mediterranean region, and elsewhere. Typical symptoms are from well-defined chocolate-coloured spotting on the leaves with streaking on the stems and petioles, to extensive blackening of the leaves, accompanied by defoliation and death. The latter symptoms, which are associated with substantial yield losses, occur after a prolonged period of mild but wet conditions. This phenomenon is commonly referred-to as the 'aggressive phase' of the disease.

Although there is some disagreement in the literature as to the relative importance of *B. fabae* and *B. cinerea;* most recent reports indicate that it is *B. fabae* which is the more damaging . Research at ICARDA has concentrated on this species, and no attention has been paid to *B. cinerea.*

Enriquez (4) has recently reviewed the literature on chocolate spot in faba beans. There are very few reports on the identification of resistance. One paper from the USSR (14) reports two cultivars, "Purple pod" and "Hmelnickie" as being resistant. In Egypt (7) in field evaluations of more than 4,000 lines, 30 showed moderate resistance with rates of infection of less than 10% . In a pot trial, the germplasm accession NEB 938 was found to be highly resistant, with an infection rate of less than 1% . Other lines showing less than 10% of infection were NEB 368, NA 67, NA 174, Sel. 23, Sel. 27 and the released cultivars Giza 1 and Giza 3. Rebaya 40 was found to be highly susceptible with an infection rate greater than 60% . These results correlated well with tests on detached leaves in which the cultivars Giza 1 and Giza 3 again showed a moderate level of resistance.

*G. Hawtin & C. Webb (Eds.): Faba Bean Improvement*
©*1982 ICARDA. ISBN -13:978-94-009-7501-9*

El-Sherbeeny and Mohamed (3) also used detached leaves in a series of studies on *Botrytis fabae*. The method used was similar to that reported by Deverall and Wood (1). Very low rates of infection were observed in NEB 519, NEB 938 and the cultivars Giza 1 and Giza 3. El-Sherbeeny and Mohamed concluded that their trials "indicate the efficiency of the detached leaf technique in testing lines for their reaction to chocolate spot (*Botrytis fabae* Sard.). Results can be determined within a few days, using only a few leaflets of the plants. The same plant can be tested with several pathogens and isolates in the same growing season". Although for specific purposes, such as those stated, the detached leaf technique appears to hold considerable promise, efforts at ICARDA have concentrated on the development of methods for screening in the field.

*Factors affecting the development of epiphytotics*

ICARDA has concentrated on factors which can be artificially manipulated in disease nurseries to ensure the induction of severe epiphytotics.

The studies reported here were conducted at the ICARDA site near Lattakia on the coast of northern Syria. This region is characterized during the faba bean growing season (October to March) by an average precipitation of about 750 mm, season mean minimum and maximum temperatures of about $11^{\circ}$ and $19.5^{\circ}C$ respectively, and mean minimum and maximum relative humidities of about 60% and 70% respectively. Isolates of *B. fabae* used in these studies were obtained from naturally-infected faba bean leaves collected from several locations along the coastal region of northern Syria.

*Date of planting:* As the date of planting was delayed, the severity of chocolate spot decreased significantly, (Table 1). The earliest planting date resulted in the greatest severity of 77.5% whereas the latest date resulted in only 45.0% infection. It was concluded that for maximum disease development, faba beans should be planted early.

Table 1.

The effect of planting date on the severity of chocolate spot in a local faba bean landrace

| Date of planting | Disease Severity, %* (Mean of 4 Reps.) |
| --- | --- |
| December 10 | 77.5 |
| December 25 | 65.0 |
| January 10 | 60.0 |
| January 25 | 57.5 |
| February 18 | 45.0 |

\* Plants with no leaves infected = 0%, plants with all leaves infected = 100%.

*Source and age of inoculum and the effects of the growing medium:* Inoculum of *B. fabae* from naturally-infected faba bean leaves, or from one-year-old refrigerated sclerotia of *B. fabae*, was found to be more virulent than that grown on potato-dextrose-agar (PDA) or faba bean-dextrose-agar (FDA) media (Table 2). A greater disease severity was also induced after inoculation with *B. fabae* grown on FDA than on PDA.

Table 2.

The effect of the source of *B. fabae* inoculum and growth medium on disease severity in faba beans.

| Source of inoculum | Growth Medium | Disease rating* (Mean of 4 Reps) |
| --- | --- | --- |
| - Sclerotia | None | 9.0 |
| - Naturally infected leaves | None | 9.0 |
| - Naturally infected leaves | FDA | 6.5 |
| - Naturally infected leaves | PDA | 4.5 |
| - Control (no inoculum) | - | 2.6 |

* Disease rating:

1 = No symptoms, 3 = few small discrete lesions.
5 = Lesions common and some coalesced lesions.
7 = Large coalesced lesions with some defoliation.
9 = Coalesced large dark areas with severe defoliation and plant mortality.

Sclerotia can be used to start infection early in the season. If highly susceptible geno-types (e.g. ILB 4, ILB 6, ILB 209 or Rebaya 40) are inoculated in this way, an early and rapid build-up of disease to epiphytotic levels can be induced. Fresh natural leaf inoculum can then be obtained from these susceptible genotypes for inoculating in the disease nursery. Alternatively, susceptible spreader rows can be inoculated early in the season, and the disease can then spread naturally into the adjacent test rows. *B. fabae* on culture medium (FDA) can be induced to produce sclerotia after about 3 to 4 weeks at room temperature. These sclerotia can then be kept in the refrigerator until next season. ICARDA is investigating a technique for using powdered sclerotia as a source of inoculum for dusting on to the test plants.

Age affects the virulence of *B. fabae* when grown on a culture medium. Last and Hamley (9) reported that conidia from young cultures (about 10 days old) produced the greatest number of lesions, and that the infectivity fell rapidly after that. Last (10) found that equal percentages of conidia from 10 and 40-day-old cultures germinated but only those from the younger culture were able to infect the faba bean. Infectivity fell by 90% within 25 days of culturing whereas the germination of conidia did not fall to that extent until 85 days. Sucrose was found to be highly effective in restoring the infectivity of old cultures.

*Age of plant:* The age of the faba bean plant has an important influence on the severity of chocolate spot. In a study on the susceptibility of plants of different ages to infection by *B. fabae*, it was found that as plant age increased from 2 to 7 weeks, there was a corresponding increase in disease severity and damage (Table 3).

Table 3.

Susceptibility of plants of different ages to infection by *Botrytis fabae*.

| Plant age (Weeks) | Disease severity* Control | Inoculated | Percentage reduction in green shoot wt. of diseased plants compared to control |
|---|---|---|---|
| 2 | 1.0 | 2.3 | 0.7 |
| 3 | 1.0 | 2.3 | 1.0 |
| 4 | 1.0 | 6.0 | 4.6 |
| 5 | 1.0 | 6.0 | 29.8 |
| 6 | 1.0 | 8.0 | 54.8 |
| 7 | 1.0 | 8.3 | 59.6 |

\* Disease severity rated as in Table 2.

El-Sherbeeny and Mohamed (3) also reported that the leaflets of plants approaching maturity were more easily infected than those from younger plants.

These results are consistent with those of Mansfield and Deverall (11), Deverall and Wood (1) and Last and Hamley (9), all of whom reported that the older (lower) leaves on the plant were more susceptible than younger ones.

*Inoculum density:* An increase in the inoculum density of *B. fabae* from 0 to 500,000 spores/ml was associated with a corresponding increase in disease (Table 4). Under the conditions of this test, an inoculum density of 400,000 spores/ml caused complete mortality of the susceptible local *major* type cultivar.

Table 4.

The effect of inoculum density of *B. fabae* on the severity of chocolate spot in a local susceptible faba bean cultivar.

| Inoculum density (Spores/ml) | %Infected plants at various disease levels* 1 | 3 | 5 | 7 | 9 |
|---|---|---|---|---|---|
| 0 | 100.0 | 0.0 | 0.0 | 0.0 | 0.0 |
| 100,000 | 0.0 | 80.0 | 20.0 | 0.0 | 0.0 |
| 200,000 | 0.0 | 0.0 | 80.0 | 20.0 | 0.0 |
| 300,000 | 0.0 | 0.0 | 10.0 | 90.0 | 0.0 |
| 400,000 | 0.0 | 0.0 | 0.0 | 0.0 | 100.0 |
| 500,000 | 0.0 | 0.0 | 0.0 | 0.0 | 100.0 |

\* Disease severity rated as in Table 2.

Several other researchers have also reported increased disease severity with increasing inoculum density. Last and Hamley (9) found that the numbers of lesions which developed were directly proportional to the concentration of inoculum. Using the detached leaf technique, Deverall and Wood (1) reported that increasing the number of spores per drop (0.0002 ml) from 10 to 500, significantly increased the number of lesions. The number and rate of spread of lesions were increased if the surface of the leaf was rubbed gently with a mild abrasive before inoculation. Similarly El-Sherbeeny and Mohamed (3) reported that increasing the inoculum density from 75,000 to 150,000 spores per ml increased the number of lesions on detached leaves. They also indicated that the increased infectivities of the larger inoculum densities were of the same order on the oldest as on the youngest leaves.

*Temperature and humidity:* Visible disease symptoms begin to develop about 24 hours after inoculation. The rate at which they develop depends on various environmental factors. In a study in moist polythene chambers, it was found that an increase in the incubation period from 0 to 120 hours, was associated with increased disease severity (Table 5). Under the conditions of this test, the temperature and relative humidity were about $18^{\circ}C$ and 98% respectively.

Table 5.

The relationship between the incubation period and severity of chocolate spot.

| Incubation Period (Hrs. After Inoculation) | Disease severity * | |
|---|---|---|
| | Chocolate spot | Control (water only) |
| 0 | 1.0 | 1.0 |
| 12 | 1.0 | 1.0 |
| 24 | 3.8 | 1.0 |
| 48 | 5.4 | 1.0 |
| 72 | 6.2 | 1.0 |
| 96 | 8.2 | 1.0 |
| 120 | 9.0 | 1.0 |

\* Disease severity rated as in Table 2.

Wilson (16) reported that the maximum air temperature for infection is about $30^{\circ}C$, with no infection occurring at $32^{\circ}C$. The optimum temperature for disease development was around $20^{\circ}C$ at which temperature infection occurred in about 8-12 h. At $5^{\circ}C$ and $1^{\circ}C$, infection occurred after about 4 days and 7 days respectively.

Relative humidity appears to be the most important environmental factor influencing the development of chocolate spot. Wilson (16) reported that no infection occurred at less than 85% RH.

Although Wilson was primarily concerned with *Botrytis cinerea*, *B. fabae* has substantially the same environmental requirements; Sardina (13) reported that the minimum RH for infection by *B. fabae* was 84-85%.

To ensure adequate levels of RH in disease screening nurseries, it may be necessary to cover plants, for example with polythene sheets, for several days after inoculation. Alternatively, it may be possible to ensure high humidity during the infection period through the use of mist or sprinkler irrigation.

*Procedure for the production of epiphytotics*

Based on local experience, and the factors already considered in this chapter, a procedure has been developed at ICARDA for producing epiphytotics of chocolate spot in the field:

a. Black sclerotia are collected from a culture of *B. fabae*, isolated the previous season from an aggressive lesion and maintained on an FDA medium in the refrigerator.

b. The sclerotia are surface sterilized in 0.5% sodium hypochlorite solution (10% clorox) for 2-3 minutes, plated on FDA medium then incubated at $22^{\circ}C$.

c. After 3-5 days of incubation, the plates are exposed to an alternating cycle of 12 h darkness and 12 h white fluorescent light at room temperature.

d. After 2 days of light treatment, the plates are left at room temperature, for about one week, until spores have formed.

e. Spores of *B. fabae* are then transferred to FDA medium and propagated as in the previous step.

f. 10-12 day-old cultures are homogenized in a Waring blender for 60 seconds.

g. Homogenates are then diluted with tap water until a spore concentration of 400,000 spores/ml is obtained.

h. The inoculum is then used after sunset to inoculate 8-10 week-old faba bean plants in the field, employing 10-15 ml of the spore suspension/plant. A knapsack sprayer is used to spread the inoculum.

i. Inoculated nurseries are covered immediately with a polythene cover, supported by metal frames (2 x 6 x 1.5m), to provide adequate humidity and a water film on the leaf surface. The nursery is sprayed on alternate evenings with water to maintain high humidity.

j. After 4-5 days of incubation, the plants are uncovered. Disease readings are taken 2, 3 and 4 weeks after inoculation.

*Identification of resistant genotypes*

In an attempt to identify *B. fabae*-resistant genotypes, the above procedure was used in 1980 under field conditions in the disease nursery site at Lattakia.

Out of 1650 germplasm pure line accessions inoculated, but without the polythene cover, 21 entries were rated 1 (without disease symptoms), and 516 were rated 3 (plants having a few small discrete lesions).

In this test 162 entries were rated 7, (large coalesced lesions with some defoliation) indicating only a moderate disease pressure in the nursery. Disease pressure was far greater when the same procedure was employed on 15 faba bean progenies, but under a polythene cover. About 80% of the entries in this nursery were rated 7 or 9 (extensive lesions with severe defoliation).

However, one entry, a selection from ILB 438, was rated highly resistant.

*Strategies for developing resistant cultivars*

The first step in any disease resistance breeding programme is the identification of sources of resistance. Using the above technique, it should be possible to identify such sources in the field. Other screening methods such as the detached leaf technique, may also prove effective in this search.

The dangers of developing cultivars with only a single resistance gene conferring vertical resistance are well known. The possibility of the build-up of new virulent pathotypes which overcome the resistance is ever present. Unfortunately, almost nothing is known about the race situation in *B. fabae* and this area will require considerable research as major gene sources of resistance are identified. The development of an international disease resistance nursery is being considered by ICARDA as a possible means of studying this situation. Such nurseries have proved of considerable value in other crops and for other pathogens.

Although major gene resistance may provide only a short-term solution, it should not be neglected. However, much work remains to be done both to study the genetics of resistance of sources already identified, such as that observed in the selection from ILB 438, and in the incorporation of such resistance into elite genetic backgrounds. It may ultimately prove impractical to develop cultivars with such resistance, and it is important that alternative strategies be considered from the outset.

Faba beans have a high level of outcrossing which offers a number of exciting possibilities for developing cultivars with stable resistance. Cultivars rated 3 to 5 under heavy disease pressure, should not be neglected as sources of resistance. It may well be that they carry a range of minor genes at different loci which can be combined, through such breeding procedures as recurrent selection, into a genetic background which is uniform enough for release, as a cultivar, to farmers.

Another possibility for combining different resistance sources (both major and minor genes) is through the development of multi-lines or, more appropriately, mixtures. The current interest in synthetic cultivars offers very interesting opportunities in this respect.

The development of cultivars resistant to *B. fabae* is still in its infancy. In the long run, it may not be possible to develop cultivars which can withstand extremely heavy disease pressures. However host plant resistance is expected to become an increasingly important component in an overall strategy for controlling the disease. The integration of resistant cultivars, fungicide control and appropriate agronomy practices into a single package should be high on the list of priorities for future research on the control of chocolate spot.

## REFERENCES

1. Deverall, B.J., Wood, R.K.S. 1961. Infection of bean plants (*Vicia faba* L.) with *Botrytis cinerea* and *B. fabae*. Ann. Appl. Biol. 49: 461-472.
2. Djerbi, M., Mlaiki, A., Bouslama, M. 1979. Food Legume diseases in North Africa. In: Food Legume Improvement and Development ed. G.C. Hawtin and G.J. Chancellor, IDRC pub. 126e. Ottawa.
3. El-Sherbeeny, M.H., Mohamed, M.A. 1980. Detached leaf technique for infection of faba bean plants (*Vicia faba* L.) with *Botrytis fabae*. FABIS 2: 44-45.
4. Enriquez, G.A. 1977. Chocolate spot on *Vicia faba* caused by *Botrytis fabae* Sardina and *B. cinerea* Pers. A literature review. CATIE, Turrialba, Costa Rica, 34p.
5. Hanounik, S.B. 1979. Diseases of major food legume crops in Syria. In: Food Legume Improvement and Development ed. G.C. Hawtin and G. J. Chancellor, IDRC pub. 126e Ottawa.

6.  Ibrahim, A.A., Nassib, A.M., El-Sherbeeny, M. 1979. Production and Improvement of grain legumes in Egypt. In: Food Legume Improvement and Development, ed. G.C. Hawtin and G.J. Chancellor, IDRC pub. 126e Ottawa.

7.  Ibrahim, A.A., Nassib, A.M. 1979. Screening for disease resistance in broadbeans (*Vicia faba*) in Egypt. FABIS 1: 25.

8.  Kaiser, W.J., Mueller, K.E., Dariush, D. 1967. An outbreak of broadbean disease in Iran. Plant Disease Rep. 51 (7): 595-599.

9.  Last, F.T., Hamley, R.E. 1956. A local-lesion technique for measuring the infectivity of conidia of *Botrytis fabae* Sard. Ann. Appl. Biol. 44: 410-418.

10. Last, F.T. 1960. Longevity of conidia of *Botrytis fabae* Sardina. British Mycological Society Transactions, 43: 673-680.

11. Mansfield, J.W., Deverall, B.J. 1974. The rates of fungal development and lesion formation in leaves of *Vicia faba* during infection by *Botrytis cinerea* and *Botrytis fabae*. Ann. Appl. Biol. 79: 77-89.

12. Moore, W.C. 1949. The significance of plant diseases in Great Britain. Ann. Appl. Biol. 25: 258-288.

13. Sardina, R.J. 1929. Una nueva especie de *Botrytis* que ataca las habas. Mem. R. Soc. Espaniola Hist. Nat. 15: 291-295.

14. Sestiperova, Z.I., Timofeev, V.B., 1965. 'An evaluation of varieties of forage and culinary broad beans for resistance to diseases in the conditions of the Leningrad Province'. Mem. Leningrad. Agric. Inst. 95: 126-133.

15. Sundheim, L. 1973. *Botrytis fabae, B. cinerea* and *Ascochyta fabae* on broad bean (*Vicia faba*) in Norway. Acta. Agr. Scand. 23(1): 43-51.

16. Wilson, A.R. 1937. The chocolate spot disease of beans (*Vicia faba* L.) caused by *Botrytis cinerea* Pers. Ann. Appl. Biol. 24: 258-288.

# 26. BREEDING FOR RESISTANCE TO FABA BEAN RUST

C.C. BERNIER and R.L. CONNER

*Department of Plant Science, University of Manitoba, Winnipeg, Manitoba, Canada*

*Uromyces viciae-fabae* (Pers.) Shroet. (Syn. *Uromyces fabae* (Grev.) de Bary et Fuckel) is known to be world wide (5). The rust is common throughout the Mediterranean region and is considered one of the most serious diseases of faba beans in Egypt (3). Crop losses as high as 50% have been reported in the North Delta region but losses normally range from 5-20% (7). Williams (14), found that in Australia, severe early rust infestations could result in yield losses as high as 45%. In Europe, in regions where winter beans are usually grown, rust has not been a problem. Since the introduction of small-seeded faba beans into Western Canada some 10 years ago, rust has been reported at several locations in the provinces of Manitoba and Saskatchewan (1) (11). In Manitoba, rust was first observed only occasionally in trial plots on some cultivars. By 1974, rust was prevalent in late sown plots with some lines showing moderate to severe infection. Although the commercial cultivars were not affected by rust, it now appeared that rust might be serious if infection started early in the season and the acreage of the crop increased. Furthermore, rust was also found to occur on native vetches throughout the region and could act as a source of early inoculum.

This rust also infects crops such as peas *(Pisum sativum* L.) and lentils *(Lens culinaris* Medic.) as well as certain wild and cultivated species of *Vicia* and *Lathyrus* (5). Studies in Europe indicate that several *formae speciales* exist within this rust (4). Similar studies with isolates from hosts native to North America have never been conducted.

Very few attempts have been made to differentiate races of *U. viciae-fabae* and identify sources of resistance in *V. faba*. In Egypt, the cultivars Giza 1 and Giza 3 which are adapted to the North Delta region, possess tolerance or moderate resistance to rust. These cultivars are the result of many years of testing and selection under conditions of natural rust infection in the Delta region (7). However, no sources of specific resistance appear to have been identified.

Hiratsuka (6) tested nine broad bean cultivars against five sources of rust but found no resistance. Kispatic (8) proposed a scheme for rating infection types of rust reactions on *V. faba*. He tested 14 cultivars against single spore isolates from three locations but found that the lack of uniformity for rust reaction in the cultivars made it impossible to distinguish races.

Race identification has been more successful in self pollinated crops such as peas and lentils. Kispatic (8) differentiated nine rust races based on their reaction on a number of pea cultivars. Sohi *et al.* (13) tested 35 lines of peas to the natural inoculum in the area near Banglore, India. Certain lines were less severely attacked than others suggesting they had some type of resistance.

*G. Hawtin & C. Webb (Eds.): Faba Bean Improvement*
©*1982 ICARDA. ISBN -13:978-94-009-7501-9*

Nene *et al.* (12) tested between 129-157 lines of lentils to rust over two seasons. A few lines were found to be either immune or to have low terminal rust severity.

Our studies at the University of Manitoba were initiated to gain a better understanding of the variability that exists for host range and race composition within this rust in Manitoba. Such information is essential to the development of stable resistance. The genetic basis of rust resistance in selected inbred lines of *V. faba* was also examined and methods for the detection of a rate limiting or slow rusting resistance in faba bean lines were evaluated (2).

*Methodology*

All experiments, with the exception of the slow rusting experiments, were conducted in the greenhouse or in growth chambers. Plants were grown in a rust-free area at 16-20°C under a 16 h photoperiod.

Most rust isolates were increased from single pustules on their original host. Other isolates used in the host range study were increased as field collections from specific localities. Standard procedures to prevent contamination of isolates were used throughout (Appendix 1) and inoculum was stored at 5°C for no longer than a month before use.

Three weeks after emergence, plants were inoculated with uredospores of a given isolate suspended in a light oil (Soltrol 170). In the host range study, plants were sprayed with the uredospore suspension, whereas in the other studies, inoculum was applied to two youngest sets of fully expanded leaves using a fine nylon brush. After drying, the plants were incubated at a 100% relative humidity for 24 h and then transferred to the greenhouse. In the host range study, plants were scored for the presence or absence of sporulation 2 weeks after inoculation. In the other studies, infection types were classified on the basis of pustule size 11 days after inoculation (Table 1) to effectively limit the spread of the rust and allow the sequential testing of more than one isolate on the same plant.

Table 1. Infection types of *Uromyces viciae-fabae* on *Vicia faba* and *Pisum sativum*.

| Infection type | Cultivar Reaction | Reaction class |
|---|---|---|
| 0 | No sign of infection | Immune |
| ; | Necrotic Flecks | Hypersensitive |
| 1 | Uredia minute, barely sporulating | Very resistant |
| 2 | Uredia small (about .5 mm in diameter) | Resistant |
| 3 | Uredia large (about 1.0 mm in diameter) | Susceptible |
| 4 | Uredia very large (>1.0 mm in diameter) | Very susceptible |

*Results*

*Host range.* Rust isolates collected from species of *Vicia, Lathyrus* and *Pisum* were used in cross-inoculation tests on a total of 20 *Vicia* spp., 10 *Lathyrus* spp., one *Lens* sp. and one *Pisum* sp. The isolates were found to have much larger host ranges than had previously been indicated. The isolates shared so many hosts in common that it was not possible to group them into *formae speciales* as proposed by Gaumann (4). Isolates from native species of *Vicia* and *Lathyrus* infected one or more of the cultivated crop species of *V. faba, P. sativum* and *L. culinaris* indicating that they could play an important role in the epidemiology of the disease.

*Pathogen variation and host resistance :* Four licenced cultivars presently grown in Manitoba and a rust susceptible line (PI 222128) were evaluated to establish their reaction to two rust isolates. The results indicate that with the exception of PI 2 22128 which is uniformly susceptible to both rust isolates, there is a definite race effect for all cultivars with a significantly greater percentage of plants being susceptible to isolate SP51 (Table 2). The heterogeneity within the cultivars is clearly demonstrated in their response to isolate SS3 but not SP51. It would seem that if isolates similar to SP51 ever become predominant, they could be very damaging in areas where these cultivars might be grown.

Seventeen selections from licenced cultivars and accessions were chosen on the basis of their resistance to rust in preliminary tests. After two to four generations of inbreeding and selection, eight lines and an additional line that did not react uniformly to SS3 were found to act uniformly to a second rust isolate, SP51. These nine inbred lines were used as differentials in race identification.

Table 2. Percentages of plants within faba bean cultivars infected by rust isolates SS3 and SP51.

| Host | % plants infected | |
| | SS3 | SP51 |
|---|---|---|
| Diana | 7.0 A | 96.0 C |
| Ackerperle | 24.0 A B | 84.0 C |
| Herz Freya | 24.0 A B | 94.0 C |
| Erfordia | 40.0 B | 96.0 C |
| PI 222128 | 98.0 C | 100.0 C |

Turkey's W procedure at 5% level of significance. Values followed by the same letter are not significantly different.

Two races were identified on the inbred lines, from amongst four isolates collected on *V. faba* (Table 3). Rust isolates from *V. americana* Muhl. behaved the same as race 1 from faba bean. Additional races were detected amongst the three rust isolates from *L. venosus* Muhl. and an isolate from *L. ochroleucus* Hook. was avirulent on all the inbred lines.

*Inheritance of rust resistance.* $F_2$ and backcross data from crosses between the nine inbred faba bean lines selected for uniform reaction against rust isolates SS3 and SP51, revealed the presence of three different dominant genes for resistance (Table 4). Resistance genes Fr 1 and Fr 2 provide resistance to SS3 while Fr 3 provides resistance only to SP51.

Table 3. Reactions of rust isolates on inbred faba bean lines.

| Rust isolate | | Inbred lines | | | | | | | | |
|---|---|---|---|---|---|---|---|---|---|---|
| | | 2N 447-7 | 2N 40-3-5 | 2N 62-6 | 2N 12-2 | Ack -11 | Erf -1-4 | D -50-7 | HF -13-4 | MB -1-2 |
| From *V. faba* | | | | | | | | | | |
| SS3 ⟋ Race 1 | | + | + | + | − | − | − | − | ± | − |
| SP80 ⟍ | | + | + | + | − | + | − | − | − | − |
| SP51 ⟋ Race 2 | | + | + | + | + | + | + | ± | − | − |
| SP17 ⟍ | | + | + | + | + | + | + | − | − | − |
| Rust from | | | | | | | | | | |
| *V. americana* | Steinbach | + | + | + | − | − | − | − | − | − |
| *V. americana* | Falcon Lk. | + | + | + | − | − | − | − | − | − |
| *L. ochroleucus* | Oak Point | *̲ | − | − | − | − | − | − | − | − |
| *L. venosus* | Haddashville | * | * | + | * | | − | − | *̲ | * |
| *L. venosus* | Killarney | + | *̲ | *̲ | *̲ | | − | − | *̲ | *̲ |
| *L. venosus* | Brokenhead Rv. | + | + | + | *̲ | * | * | * | * | * |

+ = Infection type 3-4
* = Infection type 2.
- = Infection type 0-1.

Additional heterozygous resistance genes were detected in two lines where their segregation was masked by other homozygous resistance genes. Resistance in two lines was still not homozygous after three generations of inbreeding and selection.

Table 4. Proposed genotypes of inbred faba bean lines based on their response to isolates SS3 and SP51.

| Host | Proposed genotype[a] | | | |
|------|------|------|------|------|
| 2N447-7 | fr1fr1, | fr2fr2 | fr3fr3 | |
| 2N40-3-5 | fr1fr1, | fr2fr2, | fr3fr3 | |
| 2N62-6 | fr1fr1, | fr2fr2, | fr3fr3 | |
| 2N12-2 | Fr1Fr1 | fr2fr2, | fr3fr3 | |
| Ack-11 | Fr1Fr1 | fr2fr2 | fr3fr3 | |
| Erf-1-4 | fr1fr1, | Fr2Fr2, | fr3fr3 | Rr |
| MB-1-2 | fr1fr1, | Fr2Fr2, | Fr3Fr3, | Rr |
| D-50-7 | Rr, Rr | | | |
| HF-13-4 | Rr, Rr | | | |

[a]Fr1 Fr2 - provide resistance against isolate SS3.
Fr3 - provides resistance against isolate SP51 only.
R represents unidentified resistance genes.

*Slow rusting resistance.* Also referred to as partial resistance or horizontal resistance. In this type of resistance, rust increases more slowly in the field on some cultivars than on others, even though the cultivars are susceptible. The resistance thus reduces the rate of epidemic development. To determine if slow rusting existed in faba bean cultivars, 25 accessions showing low to high rust severity were selected for further study from among other lines which were severely rusted in field tests. The lines were tested in single row plots replicated five times with rust spreader rows at the end of each plot. Rate of rust development after inoculation of the spreader rows was assessed by comparing the area under the disease progress curve (AUDPC) and terminal disease severity. The results indicate that three lines consistently behaved as slow rusters over three years of field testing, and another six lines had AUDPC values that shifted from low to intermediate. Typical results are illustrated in Fig. 1.

*DISCUSSION*

The results of these studies are encouraging. They demonstrate that the development of rust resistant cultivars should be possible. However, they must be viewed as but a first step in the long process of ensuring effective, stable resistance to rust. The resistance genes require wider testing and additional resistance genes must be sought. Seed of the resistant inbred lines will be sent to ICARDA for testing in the region as soon as sufficient seed is available. We also propose to include rust isolated or races from the Mediterranean region in our program. These isolates will be tested in growth rooms during the winter months only.

The wide host range of the rust isolates from Canada, together with the previously reported host range studies from other areas of the world (4) (9) demonstrate that considerable variability exists within *U. viciae-fabae*. Thus, as experience with rust diseases of other crops suggests, the use of single resistance genes in cultivars would likely not result in long term rust control. This problem can be alleviated in two or three different ways. Gene pyramiding, that is the addition of two to three resistant genes to a single cultivar, might be succesfull. To be effective, the genes in such complexes should not be released singly in other cultivars, in order to prevent stepwise increases in virulence which could destroy multigene resistance (10).

The use of multiline or synthetic cultivars would seem to be a better choice for a partially outcrossing crop such as faba beans. Multilines are composed of a number of lines, each carrying different resistance genes or combinations of resistance genes. Such cultivars could likely be released following a few generations of backcrossing the resistance genes into one or several agronomically suitable lines.

Long term rust control might best be provided by slow rusting resistance. We have shown that it is possible to identify slow rusting lines in small field plots by the use of area under the disease progress curve or final rust severity values. The nature of this resistance and its stability against additional races obviously requires further study in Canada as well as in the Mediterranean region. Individual plants cannot yet be assessed for this resistance either in the field or greenhouse because the characteristics or components of slow rusting in faba beans are not known. Thus, the use of this type of resistance is presently limited to population improvement by recurrent selection. We are presently attempting to identify the components responsible for slow rusting in order to allow the incorporation of this resistance into suitable cultivars by traditional plant breeding methods.

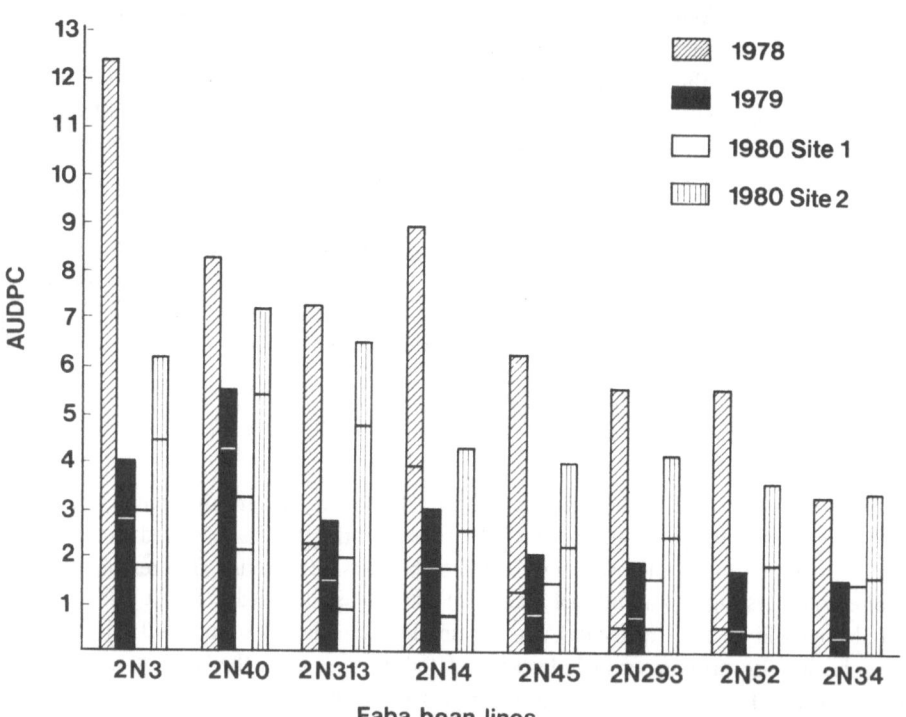

Fig. 1. Area under the disease progress curve (AUDPC) for selected faba bean lines for three years.

APPENDIX 1

*Procedures for increasing rust isolates.* Inoculation of host plants for rust increase was carried out in a plexiglass inoculation cabinet equipped with a set of fine spray nozzles. Between inoculations, these nozzles sprayed a fine mist of water from the top of the cabinet washing the air within the cabinet free of uredospores after four to five minutes. After spraying, the walls of the cabinet were hosed down. The inoculated plants were covered with a tubular polyvinyl isolation cage. Filtered air was circulated through the cage by a rubber hose attached at the base of the chamber. This slight flow of air through the isolation cages prevented contaminating spores from floating in and also prevented an excessive build-up of moisture. Rust was later collected with a cyclone spore collector or by the leaf-tap method (Browder 1971). Uninoculated check plants were always included to ensure that any possible contamination could be detected.

## REFERENCES

1. Bernier, C.C. 1975. Diseases of pulse crops and their control. In J.T. Harapiak, ed. Oilseed and pulse crops in Western Canada - A symposium. Western Co-operative Fertilizers Ltd., Calgary, Alta. pp 439-454.
2. Conner, R.L. 1981. Evaluation of resistance to rust *(Uromyces viciae-fabae)* in faba beans *(Vicia faba)* and pathogenic variability in *U. viciae-fabae*. Ph.D. Thesis, Univ. of Manitoba, February 1981, 118 pp.
3. El-Helaly, A.F. 1939. Preliminary studies on the control of bean rust. Ministry of Agriculture, Egypt. Bull. 201. pp. 19.
4. Gaumann, E. 1934. Zur kenntnis des *Uromyces fabae* (Pers.) de By. Annales Mycologici 32 : 464-470.
5. Guyot, A.L. 1957. Les Urédinées (ou rouilles des végétaux) III. *Uromyces.* Encycl. Mycol. 29 : 505-554.
6. Hiratsuka, N. 1934. Physiological studies on *Uromyces fabae* f. sp. *viciae-fabae*. The Botanical Magazine 48: 309-325.
7. Ibrahim, A.A., Nassib, A.M., El-Sherbeeny, M. 1978. Production and improvement of grain legumes in Egypt. In G.C. Hawtin and G.J. Chancellor, eds. Food Legume improvement and development. ICARDA and IDRC, Aleppo, Syria. pp. 39-46.
8. Kispatic, J. 1944. Einleitende versuche uber rassenbildung bei *Uromyces fabae* (Pers.) de Bary. Phytopathol. Z. 14: 475-483.
9. ------------ 1949. Prilog poznovanju biologije i suzbijanja bobove rdje *Uromyces fabae* (Pers.) de By. f. sp. *viciae-fabae* de By. Ann. Trav. Agric. Scient., Belgrade 1(2): 1-61.
10. Knott, D.R. 1972. Using race-specific resistance to manage the evolution of plant pathogens. J. Envir. Qual. 1: 227-231.
11. McKenzie, D.L., Morrall, R.A.A. 1975. Faba bean diseases in Saskatchewan in 1973. Can. Plant Dis. Surv. 55: 107.
12. Nene, Y.L., Kannaiyan, J., Saxena, G.C. 1975. Notes on the performance of lentil varieties and germplasm cultures against *Uromyces fabae* (Pers.) de Bary. Indian J. of Agric. Sci. 45: 177-178.
13. Sohi, H.S., Sokhi, S.S., Rawal, R.D. 1974. Varietal reaction of peas to powdery mildew *(Erisiphe polygoni)* and rust *(Uromyces fabae)*. Mysore J. Agric. Sci. 8: 529-532.
14. Williams, P.F. 1978. Growth of broad beans infected with *Uromyces viciae-fabae*. Ann. Appl. Biol. 90: 329-334.

## 27. FACTORS AFFECTING RESISTANCE TO ROOT ROT AND WILT DISEASES

G. A. SALT

*Rothamsted Experimental Station, Harpenden, Herts, United Kingdom*

Root rot and wilt limit yields of faba beans in many parts of the world. It is important to recognise them as separate diseases, especially in breeding or selecting for resistance. But this can be difficult because some of the most obvious symptoms are similar to both, viz. reduced growth and yield, pale dull leaves turning black from the margins inwards, wilting and premature death. The characteristic symptom of root rot is a black rot of the root cortex. Wilting may occur when damage to the roots has become extensive but, more usually, the plant reduces its demand for water by shedding leaves and flowers and allowing pods to dry prematurely. By contrast, true wilt is a disease of the vascular system and when wilting occurs, there may be no visible damage to the root cortex. However, root rot usually develops sooner or later.

Also confusing is that both diseases may be associated with the same *Fusarium* species, indistinguishable morphologically in culture and only differing physiologically in their host-pathogen relationship. One widely used method of distinguishing the two diseases is to split the stem longitudinally to expose the vascular elements: a brown discolouration confined to the vascular system extends a considerable distance up the stem in plants affected by vascular wilt , but it is absent or confined to the roots and basal part of the stem in those with root rot.

### Vascular wilt

Wilt of faba beans is caused usually by *Fusarium oxysporum* f.sp. *fabae,* a physiological race which is pathogenic only to *V. faba*. It has been reported in China (36), Japan (20) (34), Canada (8), Russia (10) and Egypt (18). *F. oxysporum* f.sp. *pisi* and *F. oxysporum* f.sp *lupini,* causes of serious wilts on peas and lupins respectively, can cause a mild wilt in *V. faba* (26)(21), and *F. avenaceum* f.sp. *fabae* is reported to cause a serious wilt of *V.faba*

*G. Hawtin & C. Webb (Eds.): Faba Bean Improvement*
©1982 ICARDA. ISBN -13:978-94-009-7501-9

in China (55) and Japan (34). A new species, *F. inflexum*, has been reported only once, causing wilt in Germany (32). Wilt becomes a problem where faba beans and other grain legumes are grown too frequently on the same land. Not more than once every four or five years is advised in Japan if disease is to be avoided (24).

Unlike root rot, it is not usually a problem in crops grown in good rotations, and inoculum introduced on the surface of seeds is likely to give rise to isolated infected plants only, scattered throughout the crop. Once land is infested, the disease is difficult to control and in Canada at least four years' cropping with non-leguminous crops was necessary before beans could safely be grown again (9). As there is no economic chemical control, resistant cultivars to the disease are highly desirable in countries where faba beans are widely grown and wilt disease casuses serious losses.

Unfortunately, suitable sources of resistance have not been found. However, some variation in resistance among different cultivars has been reported in Egypt, (1) (15), Poland (6) (33) and Russia (23). In Egypt, a relationship was found between resistance and the B-alanine content of seeds, roots and root exudates. In resistant seedlings, B-alanine was produced during hydrolysis of materials stored in the seed (15). It occured in the resistant cv. Klein Thuringer but not in the susceptible cv. Baali (14) and it was found in root tissue and root exudate of the resistant cv. Erfordia (1). Seedlings of the susceptible cv. Baali were severely infected when grown in infested soil, but remained healthy when the soil was drenched with a solution of B-alanine. Drops of a solution placed on the leaves prevented infection of seedlings, indicating downward translocation.

Detection of proteins associated with resistance to wilt has been attempted in Russia, using serological methods (11). Antigens from faba bean seeds resistant and susceptible to *F. oxysporum*, were used to produce anti-sera, and differences in resistance were associated with the number and position of precipitate layers in serological tests, using agar gels. No further development of this or the B-alanine work has been reported since.

## Root rot

The most widespread and damaging root rot, especially in dry areas, is a complex disease involving several species of *Fusarium*, especially *F. oxysporum, F. solani, F. avenaceum* and *F. graminearum.* Also present may be *Rhizoctonia solani,* several species of *Pythium, Phoma and Cylindrocarpon,* but the species present may vary between different localities. These fungi have a much wider host range than those which cause wilt, with the exception of *F. solani* f. sp. *fabae* which is pathogenic only to *V. faba* and reported in China (37), Japan (34) and Sudan (19).

The unspecialized pathogens are normal inhabitants of most cultivated soils where they survive as spores or sclerotia, and as weak parasites on root surfaces of several species of plants without causing visible damage (3). They can be isolated from apparently healthy roots of *V. faba*, and research on similar root rots of other legume species suggests that they colonize young roots soon after emergence and develop a passive parasitic relationship which changes to aggressive parasitism when the host plant is affected adversely by some form of stress (25). Root rot symptoms then appear as a progressive necrotic breakdown of the cortex of tap-root and laterals. Removal of the stress factors often results in some recovery, new lateral roots remaining healthy, despite the proximity of inoculum, until stress factors return. Crop rotation is of very limited value against root rot, and serious outbreaks on land that has not grown legumes for many years are common. At Rothamsted during the dry summer of 1970, black root rot was serious in field beans (*V. faba minor*) grown on part of Broadbalk Field that had grown only wheat for more than 100 years. As there is no economic chemical control and no resistant varieties, control is directed mainly towards management practices that reduce stress factors.

*Stress factors that predispose plants to root rot and wilt diseases*

The severity of root rot and wilt diseases depends very much on physical and biological factors such as high temperature, shortage or excess of water, plant nutrient deficiencies, adverse soil conditions, damage by insects and nematodes, other fungal and virus diseases and competition from weeds.

Breeding for resistance would be the preferred method of control but cannot be attempted until suitable resistant material has been found. Additional difficulties are presented by the involvement of several ill-defined pathogens and the susceptibility of the crop to so many sub-optimal environmental factors. A possible solution is to concentrate on producing varieties that are more widely adapted and so are able to withstand some of the stress factors. For example, increased tolerance of high temperature or drought or resistance to a specific pest or pathogen bred into a cultivar may well increase its potential to withstand the effects of root rotting organisms.

In experiments on field beans at Rothamsted in 1976, 1977and 1978 (27) treatments that were not directly fungicidal, such as irrigation and applications of insecticides, reduced root rot more than large dressings of benomyl in the seedbed. Stem nematode (*Ditylenchus dipsaci*) was excluded by using clean seed and sowing on sites free from the pest. Seedborne viruses and *Ascochyta fabae* were avoided by using seed free from these diseases. *Aphis fabae* was controlled by spraying with pirimicarb, and losses from soil-borne pathogens were minimized by choosing sites that had not been cropped frequently with field beans or with other grain legumes. Nevertheless the use of crop protection chemicals increased yields by an average of 38 % in 1976, 17 % in 1977 and 12 % in 1978.

In these experiments, disease was assessed visually by estimating the proportion of lateral roots and tap roots blackened in samples of ten plants lifted from each plot on several occasions during growth. Each plant was placed in one of six categories according to the appearance of the tap root. This was repeated for the lateral roots. (Table 1).

Table 1.

| | The assessment of root disease | | | | | |
|---|---|---|---|---|---|---|
| | Roots more white than black | | | Roots more black than white | | |
| Category No. | 0 | 1 | 2 | 3 | 4 | 5 |
| Condition | Healthy- Most roots white | Many more white than black | Slightly more white than black | Slightly more black than white | Many more black than white | Most roots black. Plants moribund or dead |

A disease rating (D.R.) was calculated thus:

$$\text{D.R.} = \frac{\text{Sum of (category number x number of plants in that category) x 100}}{50}$$

the denominator being the maximum possible, i.e. category 5 x 10 plants. The D.Rs. for tap roots and laterals can be meaned to a single figure but it is sometimes more informative to study them separately as laterals become blackened before tap roots or there may be a late production of new disease-free laterals from tap roots that are already black.

*Soil moisture*

Reports from temperate and tropical areas agree that water stress is a major factor predisposing plants to *Fusarium* root rot and wilt disease. In the arid climate of Northern Sudan it has been shown (13) that root rot developed less in crops which were irrigated weekly than in those watered at two or three week intervals. (Table 2). Pot experiments in China (36) showed that faba beans in wet or saturated soil suffered much less from wilt (*F. avenaceum*) than those in dry to medium-moist soil. (Table 3). Even in the comparatively moist temperate climate at Rothamsted, irrigation decreased root rot more than chemical treatments in two out of three years. The third year was wet, and irrigation had no effect. (Table 4). Severe root rot in 1976 was associated with an unusually hot dry summer and unirrigated plants senesced eight weeks earlier than during normal years.

*Temperature*

In Northern Sudan, sowings early in October when temperatures were very high, had more root rot than later sowings in cooler conditions (Table 2). In other reports, high temperatures are confounded with summer drought. At Rothamsted, for example, root rot is associated with unusually warm dry summers and, in temperate climates, sowing early in spring is recommended so that plants are well developed and deep-rooted before the onset of summer (2). However, in China (36), wilt disease was reported to be more destructive in Northern Yunnan than in the warmer southern part, but the temperatures were not given.

*Insects and parasitic nematodes*

It is logical to suppose that feeding by insect pests and parasitic nematodes renders plants more susceptible to root rot and wilt diseases by reducing plant vigour and providing infection sites for fungal invasion (22). In an attempt to obtain data on this subject a common approach is to attempt to control specific organisms in an infested soil, but results are often difficult to interpret because treatments aimed at specific target organisms may be less effective or less specific than expected. At Rothamsted, we found that application of the systemic nematicide/insecticide aldicarb to the seedbed before sowing field beans consistently decreased root rot although it is active against pests rather than fungal pathogens. (Table 5). It greatly reduced adult and larval populations of the weevil *Sitona lineatus*, the migratory nematode *Pratylenchus* spp, the aphid *Acyrthosiphon pisum* and thereby bean leaf-roll virus (B.L.R.V.). The control of any one of these might well have contributed to the decrease in root rot, but it was not possible to assess their individual effects.*S. lineatus* caused damage every year. Recent work (4) suggests that larval feeding on the root nodules may be an important cause of yield loss in faba beans, although it is not known how much of this loss if any is due to increased root rot. Experimental evidence that root rot of clover caused by *F. oxysporum* was increased when *Sitona* was present was reported by Hill *et al.* (16) from U.S.A., who studied the effects of the insect alone, the fungus alone and the two together on the amount of root rot in two experiments (Table 6). In both experiments, larval feeding increased root rot.

Table 2.

Effect of sowing date and watering interval on root disease in Northern Sudan.
(After Friegoun, 1980)
Mean %infection

| Sowing Date | 1 week | 2 week | 3 week watering interval |
|---|---|---|---|
| Oct. 3 | 40 | 46 | 53 |
| Nov. 2 | 26 | 26 | 34 |
| Dec. 2 | 7 | 12 | 11 |

SE ± 3.1

264

Table 3.

The relationship of soil moisture to the development of wilt in faba beans in China
(After Yu and Fang 1948a).

| Condition of soil* | Dry | Medium-moist | Wet | Saturated |
|---|---|---|---|---|
| % water holding capacity | 18-27 | 44-65 | 70-89 | 100 |
| % wilted plants | 72 | 78 | 13 | 3 |

*Each treatment consisted of 10 pots, each with 10 plants

Table 4.

Effect of irrigation on root disease and yield at Rothamsted

| Year | Disease rating | | Grain yield (t/ha) | | Total dry weight(t/ha) | |
|---|---|---|---|---|---|---|
| | No. irrig. | Irrigation | No. irrig. | Irrigation | No. irrig. | Irrigation |
| 1976 (dry) | 99 | 70 | 1.6 | 3.0 | 4.0 | 6.5 |
| 1977 | 52 | 34 | 4.6 | 3.5 | 10.0 | 11.6 |
| 1978 (wet) | 56 | 55 | 5.8 | 6.0 | 11.9 | 12.3 |

(After McEwen et al. 1980)

Table 5

Effect of aldicarb on root disease and yield

| Year | Disease rating | | Yield(t/ha) | |
|---|---|---|---|---|
| | no. aldicarb | aldicarb | no aldicarb | aldicarb |
| 1976 | 88 | 80 | 2.1 | 2.6 |
| 1977 | 46 | 40 | 3.8 | 4.2 |
| 1978 | 61 | 50 | 5.7 | 6.0 |

Table 6.

Vascular necrosis and survival of lucerne plants subjected to *Fusarium oxysporum* f *medicaginis* and larval *Sitona hispidula* alone and in combination. (After Hill *et al.*, 1969).

| Treatment | Number of plants alive after 59 days | Experiment 1 Disease rating (0-5) | Experiment 2 Disease rating (0-5) |
|---|---|---|---|
| Control | 8.5 | 0.2 | 1.0 |
| Insect | 6.7 | 0.7 | 1.03 |
| Fungus | 4.8 | 1.2 | 2.0 |
| Insect plus fungus | 2.3 | 2.4 | 3.5 |

There is little information on the effect of nematode damage to *V. faba* except for the stem eelworm *Ditylenchus dipsaci* and root-lesion nematodes *Pratylenchus* spp. At Rothamsted in 1977, field beans sown on land for the sixth successive year were severly affected by stem eelworm and *Fusarium* root rot. The insecticide/nematicide aldicarb decreased root rot and increased height, growth and yield (Table 7) mainly, it appears, by controlling stem eelworm and *Sitona* which was also present. Thiabendazole had comparatively little effect on the nematode or *Sitona* and so the smaller increases in yield it produced, can be attributed mainly to decreases in root rot.

In our multi-disciplinary bean experiment, stem eelworm was absent and migratory nematodes of the genus *Pratylenchus* were considered to be the most important parasitic species present. *P. fallax* and *P. pinguicaudatus* cause necrotic lesions on bean roots, resembling the earliest stages of a root rot. *P. neglectus* and *P. crenatus* also invade and reproduce in bean roots but cause necrosis only when very many are present. (7). Aldicarb effectively controlled parasitic nematodes, but the effect of this on root rot could not be separated from other effects of aldicarb. Plant parasitic nematodes often play a major role in root disease complexes of many different species of plants (30).

*Virus diseases*

The debilitating effect of virus diseases , could be another factor in predisposing plants to root disease. Increased *Fusarium* root rot has been reported in while lupin infected with bean yellow mosaic virus (29) and in peas infected with pea mosaic virus.(12). In peas, virus infection did not necessarily increase the susceptibility of the host to *Fusarium* but it increased the amount of root exudate and its content of amino-acids and thus increased the activity of the pathogen (5). In pot experiments at Rothamsted, *Fusarium* wilt in lupins *Lupinus albus* - cv. Kievski) grown in soil inoculated with a spore suspension was increased by inoculating leaves with bean yellow mosaic virus or clover yellow vein virus at flowering time (31).

Inoculation of the compost with *F.oxysporum* when seedlings were transplanted caused no root rot or wilt symptoms. However *F. oxysporum* was isolated from the cortex of roots and hypcotyls of all inoculated plants, but from the steles in the hypocotyl and stem bases only of plants that had been inoculated with both fungus and virus. In plots of field beans in 1980, root rot was more severe on plants naturally infected with pea enation mosaic or bean leaf-roll virus than in adjacent plants without virus symptoms (Table 8). The result indicated that in *V. faba* also, virus infection may increase root rot.

*Other diseases*

There is no information on the effects of other diseases on root diseases of *V. faba*, and very little on other species of plants. However if the proposition that plants that are weakened by other adverse factors are more susceptible to root disease holds good, then leaf diseases are equally likely to act in this way. Late sprays of benomyl decreased leaf areas affected by chocolate spot and rust, delayed senescence and increased yields significantly but had no effect on root rot which was already well developed at this late stage of growth. However, O'Rourke and Millar (28) found that there was a direct relationship between the amount of *Ascochyta* leaf spot of lucerne and the amount of root rot (Table 9).

The effect of parasitic weeds such as *Orobanche* on root rot could be included under this same heading of additional stress factors, but no such relationship has yet been published. It would be interesting to know whether root rot is severe on plants infected by *Orobanche* and if any reduction has been observed in experiments where *Orobanche* has been controlled.

Breeding for resistance or improved adaptation to any of these may contribute a little towards ameliorating the damage from root diseases. That improvements in adaptation to more than one factor are likely to be more than additive was suggested by our results at Rothamsted, where irrigation and soil treatment with aldicarb and benomyl each contributed a small decrease in root rot and the combined effect of all three was relatively large (Table 10).

Table 7.

Effect of thiabendazole and aldicarb on stem nematode, root rot and yield.
(After Hooper, 1978).

|  | Untreated | Thiabendazole | | Aldicarb |
|---|---|---|---|---|
|  |  | To seed-bed at 6 kg.ha$^{-1}$ | To seed at 24 g.kg$^{-1}$ | To seed-bed at 5 kg.ha$^{-1}$ |
| Root rot D.R. | 80 | 50 | 30 | 20 |
| % stems infested by *D. dipsaci* | 52 | 28 | 40 | 0 |
| Stem height cm. | 55 | 64 | 67 | 85 |
| Seed yield (t.ha$^{-1}$) | 0.44 | 0.74 | 0.88 | 1.61 |
| SED. $\pm$0.265 | | | | |

Table 8.

Effect of natural infection with viruses on root rot of faba beans in field plots.

| Leaf symptoms | Number of plants | Disease rating | |
|---|---|---|---|
| | | Tap roots | Lateral roots |
| Pea enation mosaic virus | 15 | 19 | 28 |
| Bean leaf-roll virus | 20 | 26 | 32 |
| Healthy | 35 | 16 | 18 |

Table 9.

Influence of *Ascochyta* leaf spot on root rot of lucerne.
(After O'Rourke and Millar 1966).

| | Inoculum concentration (spores ml$^{-1}$) | | | |
|---|---|---|---|---|
| | 0 | $2.5 \times 10^5$ | $5 \times 10^5$ | $1 \times 10^6$ |
| Leaf spot % area | 8 | 20 | 30 | 37 |
| Root rot (0-5) | 1.3 | 1.7 | 2.0 | 2.5 |

Table 10.

Effects of irrigation, aldicarb and benomyl to the seed bed on % disease-rating of tap roots in July 1976 and on yield. (After McEwen *et al.* 1980)

| | % Disease-rating | | | |
|---|---|---|---|---|
| | untreated | benomyl | aldicarb | benomyl + aldicarb |
| No. irrig. | 84 | 81 | 70 | 45 |
| Irrigated | 65 | 46 | 43 | 26 |
| | Yield (t./ha.) | | | |
| No irrig. | 1.35 | 1.42 | 1.93 | 1.90 |
| Irrigated | 2.82 | 2.83 | 3.26 | 3.29 |

REFERENCES

1. Abdel Rehim, M. A., Michail, S.M., Hashim, M. 1968. On the control of infection of *Vicia faba* by *Fusarium oxysporum*, f. *fabae* through the application of beta-alanine. Flora, Jena, Abt. A. 159, 135-140.

2. Al'Zhanov, Zh. Sh. 1964. Diseases of Broad Bean in the Tselinograd district. Vest. sel.-khoz., Nauki, Almo-Ata, 7, 29-31. English summary in Review of Applied Mycology 44, 2002.

3. Armstrong, G.M., Armstrong, J.K. 1948. Nonsusceptible hosts as carriers of wilt Fusaria Phytopathology 38, 808-826.

4. Bardner, R., Fletcher, K.E. 1979. Larvae of the pea and bean weevil *Sitona lineatus* and yield of field beans. Journal of Agricultural Science, Cambridge 92, 109-112.

5. Beaute, M.K., Lockwood, J.L. 1968. Mechanism of increased root rot in virus-infected peas. Phytopathology 58, 1643-1651.

6. Bojarczuk, M., Brodowska, A., Bojarczuk, J. 1972. Study of fusariosis of broad bean, *Vicia faba*. Hodowla Roslin Aklematyzacja & Nasiennictivo 16, 293-304. Ref. RAM 52, 2077.

7. Corbett, D.C.M., Webb, R.M. 1974. Root-lesion nematodes. Rothamsted Report for 1974, Pt. 1, 182.

8. Coulombe, L.J. 1957. Fusarium wilt of broad beans. (Le fletrissement fusarium des Gourganes). Rep. Quebec Soc. Prot. Pl. 38 (1956) 26-33. English summary in Review of Applied Mycology 37. 434.

9. Coulombe, L.J. 1961. Control of Fusarium wilt of broad bean, *Fusarium oxysporum*, f. *fabae*. Canadian Plant Disease Survey 41, 191-193.

10. Dunin, M.S. 1962. Methods of protecting fodder beans from diseases. Izvestiya, Timir-yazev Sel. Akad. 1962, 11-31. English summary in Review of Applied Mycology 42. 64.

11. Dunin, M.S., Abdel Rehim, M.A., Vinitskaya, O.P. 1966. Serological correlations between broad bean and the casual agent of Fusarium wilt. Sel. Khoz. Biol. 1, 265-276. Ref. in Review of Applied Mycology, 46, 1800.

12. Farley, J.D., Lockwood, J.L. 1964. Increased susceptibility to root rots in virus-infected peas. Phytopathology 54, 1279-1280.

13. Freigoun, S.O. 1980. Effect of sowing date and watering interval on the incidence of wilt and root rot diseases in faba bean. FABIS 2 41.

14. Hashem, M., Abdel Rehim, M.A. 1967. The possible role of amino-acids in resistance of plants to fungal infection. Flora Jena Abt. A. 158, 265-267.

15. Hashem, M. 1969. The mechanism of resistance of some varieties of *Vicia faba* towards infection with *Fusarium oxysporum*, f. *fabae*. Flora Jena Abt. A. 160, 164-168.

16. Hill, R.R.J. , Newton, R.C., Zeiders, K.E., Elgin, J.H. 1969. Relationship of the clover-root curculio, *Fusarium* wilt and bacterial wilt in alfalfa. Crop Science 9, 327-329.

17. Hooper, D.J. 1978. Stem nematode (*Ditylenchus dipsaci*). Rothamsted Report for 1977, Pt. 1. 179-180.

18. Ibrahim, I.A., Abdel Rehim, M.A. 1965. *Fusarium* root rot and wilt on horse bean (*Vicia faba* var *equina*) in U.A.R. Alexandria Journal of Agricultural Research 13, 415-426.

19. Ibrahim, G., Hussein, M.M. 1974. A new record of root rot of broad bean (*Vicia faba*) from the Sudan. Journal of Agricultural Science 83, 381-383.

20. Ikata, S. 1951. The disease of food crops No. 1. Rice plants and beans. Asakura Shoten Tokyo. In Japanese. English summary in Review of Applied Mycology 31, 458(1952).

21. Kadow, K.J., Jones, L.K. 1932. Fusarium wilt of peas with special reference to dissemination. Washington Agricultural Experimental Station Bulletin 272.

22. Kilpatrick, R.A. 1961. Fungi associated with larvae of *Sitona* spp. Phytopathology 51
640-641.

23. Kiselev, A. 1976. Inheritance of resistance in broad bean to *Fusarium* wilt. Trudy Tul'
skoi S. -Kh. Op. Stantsii (1976) 182-186. English summary in Review of Plant Pathology, 57, 1908.

24. Kiyoshi Kogure 1979. Broad beans in Japan: origin and development. FABIS 1, 11-14.

25. Leath, K.T., Lukezic, F.L., Crittenden, H.W., Elliott, E.S., Halisky, P.M., Howard, F.L.,
Ostazeski, S.A. 1971. The *Fusarium* root rot complex of selected forage legumes in the
Northeast. Pennsylvania State University, College of Agriculture, Bulletin 777.

26. Linford, M.B. 1928. A *Fusarium* wilt of peas in Wisconsin, Wisconsin Agricultural Experimental Station Research Bulletin 85.

27. McEwen, J., Bardner, R., Briggs, G.C., Bromilow, R.H., Cockbain, A.J., Day, J.M.,
Fletcher, K.E., Legg, B.J., Roughley, R.J., Salt, G.A., Simpson, H.R., Webb, R.M., Witt,
J.F., Yeoman, D.P. 1980. The effects of irrigation, nitrogen fertiliser and the control
of pests and pathogens on spring-sown field beans (*Vicia faba* L.) and residual effects
on two following winter wheat crops. Journal of Agricultural Science, Cambridge 96.
129-150.

28. O'Rourke, C.J., Millar, R.L. 1966. Root rot and root microflora of alfalfa as affected
by potassium nutrition, frequency of cutting and leaf infection. Phytopathology 56,
1040-1046.

29. Patil, P.L. 1974. Interaction of fungal and virus infections in white lupine (*L. albus*).
Research Journal Mahatina Phale Agricultural University 4, 118-122. From abstract in
Review of Plant Pathology 53, 4491.

30. Powell, N.T. 1971. Interactions between nematodes and fungi on disease complexes.
Annual Review of Phytopathology 9, 253-274.

31. Salt, G.A., Cockbain, A.J. 1978. Susceptibility of lupins to *Fusarium* increased by virus
infection. Rothamsted Report for 1977, Pt. 1. 220.

32. Schneider Roswhita, Dachow, J. 1975. *Fusarium inflexum* sp. nov. als erreger einer
Welkekrankheit an *Vicia faba* in Deutschland Phytopathologische Zeitschrift 82, 70-82.

33. Tomaszwwski, Z., Furgal, H. 1979. The assessment of the degree of resistance in broad
bean (*Vicia faba* minor, Beck) to ascochytosis and fusariosis under laboratory-glasshouse
conditions. Institute of Plant Breeding & Seed Production, Olsztyn, Poland. Abstract in
Review of Plant Pathology 58, 5619.

34. Yamamoto, W., Oyasu, N., Takigawa, K. 1955. Studies on the wilt disease of broad
bean. 1. Sci. Rep. Hyogo Univ. Agric., 2 53-62, 1955. English summary in Review of
Applied Mycology 37, 129.

35. Yu, T.F. 1944. *Fusarium* diseases of broad bean. 1. A wilt of broad bean caused by
*F. avenaceum* var. *fabae*, n. var. Phytopathology 34, 385-393.

36. Yu, T.F., Fang, C.T. 1948 a. *Fusarium* diseases of broad bean. II. Further studies on
broad bean wilt caused by *F. avenaceum* var. *fabae*. Phytopathology 38, 331-342.

37. Yu, T.F., Fang, C.T. 1948 b. *Fusarium* diseases of broad bean. III. Root rot and wilt
of broad beans caused by two new forms of *Fusarium*, Phytopathology 38, 587-594.

## 28. FABA BEAN PESTS IN EGYPT

ABDEL-HAKIM KAMEL

*Stored Product Pests Research, Plant Protection Research Institute, Ministry of Agriculture*

### Insect Pests

Faba bean in Egypt is liable to attack by several insect pests, either when seedlings at later developmental stages, or post harvest. More than 20 species belonging to the orders, Lepidoptera, Diptera, Hemiptera, Homoptera, Hemiptera, Heteroptera, Thysanuara and Coleoptera are encountered in faba bean fields and during storage. The most important of these are aphids, mainly the bean aphid, *Aphis craccivora* Koch and, to a certain extent, *Myzus persicae* (Sulz). Leaf miners, *Liriomyza congesta* (Becker); bruchid infesting in the field, *Bruchus rufimanus* Boh.; and bruchids causing damage in storage, *Bruchidius incarnatus* Boh., *Callosobruchus chinensis* L. and *Callosobruchus maculatus* F., are also important.

*Aphis craccivora* has a widespread distribution on faba beans. It begins infesting the plants at the beginning of November and reaches the first peak at the end of that month. After a decreasing period, the population increases to the second peak by the middle of December, after which it decreases in number for a while and then increases again until it reaches the maximum peak during March. The population then continues at this high level, causing severe injury to plants.

The number of individuals was found to be positively correlated with temperature, but not with the relative humidity. The number of annual generations under laboratory conditions was 52. The average period of one generation is five days during summer, and 15 during winter. The correlation between the generation period and either temperature or relative humidity was highly significant. *Aphis craccivora* has also been reported on artichokes, tomatoes, mandarin oranges, sweet lemons, cowpeas, cotton, clover and lentils.

In general, the aphid usually attacks the tops of plants and soon after, covers all other parts. Light infestation may cause a negligible effect to the plant while severe infestations can cause complete destruction and large reduction in the yield.

This species is a vector of the bean yellow mosaic virus which can destroy the crop. In general, infestation occurs in scattered patches in the field and subsequently may spread over the whole area if not controlled.

Initial studies revealed that there were no significant differences in the population density of *A. craccivora* resulting from different agricultural practices. They have indicated, however, that Giza 2 is more resistant, while Giza 4 was more susceptible. The Coleopter-

*G. Hawtin & C. Webb (Eds.): Faba Bean Improvement*
©1982 ICARDA. ISBN -13:978-94-009-7501-9

ous predators, especially *Coccinella undecimpunctata* L. were effective in minimising numbers of aphids.

Dimethoate 40% EC at the rate of 1,500 ppm, Malathion 57% EC at the rate of 2,500 ppm, and Tamaron 600 EC at the rate of 1,500 ppm have proved promising insecticides for the control of aphids.

*Liriomyza congesta*, the leaf miner fly, is a harmful and widespread species on faba beans in the country and throughout the growing season. It reaches its maximum in March when two or three mines can be seen on one leaflet. The larvae live in mines between the two surfaces of the leaf blades, consuming the palisade layer and consequently reducing photosynthesis, and may result in a considerable loss of yield.

The adult insect is a small fly. The female lays her eggs singly on the lower surface of the leaves. After hatching, the larvae tunnel inside causing serpentine mines which look silvery. The lower leaves are more infested than the upper ones and growing tips. The larvae pupate at the end of the mine or outside the leaf. In high infestations, the leaves become yellow and dry and may cover the plant. The number of generations in Egypt is not known.

Host plants of this species are faba beans and peas. Several natural enemies attack this pest, mainly the hymenopterous ones such as *Diglyphus* sp., *Halticoptera* sp. and *Hemiptarsenus zilahisebossi* (Erdus).

Preliminary studies indicated that *L. congesta* attacked all faba bean cultivars, but Giza 2 was the most resistant. Agricultural practices were ineffective in minimising attack.

*Bruchus rufimanus* Boh., the bean weevil. According to Kamel (3), this insect is monovoltine, monophagous and infests the green pods of faba beans. Infestation is transferred through the seeds to subsequent crops. The annual routine fumigation of beans for seed resulted in the decrease of the infestation from more than 12% to less than 1% within five years.

The faba bean, harvested in April and May, is infested with small larvae which continue to feed and pupate in late August and early September. The adult is formed during September, but stays inside the seed until it is planted in November. The beetle emerges when the seed becomes wet and the cotyledons separate. Bishara *et al.* (1) found that the beetle could remain inside the testa of the plant for about two or three months. They also found adults hiding during the daytime between petals of faba bean flowers. When the weather becomes favourable the insects fly from their shelter and begin to lay eggs. If the beans are kept in storage, about 90% of the adults die before the following March.

Irritation of the pod by the female's ovipositor causes a round pale brown area which extends inside the pod until it affects the seed. This mark persists on the dry seed and is diagnostic of infestation by this insect. The female has to choose a suitable place for laying but not all of the eggs hatch. Similarly, not all of the newly hatched larvae are able to continue their development, so external infestation does not necessarily represent an internal infestation.

In the laboratory, some beetles leave the seed 24 hours after sowing, others may delay up to 12 days.

Normally the adult is found lying inside a compact cell within the seed apparently formed by the larva from frass particles glued together. Pupae are also found within such cells. The time required from egg laying in February or March until the adult stage is about 7 months. The percentage of infestation decreases towards the South of Egypt.

*Bruchidius incarnatus, Callosobruchus maculatus* and *C. chinensis* infest faba beans in storage. Infestation begins just after harvest. The biology and behaviour of this group are similar. The eggs are laid singly on the testa of the seed and are easy to see with the naked eye.. One or more eggs are laid per seed. During summer, the incubation period lasts for about four days after which the newly hatched larvae bore directly inside the cotyledon where they remain until they are fully grown. The pupal stage and part of the adult stage are spent inside the seed after which the adult bores its way through an exit hole. It starts to repeat its life cycle. The five or more generations a year cause tremendous damage to faba beans.

In Egypt, protective measures involve sanitation in storage centres, chemical disinfection before entry of the new crop, disinfection of second hand bags, periodic inspection and spraying with 2.5% Malathion. Fumigation is needed as soon as any symptoms of insects can be found. This regime has proved to be very satisfactory in greatly reducing losses in stored faba beans.

## Storage

South of the Nile Delta, in Lower Egypt, Barheem village (Monoufia Governorate) is the only village in which faba bean is stored on a commercial basis in underground pits. This village has a storage capacity of about 4,500 tons in 5000 pits scattered in an area of about 12 ha.

When first excavated, the pit has a diameter of about three meters at the bottom and is two meters deep with a flat bottom and an entrance hole of one meter diameter. With these dimensions, it holds about six tons of beans. After being dug, the walls are well pressed and smoothed. Before the second and third storage seasons, the pit is enlarged until it reaches its full capacity of about 11 tons (3.25 in diameter and 3 m deep).

Grain can be introduced or removed through the only hole at the top. As the pits are filled, the bottoms and walls are lined with faba bean straw or fenugreek straw, and finally sealed with mud. The percentage of infestation in stored beans in such pits never increases and the testa remains white instead of turning brown as with beans exposed to the different weather and light conditions in the open air. This method of storage results also in better cooking properties for beans. Consequently, beans stored in such a way, known locally as makmoura, command 25% higher prices than beans stored in any other way.

The storage season in underground pits begins early in May, directly after harvest and might last from two months up to one year according to the market demands. Losses under such conditions never exceed 1% which results from direct contact of some beans with the walls.

## Vertebrates

As in all other countries, Egypt has a vast number of animal which share Mans's food Besides certain ticks and mites, various rodents, birds and land snails have a leading place insofar as their constant attack on crops and large scale damage to them are concerned.

*Rodents* : among the animals which are harmful to agriculture, rodents are the most important; with rats being the most prevalent as far as Egypt is concerned. The most important of the numerous species of rodents from the agricultural point of view, belong to the family *Muridae.* The six Egyptian species belong to genera *Rattus, Arvicanthis,Acomys* and *Mus.*

This group cause vast damage to nearly all types of standing crops in the field. They devour all the foodstuffs they can hold without discrimination, and seize every available opportunity to share everything we own.

The rat begins to attack after the seeds are sown. It eats some seeds immediately and stores others in its burrow; thus reducing the plant stand. Seeds which escape attack may be eaten after germination. The crop may eventually be safe, but only until near maturity. At this time, the rodent either climbs or bends the plants down, and then begins attacking the premature pods. The damage in severe infestations might reach 5%.

An integrated program for rodent control involves :
— general sanitation in the fields, including mainly the eradication of weeds in the boarders, and on the sides of water canals.
— periodical inspection to estimate the rat population.
— rodenticides and poison baits in burrows and in the pathways.
— re-treatment whenever necessary.

*Sparrows* : Until recent years, sparrows were protected by law in Egypt. But during the past few years, they have become one of the leading agricultural pests, posing a considerable threat to many crops in the fields and storehouses. This situation is mainly due to a disruption in the balance of nature as a result of the widespread use of pesticides in Egypt, and increased open-air storage. The species in Egypt is the house sparrow, *Passer domesticus.*

This bird feeds largely on grains, particularly in the open as well as in covered storehouses and granaries. It also feeds on grains and young seedlings in the field, as well as on the flowers, young pods and the premature grains inside the pods.

This species gathers in hundreds or thousands in trees, particularly during autumn when they are easier to hunt. Flocks move from place to place, depending on where food is available.

In faba bean fields, flocks of the passer sparrow may attack the young seedlings in November, and may result in a need for re-sowing. In the winter and in late March and early April, flocks of the bird attack the young premature pods. At that time losses can be high, especially at the edges of the faba bean fields near trees.

This pest needs an integrated program which includes hunting in granaries, the use of bird lime (an extract from the fruit of *Cordia mixa),* harassment, the use of chemicals in the roosting places, or chemicals which act as bird repellants, the use of poison baits and the extermination of birds in their gathering places such as branches of trees. All of these methods should be used throughout a national campaign.

# REFERENCES

1. Bishara, S.I., Haggag, Y.M.Y., Riad, A.A. 1967. Field infestation of broad beans by Bruchids in U.A.R. Agric. Res. Rev. 45 (2).

2. Hosny, M.M., Assem, M., Nasr, E.A. 1976. Agricultural insect and animal pests.

3. Kamel, A.H. 1967 A study of the possibility of eradicating the broad bean weevil *Bruchus rufimanus* Boh. (Coleoptera, Bruchidae) by fumigation. J. Stored Prod. Res. 3, 365-369.

4. --------------- 1977. Vertebrates harmful to agriculture. The Egyptian International Centre for Agriculture. Course on Plant Quarantine : 299-332 - Cairo, Egypt.

5. ---------------- 1980. Underground storage in some Arab countries. Symposium of grain storage under controlled atmosphere. Rome (May 12 to 15, 1980).

6. Saleh, M.R.A. 1968. Studies on certain species of legume aphids. M.Sc. thesis, Faculty of Agriculture, Assiut University.

SIDDIG AHMED SIDDIG

*Shambat Research Station, P.O. Box No.30, Khartoum North, Sudan*

Several insects severely limit faba bean production in Sudan. As reported by Remane (7), Hussein (2) and Siddig (8), they include : *Aphis craccivora* Koch; *Acyrthosiphon sesbaniae* Kan. Dav; *Aphis gossypii* Geov.; *Bemisia tabaci* Genn; *Empoasca lubica* Deberg; *Erythroneura lubiae* China; *Creontiades pallidus* Ramb; *Spodoptera exigua* HB; *Maruca testalis* Gey; *Caliothrips impurus* Pr; *Caliothrips sudanensis* Bagn Cam; *Bruchus elanairensis* Pic; and *Callosobruchus maculatus* F.

## A.   The lesser army worm

The lesser army worm *(Spodoptera exigua)* is a polyphagous pest that attacks plants of different families at the seedling stage e.g. faba beans, cotton, lucerne, ground nuts, castor bean, safflower, potato, sorghum, sesame, and onion.

*Damage and economic importance :* The female moth lays several hundred eggs at night, in patches of more than 20. After hatching, the larvae feed on the lower surface of young leaves, leaving the upper epidermis intact. Heavily damaged leaves may dry up and drop. The older larvae devour the foliage completely and sometimes feed on stems below ground level, often ringing them. This type of damage produces a delayed sowing date effect and hence results in reduced yields.

Furthermore, Siddig and Abu Salih (12) investigated the occurrence of the Sudanese broad bean mosaic virus in the field before the appearance of aphids and when spodoptera was infesting the crop. They demonstrated that *Spodoptera exigua* was responsible for transmitting that virus disease at that part of the season (Table 1).

*Bionomics :* The lesser army worm breeds on lucerne slowly from December to March and the population begins to increase during April with rising summer temperatures. Moths migrate in July from lucerne to infest the seedlings of different summer crops. The maximum moth flight (Table 2), which was studied by Siddig from 1967 to 1972, occurs in October; coinciding with the presence of faba bean seedlings on which it inflicts serious damage. From October onward, the population decreases with the lower nutritional value of faba beans and, in December, the pest migrates back to lucerne.

*G. Hawtin & C. Webb (Eds.): Faba Bean Improvement*
©1982 ICARDA. ISBN -13:978-94-009-7501-9

Table 1. Percentage transmission of two isolates of SBBMV between faba beans by sap transmission and insects (Hudeiba 1972).

| Isolate LM | | | Isolate AM | | |
|---|---|---|---|---|---|
| Method of transmission | | | Method of transmission | | |
| Sap Inoculation | *A. craccivora | **Spodoptera exigua | Sap Inoculation | A. craccivora | Spodoptera exigua |
| 100 | 19 | 14 | 60 | 38 | 19 |

\* 10 min. acquisition and 10 min. transmission feeding
\*\* 18 hrs. acquisition and 6 hr. transmission feeding

Magzoub and Venkatraman (3) reported *Disophrys lutea* (Braconidae) as a larval parasite in the Northern Sudan, and eight different parasites in the Khartoum area, of which *Zelomorpha sudanesis* was the most effective.

Weed control is an effective measure against the lesser army worm because many weed species are preferred hosts for egg laying (7).

Field trials at Hudeiba (10) showed that Torbidan E.C. at 4.76 litres product/ha, Endosulfan E.C. at 2.16 kg ai/ha, Lannate 90% W.P. at 536 g product/ha, and Supracide 40% E.C. at 3.57 litres product/ha were effective (Table 3).

### B. The grey cotton leaf thrip

*Caliothrips sudanensis* Bagh and Can : Host plants include faba beans, cotton, ground nuts, lucerne and other legumes and vegetables. The main alternative hosts among weeds are *Leucas urticaefolia, L. nubica, Heliotropum europacum* and a number of *Ipomoea* spp. (7).

Siddig showed that the variety BF 2/2 suffered significantly less damage by this pest than the other four cultivars which he tested; Giza, RB 29, RB 40, and IW 5.

*Damage and economic importance :* The nymphs and adults suck the cell sap of leaves whose upper surface is preferred. As a result, the mesophyll tissue is destroyed. During the day, the pest feeds inside flowers and its toxic saliva induces flowers to drop. Thus the depletion of cell sap and the shedding of flowers appreciably contribute to losses in yield.

*Bionomics :* The pest spends the dry season in association with the dark cotton thrips *Caliothrips impurus* PR in irrigated regions on lucerne and alternative weed hosts. It multiplies quickly during the second half of the rainy season (September - October) and migrates to infest faba beans and other legumes where it maintains high populations.

*Natural Enemies :* Lacewings *(Chrysopidae),* Lady birds *(Coccinellidae),* the larvae of syrphid flies *(Syrphidae)* and the anthocorid bug *Orius albidipennis* have been recorded as predators in the Sudan. The wasp *Thripoctenus* sp. is also a parasite (7).

Table 2. Monthly light trap catches of army worm moths at Hudeiba Research Farm 1967-72.

| Season | July | August | September | October | November | December | January | February | March | April | May | June | Total |
|---|---|---|---|---|---|---|---|---|---|---|---|---|---|
| 1967-68 | 1312 | 230 | 1874* | 1338 | 225 | 57 | 61 | 79 | 110 | 463 | 750 | 599 | 7098 |
| 1968-69 | 222 | 70 | 129 | 584* | 240 | 114 | 110 | 108 | 114 | 164 | 163 | 333 | 2351 |
| 1969-70 | 458 | 347 | 9882* | 6798 | 240 | 96 | 92 | 87 | 93 | 564 | 1407 | 409 | 20473 |
| 1970-71 | 390 | 98 | 122 | 635* | 98 | 16 | 12 | 56 | 84 | 93 | 105 | 696 | 2405 |
| 1971-72 | 333 | 144 | 316 | 1160* | 107 | 110 | 92 | 78 | 74 | 50 | 54 | 195 | 2713 |

*Peak light trap catches

Table 3. Evaluation of insecticides for the control of major insect pests of faba bean (Hudeiba 1974).

| Insecticides | Spodoptera damage in degrees | Thrips damage in degrees | Aphid damage transformed into $\sqrt{x+1}$ | Seed yield kg/ha |
|---|---|---|---|---|
| | S.E. = ± 1.5 | S.E. = ± 3.3 | S.E. = ± 2.1 | S.E. = ± 62.6 |
| Control | 12.9 | 47.3 | 21.3 | 2199.0 |
| Torbidan E.C. | 2.6*** | 41.7 | 8.5*** | 2210.7 |
| Fundal forte W.P. | 6.5*** | 37.5* | 17.4 | 2253.1 |
| Lannate W.P. | 4.7*** | 45.8 | 9.4*** | 2275.7 |
| Neoron E.C. | 8.1* | 49.1 | 8.9*** | 2348.8 |
| Endosulfan E.C. | 5.3*** | 41.3 | 5.2*** | 2358.6 |
| Ekatin E.C. | 13.1 | 40.6 | 1.8*** | 2410.7* |
| Nexion E.C. | 15.0 | 42.9 | 3.5*** | 2430.0* |
| Azodrin E.C. | 9.5 | 38.5 | 3.0*** | 2512.1*** |
| Nogos E.C. | 5.9** | 45.4 | 4.4*** | 2529.5*** |
| Diazinon E.C. | 4.4*** | 42.1 | 2.0*** | 2545.7*** |
| Anthio E.C. | 11.9 | 38.2 | 1.6*** | 2569.5*** |
| Supracide E.C. | 8.6* | 33.9** | 2.3*** | 2695.5*** |
| Folimat E.C. | 11.2 | 33.5** | 1.4*** | 2707.6*** |

*Control :* Siddig (13), found that Supracide 40% E.C. at 3.57 litres product/ha and Folimat 50% E.C. at 2.14 litres product/ha were effective insecticides against this pest (Table 3).

## C. The Cowpea Aphid (Aphis craccivora Koch)

Faba beans, ground nuts, lucerne, cowpeas, pigeon peas, lentils, cotton and a number of wild plants are host plants of the cowpea aphid. Siddig (14) reported **RB 29** as the least infested variety compared to four other promising varieties of faba beans, but the difference in infestation was not significant.

*Damage and economic importance :* Colonies of the cowpea aphid feed on the tips and leaves of tender shoots. Leaves curl and shoots become stunted. Moreover, heavy indirect injury is caused by transmission of the Sudanese broad bean mosaic virus (Table 1). Abu Salih *et al.* (1) reported an increasing infection with this disease from September to January as a result of an increasing infestation with the cowpea aphid during the same period. They further reported a highly significant correlation between the disease infection and yield of faba beans. Thus late sowings of the crop expose it to higher infections with this aphid species and diseases, and consequently higher losses in yield.

*Bionomics :* The cowpea aphid appears on leguminous crops, particularly field beans, during September and produces a number of parthenogenic generations in winter. The population increases from September to January and then decreases, probably with the deteriorating nutritional conditions of the crop. The pest then migrates to lucerne for the summer.

*Natural Enemies :* Lady birds, e.g. *Coccinella undecimpunctata* L., and an unidentified syrphid fly were observed as predators of the colonies of this pest (7).

*Control :* Siddig (13) reported Folimat 50% E.C. at 2.14 litres product/ha and Supracide 40% E.C. at 3.57 litres product/ha as effective insecticides against this pest (Table 3).

## D. Bruchids
*Bruchus elnairensis PIC, Callosobruchus maculatus F :* The two Bruchid spp. inflict serious damage on stored leguminous seeds in the Sudan. These include faba beans, peas, cowpeas and lentils.

Mudathir (4) reported 30% damage on faba beans, 26% on chickpeas, 19% on peas; 5% on lentils; 2% on lupins and zero on haricot beans.

Siddig (11) found that RB 29 was significantly more susceptible to bruchid damage than the other nine cultivars which he tested. Cultivars BM/9/3, RB 30 and BF/47/4 were the least damaged, although the difference in damage was not significant.

*Damage and economic importance :* The females lay their eggs in the field, on or in the pods. Hatching occurs in the store and young larvae penetrate the seed to feed inside it until pupation. When adults emerge, they eat their way out, leaving circular exit holes in the seeds. Such damage of seed can result in 30% loss of seeds in three to four weeks.

Table 4. Response of 10 cultivars of faba bean to artificial bruchid infestation (Hudeiba, 1972).

| Cultivar | Mean no. of bruchids per cultivar |
|---|---|
| | S.E. = ± 35.4 |
| BF, 2, 7 | 153 ab |
| 1W | 195 ab |
| 188 | 109 ab |
| BF/47/4 | 105 b |
| RB 29 | 229 a |
| RB 30 | 76 b |
| BF, 2, 2 | 174 ab |
| Giza 1 | 135 ab |
| BM, 9, 3 | 91 b |
| Siliam | 117 ab |

Figures followed by the same letter are not significantly different (P ≤ 0.05).

*Control :* Clean cultivation, crop hygiene and early harvesting can be useful in reducing infestation in the field.

In stores, fumigation with methylbromide is effective.

*Effect of controlling major pests on the yield of faba beans*

Siddig (13) reported that Folimat 50% E.C. at 2.14 and Supracide 40% E.C. at 3.57 litres product/ha, significantly increased the yield of faba beans when sprayed twice at flowering against Spodoptera and thrips, and twice in January against aphids.

Mudathir and Kannan (5) reported similar increases in yield as a result of Spodoptera and aphid control.

*Timing and frequency of spraying*

Siddig (13) also investigated the frequency and timing of insecticides for the economic control of the major pests of faba beans. He reported that one spray against Spodoptera and thrips 21 days from sowing followed by another one on January 7 against aphids, significantly increased the yield of the crop and were comparable to two sprays against Spodoptera and thrips followed by two sprays in January against aphids (Table 5).

*Joint control of major pests and powdery mildew*

In field trials to investigate the effect of a joint control of major pests of faba bean and powdery mildew disease, Siddig (10) found that six sprays of Sofril or Sulfinette reduced powdery mildew significantly, but did not increase the yield to any significant level. On the other hand, all treatments in which an insecticide was involved, significantly increased the yield of faba beans.

Table 5. Timing and frequency of insecticide application for the control of major pests of faba bean

| Treatments | Yield (kg per hectare) 1973-74 S.E. = ± 98.8 | 1974-75 S.E. =± 70.0 |
|---|---|---|
| Control | 1906.2 | 2129.8 |
| One early spray (Nov. 2) | 2235.7* | 2580.9*** |
| two early sprays (Nov. 2 and 16) | 2104.0 | 2585.0*** |
| One late spray (Jan. 7) | 1875.7 | 2471.9*** |
| Two late sprays (Jan. 7 and 21) | 1937.6 | 2354.8** |
| One early spray and one late spray | 2265.4 | 2721.9*** |
| Two early sprays and one late spray | 2111.7 | 2900.7*** |
| One early spray and two late sprays | 2036.9 | 2616.4*** |
| Two early sprays and two late sprays | 2288.3 | 2725.7*** |

REFERENCES

1. Abu Salih, H.S., Ishag, H.M., Siddig, S.A. 1973. Effect of sowing date on incidence of Sudanese broad bean mosaic virus in and yield of *Vicia faba* L. Ann. App. Biol. 74, 371-378.
2. Hussein, M.H. 1963. Pest survey of crops grown at Hudeiba. Ann Rep. Agric. Res. Div., Sudan, 224.
3. Magzoub, O.B., Venkatraman, T.V. 1968. Insect parasite complex of barseem army worm *Spodoptera exigua* HB (Lepidoptera, Noctuidae). Entomophaga 13, 151-158.
4. Mudathir, K. 1976. Susceptibility of different leguminous seeds to bruchid attack (unpublished).
5. Mudathir, K., Kannan, H.O. 1976-79. Chemical control of *Spodoptera exigua* and *Aphis craccivora* on broad bean (unpublished).
6. Remane, F. 1961. Pest survey of crops at Hudeiba. Ann. Rep. Hud. Res. Sta., Sudan.
7. Schmutterer, H. 1968. Pests of crops in Northeast and Central Africa. Gustav Fisher Verlag, Stuttgart, Portland, U.S.A.
8. Siddig, S.A. 1967. Pests survey in the Northern Province of the Sudan. Proc. 10th Agric. Res. Coll. Hud. Res. Sta. Sudan.
9. ---------------- 1967-72. Seasonal population changes of the lesser army worm *Stodoptera exigua* (HB) in the Northern Province of the Sudan (unpublished).
10. -------------- 1971-74. Evaluation of insecticides for the control of *Spodoptera exigua*, thrips and aphid in faba beans. Ann. Rep. Hud. Res. Sta., Sudan.
11. -------------- 1972. Susceptibility of ten varieties of faba bean to bruchid attack (unpublished).
12. Siddig, S.A., Abu Salih, H.S. 1972. Studies on Sudanese broad bean mosaic virus (SBBMV) Ann. Rep. Hud. Res. Sta. Sudan.
13. Siddig, S.A. 1978. Recommendation of pesticides for the control of pests attacking faba bean. Ann. Meet. Pests and Diseases Committee, Agric. Res. Corp. Sudan.
14. ---------------- 1980. Differential susceptibility of five promising varieties to major pests of faba bean (in press).

## 30. BREEDING FOR RESISTANCE TO APHIDS

FRED A.J. KLINGAUF

*Professor of Entomology, Head of the Institute for Biological Pest Control, Federal Biological Research Center for Agriculture and Forestry, Heinrichstrasse 243, D-6100 Darmstadt, Germany*

Most phytophagous insects attack a specific and limited group of host plants. Genuine polyphagous species, being noxious to a great number of different plants, are far less numerous. Specialization of insects to certain host plants implies a close adaptation to the specific properties of the food : (i) The insect must have mechanisms to recognise its host plant among the overwhelming number of non-hosts. Certain properties of the host may act as key stimuli, e.g. a specific shape or plant substances. On the other hand, repellents may hinder the infestation of plants. (ii) Furthermore, the insect has to be physiologically adapted to the characteristic qualities of the food.

Some modern approaches to plant protection are derived from these interrelationships between insects and plants : Attractive colours as well as attractive or repulsive chemicals are used in prognosis and control (colour traps, lures, as additives to insecticides, repellents).

So far the most important advantage taken of host plant specialization is in breeding for plant resistance. This discipline aims at using genetically fixed properties or induced changes of the plant which complicates or hinders recognition and/or diminishes nutritional quality of the host plant.

The main aphid pests of faba beans are two green and two black species which are common in dry areas. The fact that both pairs of species are not differentiated may lead to failures in controlling the pests. In contrast to the black aphids, the green ones are often of minor importance. The distinguishing feature of the apterae of the green species, *Acyrthosiphon pisum* (Harris) and *A. gossypii* Mordvilko, is the length of the cauda in relation to the length of the siphunculi. While the cauda of *A. pisum* is about four-fifths the length of siphunculi, the cauda of *A. gossypii* is only about one-half that length (Fig. 1). Both species are slim; body length being about 3.3-5.2 mm.

The apterae of the black species, *Aphis fabae* Scopoli and *A. craccivora* Koch, are more roundish with a body length of about 1.6-2.6 and 1.2-2.3 mm, respectively. The apterous adults of *A. craccivora* are shining black, whereas those of *A. fabae* are mat black. After maceration in boiling alkaline solution, *A. craccivora* shows a striking dorsal pigmentation which is less extended in *A. fabae* (Fig. 2).

Species of the genus *Aphis* are characterised by marginal tubercles of the abdominal segments I and VII. These and further tubercles are important for discriminating the alatae of *A. fabae* and *A. craccivora* from similar winged aphids. In contrast to some other *Aphis*

*G. Hawtin & C. Webb (Eds.): Faba Bean Improvement*
©*1982 ICARDA. ISBN -13:978-94-009-7501-9*

species, *A. fabae* and *A. craccivora* show only sometimes, and then very little, additional tubercles on segments II-IV. Like the apterae, alatae of *A. craccivora* possess more pigments in the siphunculi region than *A. fabae* (Fig. 3).

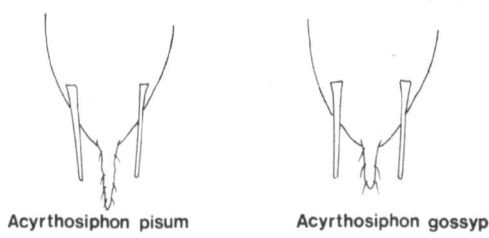

Acyrthosiphon pisum          Acyrthosiphon gossypii

Fig. 1.    Identification of *Acyrthosiphon pisum* (Harris) and *A. gossypii* Mordvilko (apterae) by different length of the siphunculi

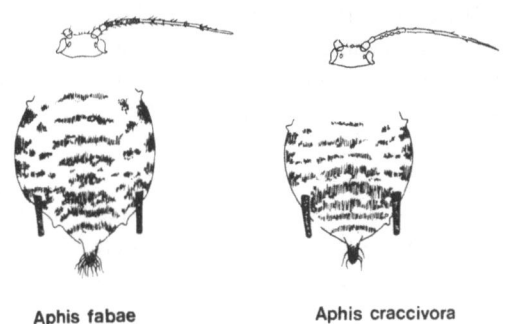

Aphis fabae          Aphis craccivora

Fig. 2.    Identification of *Aphis fabae* Scopoli and *A. craccivora* Koch **(alatae)** by different pigmentation **patterns, length** of cauda and siphunculi, number of caudal hairs, and occurrence of hairs on antennae (20).

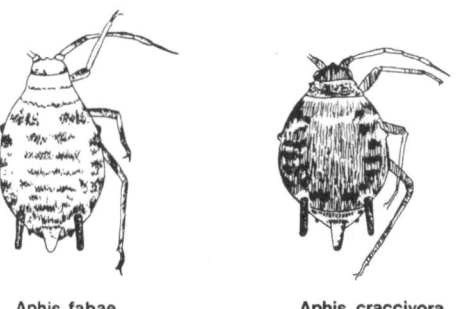

Aphis fabae          Aphis craccivora

Fig. 3.    Identification of *Aphis fabae* Scopoli and *A. craccivora* Koch (apterae) by different pigmentation after maceration (19).

Furthermore, in winged *A. fabae* the relative length of siphunculi is 0.1-0.125 of body length and 1.1-1.5 of cauda length. Other *Aphis* species with similar length of siphunculi have more tubercles, so that both characteristics are necessary for comparison. In winged *A. craccivora,* the siphunculi measure 0.17 of body length and 1.5 of cauda length. *A. fabae* has more than 11 caudal hairs (like some other *Aphis* species with more tubercles), whereas *A. craccivora* has fewer caudal hairs. At 40 x magnification, there are additional signs for identification, since the antennae of *A. fabae* have distinct hairs in contrast to those of *A. craccivora* (Fig. 3) (19) (20). Besides the above mentioned species some other aphids may occur on faba beans, e.g. the widely distributed green peach aphid, *Myzus persicae.*

*A. pisum* is widely distributed in moderate and warm climates. *A. gossypii* occurs only in warmer climates but seems to be far less frequent in the Near and Middle East than *A. pisum.*

Due to their normal high frequency and stronger noxious effects to plants, both black species are regularly more important than the two *Acyrthosiphon* species. Likewise both black species show somewhat different climatic preferences. In cooler regions *A. fabae* dominates. In the warmer regions of Europe and Asia both species are found, sometimes in mixed colonies, e.g. in the Jordan Valley and in Syria (Fig. 4). In dry and hot climates, *A. craccivora* becomes more dominant. Thus all colonies of black aphids on faba beans and other legumes found in spring 1980 in the State of Bahrain and in Oman*consisted of *A. craccivora.* The overlapping pestographs of *A. fabae* and *A. craccivora* (Fig. 5) clearly show this shifting of dominance of the two species from cooler to warm and dry regions.

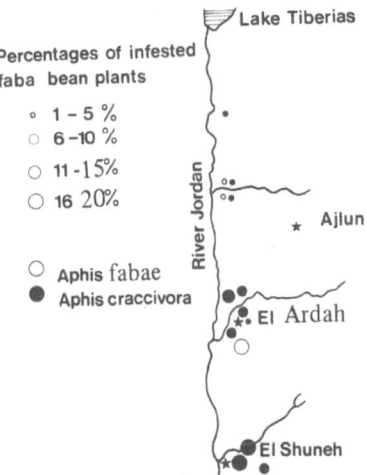

Fig. 4.    Distribution of *Aphis fabae* and *A. craccivora* on faba beans in the Jordan Valley, and percentages of infested plants, March 1980.

*Observations in Oman by Dr. M. Blaeser, personal communication.

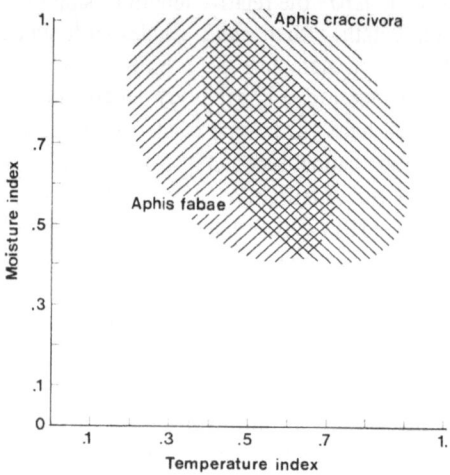

Fig. 5. Pestographs of *Aphis fabae* and *Aphis craccivora (A. craccivora* after Gutierrez *et al.* (1974)), in : Climate and Food : climatic fluctuation and U.S agricultural production. The Nat. Res. Council, Nat. Acad. Sci., Washington, D.C. 1976.

The behaviour of a migratory winged aphid is presented in (Fig. 6 I) (17). The time between last moult and start of flight, the teneral period, is characterised by the resting mood. In this stage, the aphid become ready for flight. The newly winged aphids react positively to the shortwave light of the sky and take off under suitable meteorological conditions (17) (5) (6). In this flight mood the aphid is unable to recognise a host plant. Aphids artificially placed onto a plant walk around and usually try to take off again, regardless of whether it is a host or non-host. Moericke calls this phase of migration distance flight which sometimes carries aphids over hundreds of kilometers with the help of the wind. With increasing duration of flight and/or passive drifting, a change of mood takes place. The aphids now react to the longwave light reflected from brown soil and contrasting green plants. They now fly short distances over the vegetation and become increasingly ready for alighting on plants, thus indicating their attack mood and the beginning of the attack flight. In this phase, aphids may visit different plants, taste them and then resume flying.

After alighting, aphids show a characteristic probing behaviour during which only the surface (cuticle) or at most the epidermis of a leaf is touched by the labium and the stylets. As shown in Fig. 6 II the labium in this stage is not shortened as it is during piercing. Non-hosts are often left after a series of probes and walkings on the leaf. Only after the aphid has recognised positive stimuli (e.g. from a suitable host and also some non-hosts) and the settlement mood has reached a certain level, the aphid tries to pierce the leaf tissues towards the sieve tubes (Fig. 6 III).

According to experiments with *Megoura viciae* (1) and with *A. fabae* (2), the phloem is reached after 8 min. at the earliest. After a penetration time of 1 hour, about 60% of *M. viciae* have pierced the phloem, whereas some individuals need more than 3 hours. The stylet tracks are visible by a salivary canal formed during penetration. The track in the

inner tissues is often branched like a bush which illustrates the aphids' difficulties in finding the right way. Sometimes the aphid leaves the initial point of piercing and moves on to
another, perhaps better place. The acceptance of a host plant depends on the possibility
of reaching the phloem and on the qualitative and quantitative properties of the phloem
sap. During settlement, offspring are produced. If the aphid is still capable of flying, it may
enter a second attack flight and settlement (Fig. 6 I)).

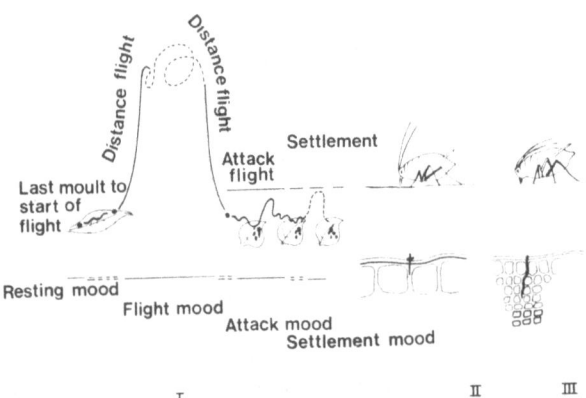

Fig 6(I)Diagram of the behavioural stages of the winged aphids; (II) probing aphid and
salivary flange; (III) piercing aphid with shortened labium and salivary canal reaching the phloem. (17).

Not all the winged aphids start for a long distance flight after moulting. Often a short
migration period leads them only to adjacent plants or fields, or the aphids may even remain on its host. Likewise the unwinged aphids remain settled or, particularly young adults
and $L_4$ of large *A. pisum* colonies, may show a migration mood (9). This finally leads them
to young recently unfolded leaves of the same or other plants nearby, whereas the older
and the very young leaves of faba beans are avoided (Fig. 7). Generally similar phases of
the cycle described above are passed. For further details of flight behaviour and penetration of aphids see reviews by Kring (1972) and Pollard (1973).

Fig. 7.   Upward movement of aphid colonies during growth of the faba bean (9).

The different phases of infestation and separation may be described by a cycle (Fig. 8) following a scheme given by H.J. Müller (21). The cycle may start with the exploitation of the host after settlement. This period ends with the exhaustion of the host plant, lack of essential food substances, water shortage and by increasing interference due to the enlargement of the colonies. Wheareas well-conditioned colonies of aphids under preferred food and weather conditions mainly consist of apterae, the absence or lack of certain amino acids and the irritation caused by crowding, stimulate the formation of increasing numbers of alate individuals, which start a new cycle of dispersion.

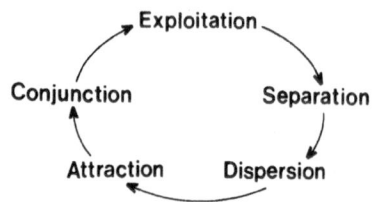

Fig. 8.    Cycle of different phases in aphid - host - plant relationships (21).

The main stimuli for dispersion are of an endogenous nature, caused by the lowered threshold for locomotion during separation. Besides meteorological factors (light intensity, wind speed, temperature), shortwave light increases flight activity.

Aphids, ready to alight, react positively to longwave light (attraction). Whether or not this is repellency of shortwave light or a greater response to longwave light is not clear yet : yellow and to a lesser extent green colours attract the parasites so that they can be caught in yellow traps, which are widely used for prognosis. On the other hand, white surfaces that reflect shortwave light or an aluminium surface both hinder alighting. This is sometimes applied to protect crops against aphids transmitting non-persistent viruses, where chemicals fail.

During attack flight hosts as well as non-hosts are visited. Only in some cases aphids were found to favour the hosts by specific odor or slight colour variations in tint or hue (18) (24) (26).

In the phase of conjunction, probing and penetration behaviour is stimulated furthermore by yellow and to a lesser extent green. Beyond colour, chemical properties of the plant surface (probing) and the inner tissues (penetration, sucking) are of primary importance. Besides other chemicals, probing behaviour is elicited by the composition of the waxy leaf surface. Nöcker–Wenzel et al. (23) identified about 20 different substances in faba bean waxes, where n-alkanes apparently play a prevailing role in host selection of some pests. The homologous n-alkane compounds with a C-chain of $C_{27}$, $C_{29}$ and $C_{31}$ are most abundant. Artificial compositions of these n-alkanes act as arrestants, whereas n-$C_{29}$ $H_{60}$, the main n-alkane component of Brassica crops, was shown to be deterrent (Fig. 9).

The waxes of        plant species consist mainly of the same components but differ in quantities. The patterns of waxy components of related plants are usually similar so that chemotaxonomists may use them as "finger prints". This specific composition supports the hypothesis of the role of waxy components in host selection during probing behaviour. They may act together with small quantitites of primary and secondary plant substances. In A. fabae , the phenolic amino acid L-Dopa and different phenols as secondary plant substances promote conjunction and have an arrestant effect (Table 1) (4) (3). Physical properties of the plant surface (wall thickness, hairs) may also influene the acceptance of a plant as host.

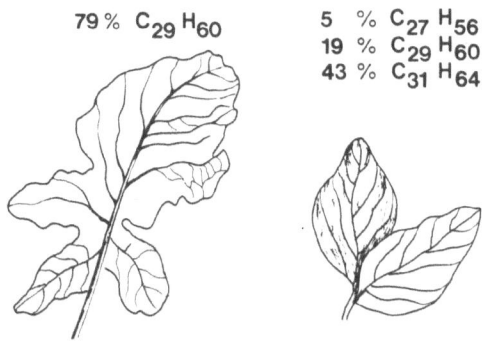

79 % $C_{29}H_{60}$

5 % $C_{27}H_{56}$
19 % $C_{29}H_{60}$
43 % $C_{31}H_{64}$

Brassica n. napus          Vicia faba

Fig. 9.  Percentages of the main n-alkanes in the alkane fractions of the cuticle waxes of faba beans and rape (alkane finger print). Some insect pests associated with legumes prefer the n-alkane pattern of *V. faba* (12).

Table 1. Phenolic compounds having different degrees of effectiveness as arrestants (3).

| | |
|---|---|
| very high effectiveness | Amino acids<br>L-Dopa (L-B-3,4-dihydroxyphenylalanine)<br>L-Ornithine<br>L-Canavanine |
| High effectiveness | Phenols<br>Phenol<br>Hydroquinone<br>Phenolether<br>Guaiacol<br>Phenolic acid<br>3-(3,4-Dihydroxyphenyl)-propionic acid |
| medium effectiveness | Phenols<br>Pyrocatechol<br>Phloroglucinol (1,3,5-Trihydroxybenzene) |
| low effectiveness | Phenols<br>Resorcinol<br>Pyrogallol (1,2,3-Trihydroxybenzene)<br>1,2,4-Trihydroxybenzene<br>Phenolamines<br>Tryamine<br>Dopamine (3-Hydroxytyramine)<br>2,4,5-Trihydroxyphenylamine<br>3,4,5-Trihydroxyphenylamine<br>Flavonoids<br>Apigenin<br>Luteolin<br>Quercetin |

The conjunction is promoted if the aphid visits non-hosts before alighting on a host during attack flight (Fig. 10). The stop on a non-host apparently decreases the threshold of host-specific stimuli (8).

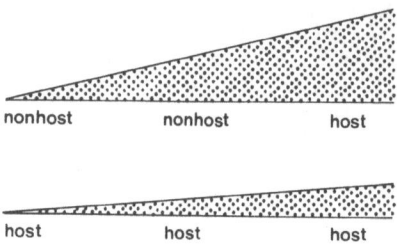

Fig. 10. Differences in the increasing settlement mood during attack flight due to the mode of sequences of visited non-host and host plants (8).

It is not known which stimuli direct piercing. Besides physical properties of the tissue, a gradient from acid to alkaline from outside to inside may support penetration (Fig. 11) (10).

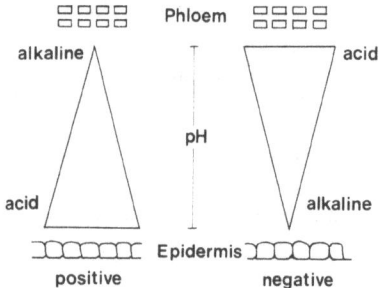

Fig. 11. Preference of *Acyrthosiphon pisum* to a pH-gradient reaching from light acid (plant surface) to light alkaline (phloem sap).

The exploitation depends on the compositon of the phloem sap, the sugar and amino acids of which have a positive synergistic effect (15) (16). Amino acids without functional groups are more attractive for *A. pisum* than those with a phenolic or mercapto-component (11).

In plant breeding, it is impossible to test the resistance of a plant line by ordinary infection tests as is done with pathogens. One and the same aphid (during the sequence of different phases) can behave like another individual. As mentioned above, aphids during flight or walking mood will not react to a plant be it a host or not. Forcing aphids in contact with a plant can lead to a settlement, though there is no attraction by specific stimuli. Thus an unattractive plant may be classified as a suitable host. The right conditioning of experimental aphids and adequate tests are essential in resistance experiments. Field experiments, though time consuming and depending on natural infestation, offer the best conditions.

Chances for resistance breeding apparently exist in altering the stimuli for attraction, conjunction and perhaps exploitation (21). Since during exploitation the aphids require nearly the same compounds being also essential for the quality of yield (e.g. amino acids for protein synthesis), this phase seems not to be generally suitable for pest management. For example the old German broad bean variety "Rastatt" shows a rather high degree of resistance, apparently resulting in antibiosis. This variety has a lowered amino acid production. Its yield is nearly the same as in other varieties, but the period of ripening is prolonged.

Better prerequisites are offered by varieties with a short vegetation period. This is usually correlated with a faster hardening of the tissues and therefore with increased resistance as the aphids prefer soft tissues (Fig. 7). Moreover, a short growing period reduces the infestation period especially if the aphids arrive later in season. This implies an indirect mode of plant resistance based on partial coincidence between suitable plant stages and the pest infestation period. Plants which resist cool weather in springtime, allow an early seeding which shortens the infestation period, provided the variety does not need an elongated growing time (Fig. 12).

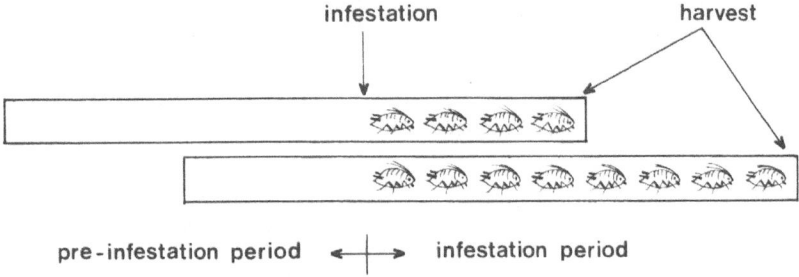

Fig. 12. Different length of infestation periods due to seeding time.

Suitable ways of interfering with the phases of attraction and conjunction, both being connected with token stimuli not important for human food (21), include the formation of thick white cell walls, intensification of grey leaf colour, and attenuation of arrestants. Faba bean varieties which have a low orthodiphenol content showed a decreased infestation (3). A special mechanism of resistance to aphids in alfalfa wasdescribedby McMurtry and Standford (14). With the highly resistant variety C 84, the spotted alfalfa aphid *Therioaphis maculata* often failed to reach the phloem. Many salivary sheaths showed blind endings within the leaf tissues some distance from the sieve tubes. Since the plants are attractive and elicit normal probing behaviour, the resistant clone is acting as a trap, causing the aphids to die of starvation. (For further behavioural differences see 1, 22).

The complex interrelationships between host and parasite, briefly demonstrated here in successive phases, may exemplify that acceptance or rejection of a plant by the aphid results from an intimate interaction of different stimuli. The degree of plant resistance or susceptibility results from the present status of balance between both competitors. Thus, under biological aspects a shifting of the balance in favour of the plant seems possible, and should be aimed-at. Since classical biological control methods of aphids often are not effective enough, the breeding for resistance is an important biological alternative.

Acknowledgement

I am grateful to Dr. A.M. Huger and Dr. E. Bode for their help in preparing the manuscript.

REFERENCES

1. Ehrhardt, P. 1963. Untersuchungen über Bau und Funktion des Verdauugstraktes von *Megoura viciae* Bckt. (Aphidae, Homoptera) unter besonderer Berücksichtigung der Nahrungsaufnahme und der Honigtauabgabe. Z. Morph. Ökol. Tiere 52, 597-677.

2. Hennig. E. 1959. Untersuchungen am Saugakt von *Aphis fabae* Scop. mit $P_{32}$. Naturwissenschaften 46, 410-411.

3. Jördens, D. 1977. Zur Wirkung pflanzlicher Oberflächen-Substanzen auf den Primärbefall durch die Schwarze Bohnenblattlaus, *Aphis fabae* Scop. (Homoptera : Aphididae). Diss. Faculty of Agriculture, Univ. Bonn, 130 S.

4. Jördens, D., Klingauf, F. 1977. Der Einfluss von L-Dopa auf Ansiedlung und Entwicklung von *Aphis fabae* Scop. an synthetischer Diät. Med. Fac. Landbouww. Rijksuniv. Gent, 42/2, 1411-1419.

5. Kenedy, J.S., Booth, C.O. 1963a. Free flight of aphids in the laboratory. J. Exp. Biol. 40, 67-85.

6. ---------------------------------- 1963b. Co-ordination of successive activities in an aphid. The effect of flight on the settling responses. J. Exp. Biol. 40, 351-369.

7. Kindler, S.D., Staples, R. 1969. Behaviour of the spotted alfalfa aphid on resistant and susceptible alfalfas. J. Econ. Entomol. 62, 474-478.

8. Klingauf, F. 1972. Modellversuche zum Einfluss der kurzen Pflanzenbesuche auf das Wirtswahlverhalten während des Befallsfluges von Aphiden. Z. angew. Entomol. 70, 352-358.

9. -------------- 1976. Die Bedeutung der "Stimmung" im Leben phytophager Insekten am Beispiel des Wirtswahl-Verhaltens von Blattläusen. Z. angew. Entomol. 82, 200-209.

10. Klingauf, F., El-Sayed, A.M.K. Stechborsten-Penetration bei *Acyrthosiphon pisum* (Harris) und *Myzus persicae* (Sulz.) (Homoptera, Aphidina) an Blattmodellen mit Saccharose- und pH-Gradienten. (in preparation).

11. Klingauf, F., Nöcker-Wenzel, K. 1972. Einfluss von Aminosäuren auf das Wirtswahlverhalten von *Acyrthosiphon pisum* (Homoptera : Aphididae) unter besonderer Berücksichtigung ihres chemischen Aufbaus. Ent. exp. appl. 15, 274-286.

12. Klingauf, F., Nocker-Wenzel, K., Röttger, U. 1978. Die Rolle peripherer Pflanzenwachse für den Befall durch phytophage Insekten. Z. Pflanzenkrankh. Pflanzensch. 85, 228-237.

13. Kring, J.B. 1972. Flight behaviour of aphids. Ann. Rev. Entomol. 17, 461-492.

14. McMurtry, J.A., Standford, E.H. 1960. Observations of feeding habits of the spotted alfalfa aphid on resistant and susceptible alfalfa plants. J. Econ. Entomol. 53, 714-717.

15. Mittler, T.E., Dadd, R.H. 1964. Gustatory discrimination between liquids by the aphid *Myzus persicae* (Sulzer) . Ent. exp. appl. 7, 315-328.

16. ----------------------------------- 1965. Differences in the probing responses of *Myzus persicae* (Sulzer) elicited by different feeding solutions behind a parafilm membrane. Ent. exp. appl. 8, 107-122.

17. Moericke, V. 1955. Über die Lebensgewohnheiten der geflügelten Blattlause (unter besonderer Berücksichtigung des Verhaltens beim Landen). Z. angew. Entomol. 37, 29-91.

18. -------------- 1969. Hostplant specific colour behaviour by *Hyalopterus pruni* (Aphididae). Ent. exp. appl. 12, 524-534.

19. Müller, F.P. 1969. Aphidina - Blattläuse, Aphiden. In: Stresemann, E. : Exkursionsfauna von Deutschland.Insekten - Zweiter Halbband. Wirbellose II/2. Volk und Wissen, VEV Berlin, 51-141.

20. -------------- 1970. Zur Kenntnis der Schwarzen Bohnen- oder Rübenblattlaus vom Gesichtspunkt der Prognose und des Warndienstes. Nachrichtenbl. Deutsch Pflanzenschutzd. DDR, 24, 96-100.

21. Müller, H.J. 1965. Das Beziehungsgefüge zwischen Blattläusen und (landwirtschaft-lichen) Kulturpflanzen als Beispiel eines Zyklus autokologischer Phasen. Der Züchter 35, 14-24.

22. Nielson, M.W., Don, H. 1974. Probing behaviour of biotypes of the spotted alfalfa aphid on resistant and susceptible alfalfa clones. Ent. exp. appl. 17, 477-486.

23. Nöcker-Wenzel, K., Klein, W., Klingauf, F. 1971. Isolierung von Oberflächensubstan-zen aus *Vicia faba* L. im Rahmen von Untersuchungen zur Insekt-Wirtspflanzen Beziehung. Tetrahedron Letters No. 46, 4409-4412.

24. Petterson, J. 1970. Studies on *Rhopalosiphum padi*. I. Laboratory studies on olfacto-metric responses to the winter host *Prunus padus*. Lantbruks-Hoegsk. Ann. 36, 381-399.

25. Pollard, D.G., 1973. Plant penetration by feeding aphids (Hemiptera, Aphidoidea) : a review. Bull. Entomol. Res. 62, 631-714.

26. Tamaki, G., Butt, B.A., Landis, B.J. 1970. Arrest and aggregation of male *Myzus persicae*. Ann. Entomol. Soc. Amer. 63, 955-960.

1. Mohr, H.F. 1962. Die Fragmentierung zwischen Blödsinn und Sinn vi...ulla...
Ihnen Asthmapillen als Beispiel einer Zwangsneurose bei Nachlassen. Ne...chen. Dtsch. Z....
81, 17-24.

2. Nelson, M.W., Poon, H. 1976. Feeding behaviour at the type of the spotted. Bull...
...spine correlation and susceptible of while signal. Zool. Soc., xxi, 13, A15-A55.

3. Junker-Waucik, R., Kleij, W., Klingel, H. 1971. Bewegung und Verhaltensweise...
...tier. Inf. Doc. 1 ... 557 ...

31. HOW YIELD STABILITY CAN INFLUENCE FARMERS' DECISIONS TO ADOPT NEW TECHNOLOGIES: THE CASE OF FABA BEAN PRODUCTION IN EGYPT.

DAVID NYGAARD * AND ABDEL-MAWLA M. BASHEER**


*Farming Systems Program, ICARDA, P.O.Box 5466, Aleppo, Syria
**Agricultural Economics Research Institute, Agricultural Research Center, Dokky, Cairo, Egypt

Agricultural production is not stable; it is risky and uncertain. Instability is one factor that inhibits farmers' adoption of new agricultural techniques. Yet, unlike other factors that affect the adoption process, it is not as easily understood by agricultural researchers, extension workers and policy makers who are trying to encourage the adoption of new techniques to improve agricultural productivity. This paper aims to (i) clarify the concept of risk and uncertainty, (ii) illustrate how it affects farmers' decisions, and (iii) show the implications it has for developing new techniques and recommendations for farmers.

The concept presented here is generally applicable to farmers anywhere who are confronted with a decision regarding an agricultural activity. While irrigated crop production tends to be more stable than rainfed crop production, yields can still be surprisingly variable.

Faba bean production in Egypt is affected by many things in addition to water availability. If we take into account the impact of risk and uncertainty early in the research process, the research results and subsequent recommendations made to farmers are more likely to be adopted. There is no question that the faba bean crop is important to individual Egyptian farmers and to Egyptian agriculture. Indeed, its importance is going to increase.

The first section of this chapter describes how risk and uncertainty affect the decision-making process. It discusses how certain characteristics of agriculture and of farmers, influence decisions on different inputs such as the choice of a seed variety and the amount of fertilizer to use.

Section two describes faba bean production in Egypt in the light of production determinants and farmer characteristics that are shown to be important in the first section. Examples from studies of farmers' behavior in other developing countries are used to support incomplete aspects of the Egyptian production picture.

The third section describes the implications of the concept on faba bean research and production in Egypt. Suggestions are made for (i) understanding farmers, (ii) researching new technology, (iii) encouraging the diffusion of new technology, and (iv) developing a supporting infrastructure and institutional system to encourage high levels of productivity.

G. Hawtin & C. Webb (Eds.): Faba Bean Improvement
©1982 ICARDA. ISBN -13:978-94-009-7501-9

## THE CONCEPT OF RISK AND UNCERTAINTY

Not long ago, production scientists focused on research efforts that increased yields, and defined optimal input quantities as those where yields were maximized. Economists have encouraged production scientists to consider profit-maximizing input quantities instead. Now a new body of research suggests that they strive to not only maximize profit but simultaneously minimize risk. This chapter explains the difference between these decision/rules and shows how incorporating risk into the decision-making process affects input use and the choice of new techniques.

### Input Use

Fig. I illustrates both yield- and profit-maximizing input quantities for phosphate on faba bean production. $X_1$ units of $P_2O_5$ will give the maximum amount of grain. However, since $P_2O_5$ is costly and faba beans can be sold for income, a profit maximizer will use fewer units of $P_2O_5$, i.e., $X_2$. This choice depends on the purchase price of phosphate relative to the sales price of beans. Using economic theory, one can show that profit is maximized where the price line, i.e., the price of the input divided by the price of the product, is tangent to the response curve. Currently this ratio in Egypt is 0.19 EL/kg /0.16 EL/kg. at official prices, and 0.33 EL/kg / 0.206 EL/kg at market prices. At this optimum, the increase in yield attributed to the last added unit of the input is just enough to pay for that input. For example, if bean prices increase, the quantity of $P_2O_5$ required to maximize profit will also increase. Higher faba bean prices induce yield increases.

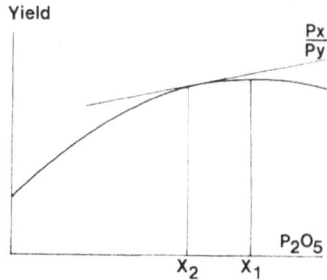

Fig. 1.

The several theories that incorporate risk and uncertainty into the decision-making process include safety-first, focus-loss, lexicographic ordering, and expected utility (15) (17). Results are not radically different from one theory to another, and we only refer to the expected utility theory in this chapter.

The two necessary components for this theory to handle risk and uncertainty are (i) a personal valuation of consequences as reflected by a person's utility function and (ii) personal beliefs about the occurence of uncertain events as reflected in a person's subjective probabilities of outcomes (4). If farmers are uncertain about yields and are risk averse, they will alter their input use from profit-maximizing quantities.*

---

* For simplification we will assume prices are fixed. This is, of course, not always the case and price instability affects decisions as well.

For example, let us assume that such a farmer needs to make a decision on how much phosphate to use on his faba bean plot. Fig. 2 shows that the impact of unstable production is to reduce the optimal amount of $P_2O_5$ used from $X_2$ to $X_3$. Note that the impact is the same as if the price of faba beans were reduced. This effect is attributed to $\theta V$, where $\theta$ measures the farmer's aversion to risk and V measures his perception of the variability in bean yields.

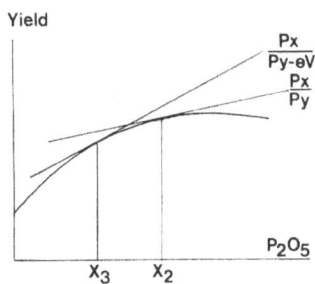

Fig. 2.

The implication of this theory with respect to recommendations to farmers in this case recommending economic optimal amounts of $P_2O_5$, is profound. Two farmers, perhaps even neighbors, may have different perceptions of $\theta$ and V, i.e. , different degrees of risk aversion and perceptions of yield variability, and will consequently use different amounts of phosphate.

*New techniques*

The theory is easily applied to decisions about adopting new techniques. Assume that we have two varieties of faba beans: $V_1$, the traditional variety and $V_2$, a new, perhaps high-yielding, variety. $V_1$ will do all right at low levels of inputs but does not respond to increasing input use, for example, phosphate. On the other hand, $V_2$ does well at high levels of phosphate but does not do well at traditional levels. See Fig. 3.

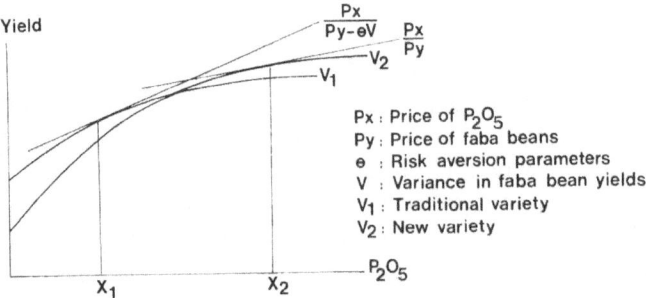

Fig. 3.

A profit-maximizing choice given the price of phosphate, Px, and the price of beans, Py, suggests that farmers would choose $V_2$ and use $X_2$ units of $P_2O_5$. However, if risk aversion and/or yield instability are great enough, a farmer will grow the old variety $V_1$ and use $X_1$ units of $P_2O_5$. Again risk aversion and yield uncertainty have the same effect as a decrease in the price of faba beans. In this case, the effect is strong enough to discourage the adoption of a new variety.

This was shown with wheat varieties in Tunisia (16) and there is tentative evidence that this may be the case with faba beans Giza 2 and Giza 4 in Menia Province, Egypt. In some faba bean on-farm trials last year, the average yield for traditional varieties at low $P_2O_5$ levels was 2.5 t/ha. However we are not sure whether Giza 4 can outyield Giza 2 at high levels of fertilization.

It is more difficult to explain the effect of variety choice if we have a neutral technological change. This is shown in Fig. 4 where the new variety has a parallel but higher response surface to the traditional variety. In this case, there is an increased likelihood that adoption will occur.

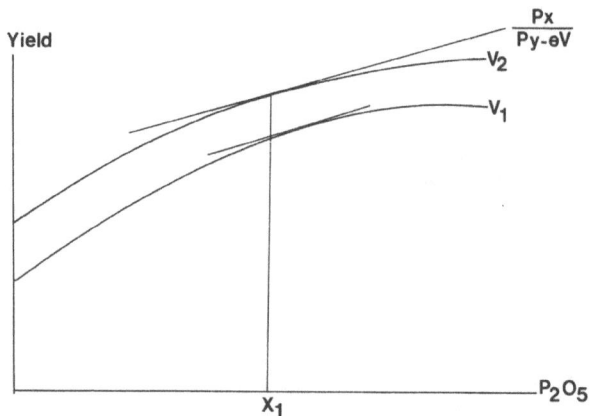

Fig. 4.

However, if a farmer even perceives that the variability in the new variety is greater than in the traditional variety, he may not switch, despite its higher yield. Risk aversion steers him away from the risky choice. It will be interesting to see if this is the case with Giza 1 and Giza 3 in this year's on-farm trials in Kafr Es-Skeikh.

It is worth mentioning at this point that the new technology presented in Fig. 4 is preferable to that in Fig. 3 for the simple reason that the farmer is not required to change his phosphate levels if he adopts $V_2$. Innovation in itself creates uncertainty and simple changes require less learning and faster adjustment to economic optimum (7).

Also while it has been shown that the risk-averse producer, facing uncertain output, reduces his input use from profit-maximizing levels, this need not always be the case. Some inputs, namely disease and pest control, can actually be used at greater levels by such producers in efforts to stabilize yields (9). Therefore, if a farmer has to make a decision regarding the mix of two inputs, e.g., phosphate and pesticide application, he may decrease $P_2O_5$ and increase the pesticide application in comparison with profit maximization levels.

Obviously, all possible variations in farmer behavior have not been covered in this section, but two of the most important effects have been discussed. As stated earlier, the model, based on the theory of expected utility, is only applicable if yields are unstable and farmers are risk averse. The next section discusses whether these assumptions are true for faba bean producers in Egypt.

## YIELD INSTABILITY AND RISK-AVERSE FARMERS

### Yield Instability

Fig. 5 and Table I show that there is a great variability in yields in Egypt at the national level. This variability was greater during 1960-69 than in 1970-80 yet the range in yields during the last 10 years was still 350 kg/ha. These national averages may mask the variability on farmers' fields.

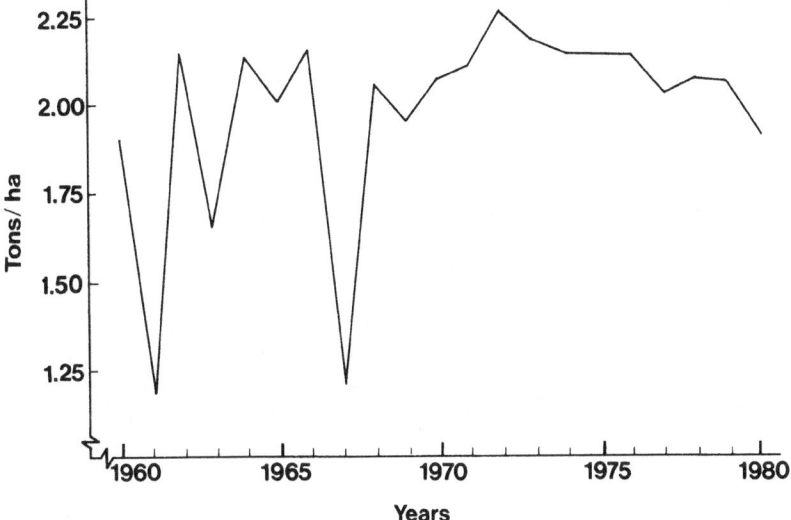

Fig. 5. The average yield of faba beans in Egypt, 1960-80.

Source : Data bank, Central Laboratory for Design and Statistical Analysis Research, Agricultural Research Center, Cairo, Egypt: Unpublished.

Fig. 6 shows the variation for Menia and Kafr Es-Sheikh Provinces, where the Nile Valley Project has on-farm trials. The standard deviation of yields during the past 20 years in Menia is 368 kg/ha compared with 290 kg/ha for the country as a whole.

302

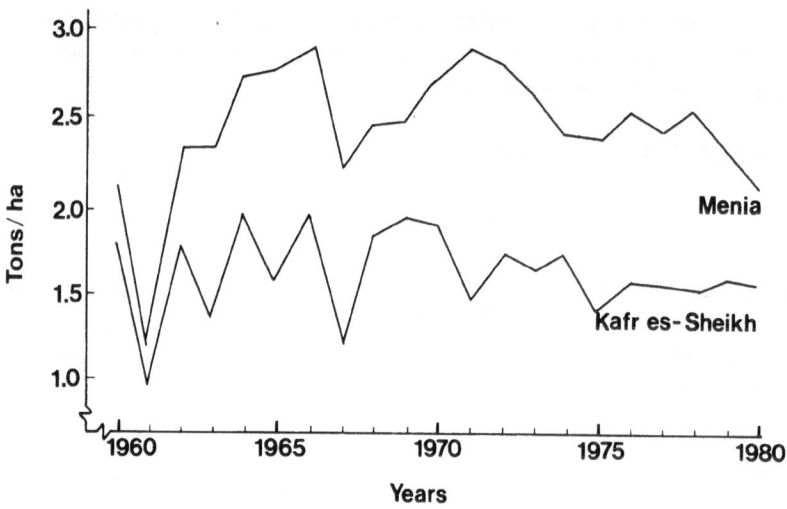

Fig. 6. The average yield of faba beans in Menia and Kafr-es-sheikh Provinces, Egypt, 1960-80.

Source : Data Bank, Central Laboratory for Design and Statistical Analysis Research, Agricultural Research Center, Cairo, Egypt: Unpublished.

However, farmers' perceptions of yield variability are more important to the decision-process than actual yields. Unfortunately, no information exists about such perceptions among Egyptian faba bean producers. Nevertheless, there is evidence elsewhere that perceptions differ from reality.For instance in Tunisia, it was found that wheat farmers perceived (i) yields to be more variable than they actually were, (ii) the variability to be larger for new varieties than old and (iii) the variability to decrease as the farmers gained experience with these varieties (14).

The most important source of yield instability for Tunisian wheat producers is irregular rainfall whereas Egyptian faba bean production is under irrigation, and this considerably decreases climatic uncertainty. Yet farmers in Menia Province last year suggested that yields were variable and this was caused by diseases, pests, problems with the irrigation system, and, most importantly, variable temperatures at the flowering and pod-setting stages.

Table 1.

## FABA BEANS YIELD IN EGYPT
1960 - 1980

| Year | Menia | Yield t/ha Kafr Es-Sheikh | Egypt |
|---|---|---|---|
| 1960 | 2.104 | 1.779 | 1.907 |
| 1961 | 1.202 | 0.942 | 1.188 |
| 1962 | 2.321 | 1.835 | 2.136 |
| 1963 | 2.321 | 1.362 | 1.650 |
| 1964 | 2.708 | 1.953 | 2.117 |
| 1965 | 2.725 | 1.585 | 2.001 |
| 1966 | 2.868 | 1.943 | 2.162 |
| 1967 | 2.193 | 1.174 | 1.202 |
| 1968 | 2.461 | 1.836 | 2.046 |
| 1969 | 2.491 | 1.936 | 1.955 |
| 1970 | 2.744 | 1.899 | 2.072 |
| 1971 | 2.881 | 1.251 | 2.120 |
| 1972 | 2.820 | 1.735 | 2.266 |
| 1973 | 2.659 | 1.659 | 2.192 |
| 1974 | 2.399 | 1.735 | 2.146 |
| 1975 | 2.402 | 1.400 | 2.146 |
| 1976 | 2.540 | 1.570 | 2.146 |
| 1977 | 2.426 | 1.539 | 2.037 |
| 1978 | 2.525 | 1.509 | 2.064 |
| 1979 | 2.311 | 1.583 | 2.056 |
| 1980 | 2.102 | 1.558 | 1.913 |
| Mean | 2.438 | 1.609 | 1.977 |
| Standard Deviation | 0.368 | 0.271 | .290 |

Source: Data Bank, Central Laboratory for Design and Statistical Analysis Research, Agricultural Research Center, Cairo, Egypt: Unpublished.

### Risk Aversion

Again, while there is a shortage of studies in Egypt that have estimated farmers' risk aversion or utility preferences, there is an impressive list of studies in other countries that show farmers to be risk averse. Examples include India (3), Kenya (22),Tunisia(16) and Mexico (13). CIMMYT scientists believe this to be a general rule for agricultural producers in developing countries. They assume farmers to be income-seeking risk-averters and that they approach technological change from this perspective (21).

Research has been done to determine characteristics that are identified with risk-averse farmers. In Mexico, off-farm income, land holdings and affiliation with farmer organisations were inversely related to risk aversion (13). In Tunisia, age was found to be positively correlated with risk aversion while farm size and good soil were significant explanatory variables of risk aversion that were negatively related. Furthermore, risk averse farmers were slower adopters of high yielding varieties (14). It should be noted that no differences in risk aversion were found in India among groups such as big and small farmers, although these farmers were risk averse (3).

Research on this subject should be encouraged in Egypt. Knowledge about the degree of risk aversion among Egyptian farmers could have a significant impact on efforts to improve agricultural production. However until that research is done, there is ample evidence to suggest that risk and uncertainty are important and they must be accounted-for, although the degree of risk aversion is perhaps still an open question. The importance of the subject to researchers who are striving to increase faba bean yields on farmers' fields is clearly great.

## IMPLICATIONS

This section describes how this concept of risk and uncertainty should affect the approach that researchers, policy makers and extension personnel use in their attempts to increase faba bean productivity. The discussion is divided into four categories. The first aims at improving our understanding of farmers. The second suggests changes in research priorities. Third, new ideas important to diffusing new technology are discussed. Finally, the importance of infrastructure development is emphasized.

### Understanding Farmers

First the theory of expected utility (risk and uncertainty) encourages us to recognize that farmers are different from each other. They have different perceptions, different attitudes, different goals and different circumstances. We know that faba bean yields are variable and we believe that many farmers perceive yields to be even more variable. The way farmers respond to this variability depends on their aversion to risk.

Farmers' goals may include, among other things, profit maximization, risk minimization, and/or having enough food for the family. The success with which they achieve these goals depends on the size of their farms, the number of children, the quality of the soil and many other attributes, all of which vary among farmers. The fact that many differences exist implies that there is no single technical solution to production problems. Farmers will not respond in the same way to new alternative technologies. Therefore, we need to think of a broader range of solutions and be aware of who may or may not benefit from technological change.

*Research*

The implications for breeding efforts are to give higher priority to yield stability, disease resistance and salt tolerance. Saline soils posed problems last year in the trials in Menia Province. While these objectives are almost always included in breeding strategies, the issue is one of setting priorities.

Agronomic research can also aim to reduce yield variability. Pest control and herbicide application are just two examples. Improved management techniques, e.g., timely application of fertilizer, can reduce variability as well as increase yields. Such improvements have the added advantage of being low in cost which is important for risk-averse farmers. Delaying decisions to use costly inputs is another effective technique agronomists should consider.

New technology adds uncertainty to the production process. This uncertainty decreases as farmers gain more experience in using the technique. Therefore, research should concentrate on simple techniques which require less adjustment. New techniques also need to show sizeable potential increases in yield before they will be adopted. The yield increase needs to be big enough to overcome the inherent increase in uncertainty that is often associated with new techniques.

*Diffusion*

Developing new technologies and encouraging farmers to adopt them, are two separate process and here lie the implications for extension workers. First, a specific recommendation, for example 70 kg of $P_2O_5$ in Menia, will not be suitable for all farmers. These producers, themselves, are known to experiment with fertilizer levels to meet their own goals. Thus an estension objective could be to increase the use of phosphate by demonstrating a range of fertilizer levels rather than one specific level. This is one reason why credit programs that force fertilizer levels on farmers have a low probability of success.*

Farmers seldom adopt a complete technological package immediately, and demonstrations should show the effect of the various pieces of the package (12). Complex packages increase uncertainty which farmers want to avoid. Similarly, demonstrations need to take place under farmers' conditions. This is one of the focuses, and I believe successes, of the on-farm trials in the Nile Valley Project both in Egypt and the Sudan. A broad range of environmental conditions, exposing more farmers to the results under their conditions, will increase adoption rates.

Finally, we must realize that each farmer has a production-consumption system that reduces the total risk he faces. A risk-averse farmer is interested in reducing his income variability and variable faba bean yields are only one factor he takes into consideration. The farmer's cropping mix, inter-cropping practices, green or dry bean production, and straw as well as grain production, are all strategies that the farmer uses to try to reduce income in-

* An example is a credit program for seed purchases where seed is made available only if the farmer takes a complete production package including a certain amount of fertilizer.

stability.

*Infrastructure*

Infrastructure development is a frequently neglected area in development programs, and there are lessons for infrastructure development as well. Timely input availability is essential. For example, consider the case of water in Egypt. The irrigation system is not under the control of individual farmers and the value of fertilizer may depend on timely irrigation. If the farmers have irrigation problems, their strategy to avoid risk will be to decrease the fertilizer input or perhaps not use fertilizer at all. Similarly a good marketing system will improve distribution and stabilize prices.

With respect to institutions, the extension service will also be a key to increasing productivity and should concentrate on improving a farmers' perceptions, particularly of a new technology, so as to reduce the uncertainty associated with change. Farmers want to understand new technologies before they adopt them.

Another institution that should not be overlooked is an insurance scheme. Guaranteeing a certain yield is, of course, one way to remove production uncertainty. Insurance is expensive. There have not been many successes of such schemes in developing countries, and they are difficult to administer. Nevertheless, insurance should not be left off the list of options, and could possibly be introduced on a pilot basis in selected small, well-administered localities.

## ACKNOWLEDGEMENT

We wish to acknowledge the help of Abdel Bari Salkini who collected and processed the micro-data used in this paper and Andree Rassam who developed several of the tables and graphs.

REFERENCES

1. Anderson, Jock R., Dillon, John L., Hardaker, J. Brian. 1977 Agricultural Decision Analysis. Ames, Iowa: The Iowa State University Press, 1977.
2. Benito, Carlos A. 1976. Peasants' Response to Modernization Projects in Minifundia Economies 1976. American Journal of Agricultural Economics. 58(May, 1976): 143-151.
3. Binswanger, Hans P. 1977 Risk Attitudes of Rural Households in Semi-Arid Tropics. International Crops Research Institute for the Semi-Arid Tropics.
4. Dillon, J.L. 1971. An Expository Review of Bernoullian Decision Theory in Agriculture: Is Utility Futility ? Review of Marketing and Agricultural Economics. 39 (March, 1971): 3-80.
5. Fitch, James B., Nordblom, Thomas L. 1977 , Agricultural Decision-Making in the Dry-lands. Proceedings of an International Symposium on Rainfed Agriculture in Semi-Arid Regions. Ed. Glen H. Cannel. Riverside, California: Consortium for Arid Lands Institute, 1977, 273-94.
6. Halter, A. N., Dean, G.W. 1971, Decision Under Uncertainty with Research Applications. Cincinnati: South-Western, 1971.
7. Hiebert, L. Dean. 1974. Risk Learning and the Adoption of Fertilizer Responsive Seed Varieties. American Journal of Agricultural Economics. 56(November, 1974):764-768.
8. Huffman, Wallace, E. 1974. Decision-Making: The Role of Education. American Journal of Agricultural Economics. 56 (February, 1974): 85-96.
9. Just, R.E., Pope, R.D. 1976 On the Relationship of Input Decisions and Risk. Giannini Foundation of Agricultural Economics. March, 1976.
10. Lin, William, Dean, C.W., Moore, C.V. 1974. An Empirical Test of Utility vs Profit Maximization in Agricultural Production. American Journal of Agricultural Economics. 56 (August, 1974): 497-508.
11. Magnusson, G. 1969. Production Under Risk: A Theoretical Study. Uppsala: Almquist and Winksells, 1969.
12. Mann, Charles K. 1977. Packages of Practices: A Step at a Time with Clusters. Paper presented at the Meeting of the American Agricultural Economics Association. 1977
13. Moscardi, E. R., de Janvry, Alain. 1977. Attitudes Toward Risk Among Peasants: An Econometric Approach. American Journal of Agricultural Economics. 56 (November, 1977): 730-738.
14. Nygaard, David F. 1979. Risk and Allocative Errors due to Imperfect Information: The Impact on Wheat Technology in Tunisia. Ph.D. Thesis, University of Minnesota, 1979.
15. Pyle, D.H. and Turnovsky, S.J. 1970. Safety First and Expected Utility Maximization in Mean-Standard Deviation Portfolio Analysis. Review of Economics and Statistics. 52 (February, 1970): 75-81.
16. Roe, Terry, Nygaard, David. 1980. Wheat, Allocative Error and Risk :Northern Tunisia. Bulletin. Economics Development Center. University of Minnesota Press. March, 1980.
17. Roumasset, James A. 1976. Rice and Risk. Amsterdam: North Holland Publishing Company, 1976.
18. Salkini, Abdel Bari, Nygaard, David, Basheer, Abdel Mawlah, Abdel Aziz, Mustafa El-Sayed. Nile Valley Project Report II. Agro-Economic Survey. ICARDA, Aleppo, Syria. (Forthcoming).
19. Scandizzo, Pasquale L. , Dillon, John L. 1976. Peasant Agriculture and Risk Preferences in Northeast Brazil: A Statistical Sampling Approach. Paper presented at the CIMMYT Conference. Mexico, 1976.

20. Welch, F. Education in Production. 1970 Journal of Political Economics. 78 (January - February, 1970): 35-59.
21. Winkelmann, Donald. 1976. The Adoption of New Maize Technology in Plan Puebla, Mexico, Mexico City: CIMMYT, 1976.
22. Wolgin, Jerome M. 1975. Resources Allocation and Risk: A Case Study of Small-Holder Agriculture in Kenya. American Journal of Agricultural Economics. 57 (November, 1975): 622-30.

## 32. THE ROLE OF FABA BEANS IN THE EGYPTIAN DIET.

GAMAL N. GABRIAL

*National Research Centre, Dokki, Cairo, Egypt*

As economic reasons prevent an increase in the use of animal proteins, an obvious alternative is to augment the diet with legumes which, as a class, are relatively rich in proteins when compared to most other foods of plant origin. Furthermore, legumes are already in common use in the daily diet. In the towns, meat is eaten mainly by people who can afford it, and legumes go to the lower income groups. In rural areas, the farmers sell their milk, butter, cheese, eggs and poultry to buy larger quantities of cheap foods such as bread and faba beans to feed a larger number of children.

Even when people can afford it, animal protein is generally restricted to the one market day of the week.

As faba beans occupy a prominent position in our national diet, stimulation of local production would be a sounder proposition than using imports to counter malnutrition.

In 1978, faba beans provided 250,000 of the 376,000 tons of pulses, nuts and seeds which Egypt produced (Table 1). The production of faba beans alone was more than half that of meat (437,000 tons). Table 2 shows that 81.6 g of the daily per capita protein intake of 94.9 g was of vegetable origin, the major proportion of which was from cereals. About 14 g of faba beans are consumed daily per caput, which accounts for about 3 g of protein.

Table 1. Production and consumption of faba beans and other food items (1000 tons)

| Types of food | faba bean | Pulses, nuts and seeds | Meat | Fish |
|---|---|---|---|---|
| Total production | 250 | 376 | 473 | 110 |
| Exported | –– | 15 | 2 (sheep) | –– |
| Imported | 16 | 81 | 59 | 68 |
| Left for human consumption (after accounting for animal feed, waste, seed use etc.) | 197 | 460 | 494 | 162 |

*G. Hawtin & C. Webb (Eds.): Faba Bean Improvement*
©*1982 ICARDA.* ISBN *-13:978-94-009-7501-9*

Table 2. Average food consumption in Egypt, 1978.

| Type of Food | g/caput/day |
| --- | --- |
| Cereals | 620.0 |
| Starch food (other) | 35.9 |
| Sugar and sugar syrups | 81.6 |
| Pulses nuts and seeds | 23.8 |
| Vegetables, fresh | 279.2 |
| Fruit | 136.6 |
| Meat | 28.2 |
| Eggs | 4.4 |
| Fish | 11.2 |
| Milk and cheese | 187.8 |
| Vegetable oils | 32.3 |
| Total food intake | 1503.3 |
| Vegetable protein | 81.6 |
| Animal protein | 13.3 |
| Calorie intake | 3503.0 |

Popular ways of preparing faba beans in Egypt are :

*Fool akhdar* : The green immature seeds of broad bean are eaten while fresh with bread and cottage cheese (Gibna Arish), at breakfast or at lunch.

*Fool matbookh* : The green immature pods of broad bean are either boiled in water to which salt and cumina are added, or are cooked in tomato sauce prepared from fried onion, fat and tomato juice.

*Fool medames* : (Stewed broad bean) : The dried seeds are stewed in about twice the volume of boiled water on a flame which is only high enough to keep the contents gently boiling for 10 to 12 hours until they become soft. Medames is served at breakfast after adding salt, cotton seed oil and lemon juice.

*Fool nabet* : The dried seeds are soaked in water for about 12 hours followed by germination for three days. The germinated beans are cooked in boiling water for about one hour and served after adding some fried garlic to its soup (Nabet soup).

*Bisara* : The decorticated dried seeds are soaked in water overnight and cooked on a mild flame with small pieces of onion and dried peppermint. Water is added several times to prevent drying. The cooked beans are blended with a small amount of water for two minutes after adding a mixture of salt, spices and chorcorus. Cooking is continued for a few minutes. Onion and garlic , fried in oil, are added to the cooked mixture and the meal is served in shallow dishes.

*Taamia (Falafel or bean-cakes)* : The decorticated dried beans are soaked in water for 12 hours. Surplus water is then drained off. Small amounts of garlic, carrot, spring onion, coriander (green and seeds), are added for flavouring. The mixture is then crushed into a thick paste. When ready, the mass is removed from the mortar and allowed to stand for some time. Fermentation is accompanied by the production of gas. The paste is finally cut into small pieces and fried in deep boiling cotton seed oil until the surface turns uniformly brown. It is usually served with wheat bread, tomato slices and green leaf vegetables. Morcos (18) found that the net dietary protein energy % of a diet containing bean cakes was 8.6 when he calculated it on the basis of following ingredients : beans *(Vicia faba)*, 60 g; cottonseed oil, 6 g; wheat flour (85% extraction), 65 g.

This meal costs 5 piasters (U.S. 7 cents) and provides 462 calories. Morcos stated that it meets the needs of infants, toddlers, children and adults, if it is fed in quantities which satisfy their energy requirements.

The Nutritive Value of Faba Beans

*Amino acids content*

The amino acids content of faba bean proteins are expressed either on the basis of mg amino acid per g nitrogen or g amino acid/100 g protein (16 g nitrogen). Faba bean proteins, like most other leguminous proteins, are deficient in the sulphur-containing amino acids, which are lower than those recommended by FAO for human consumption.

Morcos and Boctor (19) found that the Egyptian faba bean contains a high percentage of proteins. Riad (24) compared the amino acids content of the proteins of mature, immature seeds and green pods of two faba bean cultivars, Baladi and Roomi. He found that all the samples studied were markedly deficient in the sulphur-containing amino acids (cystine and methionine), and in tryptophan, which agrees with the results of Patwardhan (21). Riad (24) noticed that the lysine, phenyl alanine and threonine contents of the two varieties exceeded the values of the provisional amino acids pattern recommended by FAO. Only the green pods and immature seeds contained small amounts of isoleucine and leucine. He also found that Roomi was distinguished by its high content of amino acids as compared to the Baladi cultivar. The only exception was that the dry seeds of Baladi contained higher contents of lysine than the rest.

*Biological value*

Towards the end of the 19th century, it was realized that proteins of various legumes had different biological values; the nitrogen balance and rat growth method being used for assessing the nutritive value of dietary proteins. The former, with limitations, can furnish data on the biological value of Mitchell(17) and the digestibility and net protein utilization of Miller and Bender (16). The rat growth method of Osborne, Mendel and Ferry 1919, has been used by many workers because of its ease, in determining the protein-efficiency ratio.

In an attempt to raise the nutritive value of bread by supplementing Baladi bread with faba bean, we studied six diets (10). The protein content of the first was from faba bean only, while the protein content of the other five was a mixture of 50% protein from bread and 50% protein from either raw bean, soaked bean, germinated bean, germinated and cooked bean, and autoclaved bean. When fed to rats, the NPV and BV values of all diets in which baladi bread was supplemented with raw or treated faba bean were higher than those of the diet prepared from broad bean alone (Table 3). Fig. 1 shows the increase in body weight of rats fed on bread supplemented with all treatments of faba bean which was higher than those fed on faba bean alone except those which were fed bread supplemented with autoclaved faba bean.

In a series of food mixtures formulated in the Nutrition Institute, Cairo, and composed of different proportions of faba beans, lentils, chickpeas, rice and wheat flour, the protein content varied from 15.4 to 23.3% , their NPU ranged from 28.05 to 90.07, and

312

their PER varied from 1.36 to 3.02. The most important was Arabeana (3) (7) which was formed from a mixture of 23.5% wheat flour, 23.5% faba bean, 23.5% chickpea flour and 29.5% sugar. It gave promising results when treated for acceptability, tolerance and effect on the growth of children, (2).

Table 3. Nutritive value of raw and prepared faba bean supplemented with bread.

| Diet | % bread protein | % faba bean protein | Net protein utilization | True digestibility | Biological value |
|---|---|---|---|---|---|
| Raw bean | — — | 14.33 | 41.1 | 78.7 | 52.2 |
| Raw bean + bread | 7.22 | 7.17 | 52.7 | 78.9 | 66.8 |
| Soaked bean + bread | 6.98 | 6.93 | 51.8 | 78.9 | 65.8 |
| Germinated bean + bread | 7.01 | 6.96 | 45.0 | 76.5 | 68.8 |
| Germinate then boiled bean + bread | 6.98 | 6.93 | 46.8 | 77.5 | 60.8 |
| Autoclaved bean + bread | 6.47 | 6.41 | 41.8 | 69.5 | 60.1 |

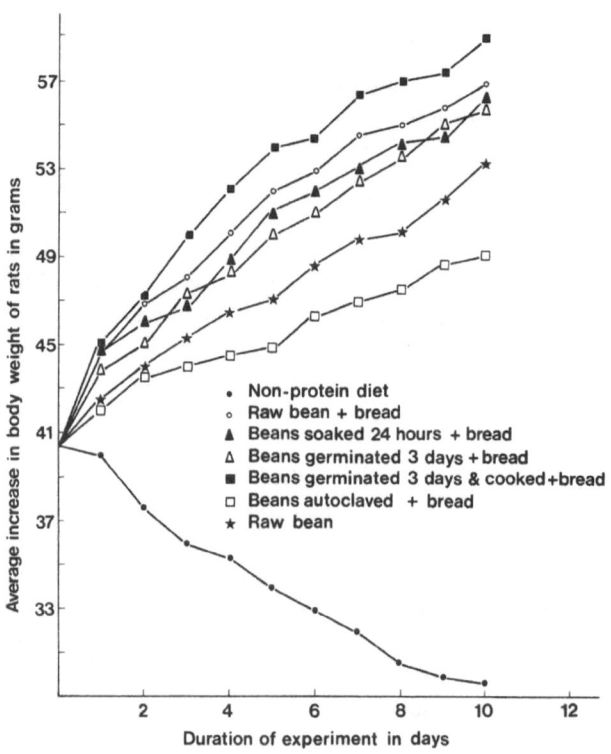

Fig. 1    Increase in body weight of rats on diets other than non-protein.

Effect of Processing on the Nutritive Value of Faba Bean.

## Soaking

Soaking in water is usually the first step in preparing faba beans for consumption. When soaked in their skins, seeds do not lose any nutrients. According to Platt (22), soaking constitutes a first stage in germination, and is important in making the constituents of the seed-stores more assimilable when eaten.

## Germination

Wiltshire (26) used germinated faba bean to cure cases of scurvy during World War I. The development of the antiascorbic factor in germinated legumes and other seeds was demonstrated before vitamin C or ascorbic acid had been identified, and chemical tests of the vitamin were available. Germination of faba beans for four days reduced the protein content and led to the increase in most of the essential amino acids needed for human consumption (12). The starch content decreased after germination, and the reducing and non-reducing sugars increased. El-Nahry et al. (8), found that the germination of faba beans showed a significant loss of the amino acids, aspartic acid and alanine. Darwish et al. (6), indicated that germination slightly improved the protein efficiency ratio of faba bean and caused a slight increase in both total serum proteins and albumin/globulin ratio on the rats' blood serum.

## Decortication

The removal of the fibrous husk improves the digestibility but sometimes leads to the loss of thiamine. Usually the germ is removed during dehusking.

## Heating

Most of the toxic factors can be removed, partially or completely, by the proper application of heat. Lepkovsky (15), stated that cooking faba beans increased their digestibility. Hussein, Gabrial and Morcos (14), noticed that during stewing, the essential amino acid lysine was the most labile. Moderate degrees of heat in preparation and cooking are beneficial. Boiling the faba bean apart from possible effects on protein digestibility, may involve some loss in water-soluble nutrients particularly when it is prolonged and excess water is used or when water is changed several times.

Our studies on the effect of processing faba bean on the availability of lysine, revealed that stewing faba beans was accompanied by a loss of its lysine content, either by biological inactivation and/or by degradation of the amino acid. Under the conditions applied for stewing the bean, the Millard-reaction can take place since the temperature and moisture content are high, and atmospheric oxygen is present in a limited volume under mild pressure. Darwish (6) showed that boiling and stewing slightly improved PER values of faba beans. El-Nahry et al. (8) found that both boiling and stewing the germinated beans caused a significant loss in the amino acids lysine, arginine, aspartic acid, alanine, methionine and tryptophan.

*Storage*

Faba beans, like most dry seeds, keep well for months or even years when properly stored. On the other hand, they deteriorate rapidly when exposed to high temperature, moisture, and insect infections. Losses and changes in nutritive value are related to conditions of storage. The two essential factors for safe storage are cleanliness and moisture control. Abd-El-Dayem (1) said that the dehydrated products he had prepared from autoclaved faba bean and other vegetables had a long shelf life because of their low moisture content.

Antinutritional factors

Antinutritional factors which have been reported in faba bean include trypsin inhibitors (25) (13) (4), hemaglutinin (10), and factors that cause favism (11). These harmful factors may either decrease the nutritive value of the protein or may even have a toxic effect.

REFERENCES

1. Abd-El-Dayem, H.H.M. 1976. M.Sc. Thesis, Fac. Agric., Al-Azhar Univ.
2. Abdou, I.A. 1968. Report on human resting of food mixtures imported from Tunisia and Algeria, submitted from Nutr. Inst. to the Ministry of Publ. Health, Egypt.
3. Abdou, I.A., El-Nahry, F.I. 1967. Bulletin of Nutrition Inst., Cairo, Egypt. (N.I.C.E.) Vol. III No. 1.
4. Anantharaman, K. 1970. Proc. Nutr. Soc., 29, 59.
5. Aykroyd, W.R., Doughty, J. 1964. Legumes in human nutrition. Food and Agriculture Organisation of the United Nations, Rome, No. 19.3.
6. Darwish, N.M., El-Nahry, F., Tharwat, S. 1976. Qual. Plant-PL-Fds. Hum. Nut. XXV, 3/4, 331.
7. El-Nahry, F.A.I. 1971. Ph.D. Thesis, Fac. Sc., Ain Shams Univ.
8. El-Nahry, F.A.I., Darwish, N.M., Nada, E. 1977. Qual. Plant-PL-Fds. Hum. Nut. XXVII 2, 151.
9. Food and Agricultural Organisation 1957. In Protein Requirements, FAO, Nutr. Stud., No. 16.
10. Gabrial, G.N. 1971. Ph.D. Thesis, Fac. Sc., Cairo, Egypt.
11. Hanna, M.B., El-Sadr, M.M. 1967. First Arab Confer. of Physiol. Sci. Cairo.
12. Hegazi, S.M. 1974. Zeitshrift fur-Ernahrungswissen shaft, 13, 4, 200.
13. Hochstrasser, V.K., Schwarz, S., Illchmann, K., Werle, E. 1968. Hoppe-Seyler's Z. Phys. Chem. 349, 19449.
14. Hussein, L., Gabrial, G.N., Morcos, S.R. 1974. J. Sci. Fd. Agric., 25, 1433.
15. Lepovsky, S., Bingham, E., Pencharz, R. 1959. Poultry Sc., 38, 1289.
16. Miller, D.S., Bender, A.E. 1955. Brit. J. Nutrition, 9, 382.
17. Mitchell, H.H. 1923-1924. J. Biol. Chem. 58, 873.
18. Morcos, S.R. 1966. J. Arab Vet. Med. Assoc., 26, 315.
19. Morcos, S.R., Boctor, A.M. 1959. Br. J. Nutr., 13, 163.
20. Osborne, T.B., Mendel, L.B., Ferry, E.L. 1919. J. Biol. Chem. 37, 227.
21. Patwardhan, V.N. 1962. Amer. J. Clin. Nutr., 11, 12.
22. Platt, B.S., 1956. Chem, and Ind., 18, 834.
23. Production and consumption sheet given by the Egyptian Ministry of Agric., Dept. Stat., 1978.
24. Riad, W.Y., 1970. M.Sc. Thesis, Fac. Agric., Ain Shams Univ.

25. Sohonie, K. Huprikar, S.V. and Joshi, M.R. 1958. J. Sc. and Indust. Res. (India), 18C, 95.
26. Wiltshire 1918 cited from Aykroyd, W.R. and Doughty, J. (1964), Food and Agriculture Organisation of the United Nations, Rome (1964), No. 19, 56.

# 33. FABA BEANS AND THEIR ROLE IN DIETS IN SUDAN

ABDALLA EL MUBARAK ALI, GASM ELSEED EL AMIN AHMED AND
EL KHATIM BALLA EL HARDALLOU

*Food Research Centre, Shambat. Khartoum North, Sudan*

Higher meat prices, during recent years, and the need for protein-rich foods have led people to shift their consumption to certain grain legumes, especially *Vicia faba*.

The production of *Vicia faba* is heavily concentrated in the Northern, Nile and Khartoum provinces and, to a lesser extent, other parts of Sudan.

Grain legume consumption is heavily concentrated in the capital and large cities of the Sudan where per capita income is relatively high and where industrial areas exist. These two aspects have resulted in the increase of per capita consumption of food legumes, particularly *Vicia faba*, which have become significant in popular dishes. The demand has become higher and consequently prices of the crop have gone up very fast, although both area and production have increased.

Most city people take stewed faba beans for breakfast and supper as well as in sandwiches at any time of the day. Many additives such as oil, spices, and cheese are usually considered an adjunct of faba bean dishes. Faba beans, especially crushed seeds, are widely used in Sudan in the popular food "Taamia", a breakfast and supper delicacy. The widespread consumption of faba beans may be because they are so easy to prepare at home, and store, as well as their acceptable taste and flavour. Their popularity is expected to increase.

As the faba bean is grown in Sudan mainly for human consumption, the hard seed problem affects the cookability and market price of the product, and consequently a lot of research has been done on this problem.

Studies on the factors affecting the nutritional value of faba beans have shown that the balance of amino acids in the seed is good, except for the sulphur amino acids, in particular methionine.

Lysine, arginine, threonine and methionine content were affected by germination, cooking or baking, while the contents of valine and leucine - isoleucine increased slightly and those of cristidine and tryptophan decreased. It is concluded that 100 g. of germinated, cooked or baked faba bean would supply, with the exception of methionine and threonine, a considerable proportion of the daily requirements for essential amino acids. Cooking also

*G. Hawtin & C. Webb (Eds.): Faba Bean Improvement*
©1982 ICARDA. ISBN -13:978-94-009-7501-9

had a marked effect on the phosphorous content but only a slight effect on ash, fat, protein, fibre and starch (3). Cooking quality was positively correlated with lightness in the colour of the hull; 100 seed weight; specific gravity; and hydration co-efficient (1).

Table 1.

Proximate analysis (%) and percentage of hard seeds for six *Vicia faba* cultivars grown at Aliab in Northern Sudan.

| Variety | Moisture | Protein | Crude fibre | Fat | Ash | Carbohydrate | Hard seed % |
|---------|----------|---------|-------------|-----|-----|--------------|-------------|
| NEB 424.5 | 8.80 | 28.44 | 6.90 | 0.80 | 2.75 | 52.31 | 20.9 |
| NEB 423 | 8.84 | 31.34 | 5.16 | 0.76 | 3.03 | 50.87 | 7.8 |
| NEB 424 | 8.63 | 28.90 | 5.00 | 1.36 | 2.65 | 53.46 | 7.8 |
| NEB 425 | 8.28 | 30.41 | 6.12 | 1.12 | 2.88 | 51.19 | 7.8 |
| 188 XG −1 | 8.74 | 27.79 | 6.00 | 1.35 | 2.82 | 53.24 | 14.3 |
| H. 72 | 8.68 | 27.84 | 7.30 | 0.70 | 2.68 | 52.78 | 14.8 |

Faba beans contain 7.1 to 11 % cellulose ; 25 to 30 % azotic substances and 45 to 48 % non-azotic substances. Free sugars include sucrose, glucose, fructose, raffinose and mattotriose tetraose and pentaose have been detected (4). Elimination of condensed tannins from the common cultivars of faba beans through breeding would not only increase their nutritional value but would also eliminate compounds which are responsible for darkening the seed coat (3). The latter effect would greatly improve the organoleptic quality of faba beans, thereby enhancing their value as human food.

As milk production in Sudan is inadequate, very little of it being available for preschool and school children, malnutrition is a common feature in this vulnerable group. The need for protein is acute in regions where root cops such as cassava are the staple food. Formulated weaning foods are absent.

From our preliminary studies, we conclude that faba beans can provide a good protein supplement to produce weaning foods of high nutritive value.

REFERENCES

1. El-Tabey Shehata, A.M., Youssef, M.M. 1979. Relationship between physical properties and cooking quality of faba bean (*Vicia faba*) FABIS No. 1 June 1979.
2. Hussein, M.A., Youssef, K.E. 1977. Effect of thermal treatment on the essential amino acids of *Vicia faba* L. F.S.T.A. Vol. 8.
3. Marquardt, R.R. 1979. Factors affecting faba bean utilization FABIS No. 1, June 1979.
4. Pritchard, J.P., Drybargh, A.E., Wilson, J.B. 1973 Carbohydrates of Spring and Winter field beans (*Vicia faba* L.) J. of the Sc. of Food and Agric. Vol 24   No. 6
5. Ministry of National Planning, Khartoum, Sudan, 1977.

# 34. PROTEIN QUANTITY AND QUALITY IN *VICIA FABA*

## JAN SJÖDIN

*Svalöf AB, S-268 00 Svalov, Sweden*

The nutritional contribution of faba beans might be improved through increased yields, increased protein concentration or improved protein quality. As a minimum we should try to increase any of these three factors without decreasing the other two. Ideally we would make the greatest improvement if we could increase crop yield, protein quantity and protein quality. The usefulness of such an improved faba bean would be still greater if we could reduce the amounts of certain antinutritional substances that are known to have a negative influence.

However seed yield, protein content and protein quality are more or less closely related. This chapter describes data regarding these relationships and their implications in breeding programs.

### Variation of Protein Content in Vicia faba

Table 1 shows variation of protein content both between and within different populations of faba beans. During five years, Picard (12) found the variation among populations of different origin to be mostly between 26 and 41%. A few populations from the Middle East and Ethiopia showed a few per cent units lower protein content. Bond and Toynbee-Clarke (3), and Bond (5) found about the same range but observed that winter beans on average, had lower protein content than spring beans. Other authors have obtained about the same range of variation, and all concluded that a great part of the differences between populations were due to heritable factors. With protein content under genetic control there are good possibilities for change by selection.

Table 1. Variation in protein content between different populations of *Vicia faba*

| Author | Range of protein content in % |
|---|---|
| Picard (1977) | 23 - 41 |
| Bond (1977) winter beans | 22 - 28 |
| Bond (1977) spring beans | 26 - 36 |
| Fröhlich *et al.* (1974) | 28 - 37 |
| Griffiths and Lawes (1978) | 23 - 38 |

*G. Hawtin & C. Webb (Eds.): Faba Bean Improvement*
*©1982 ICARDA. ISBN 90 247 2593 3. Printed in the Netherlands.*

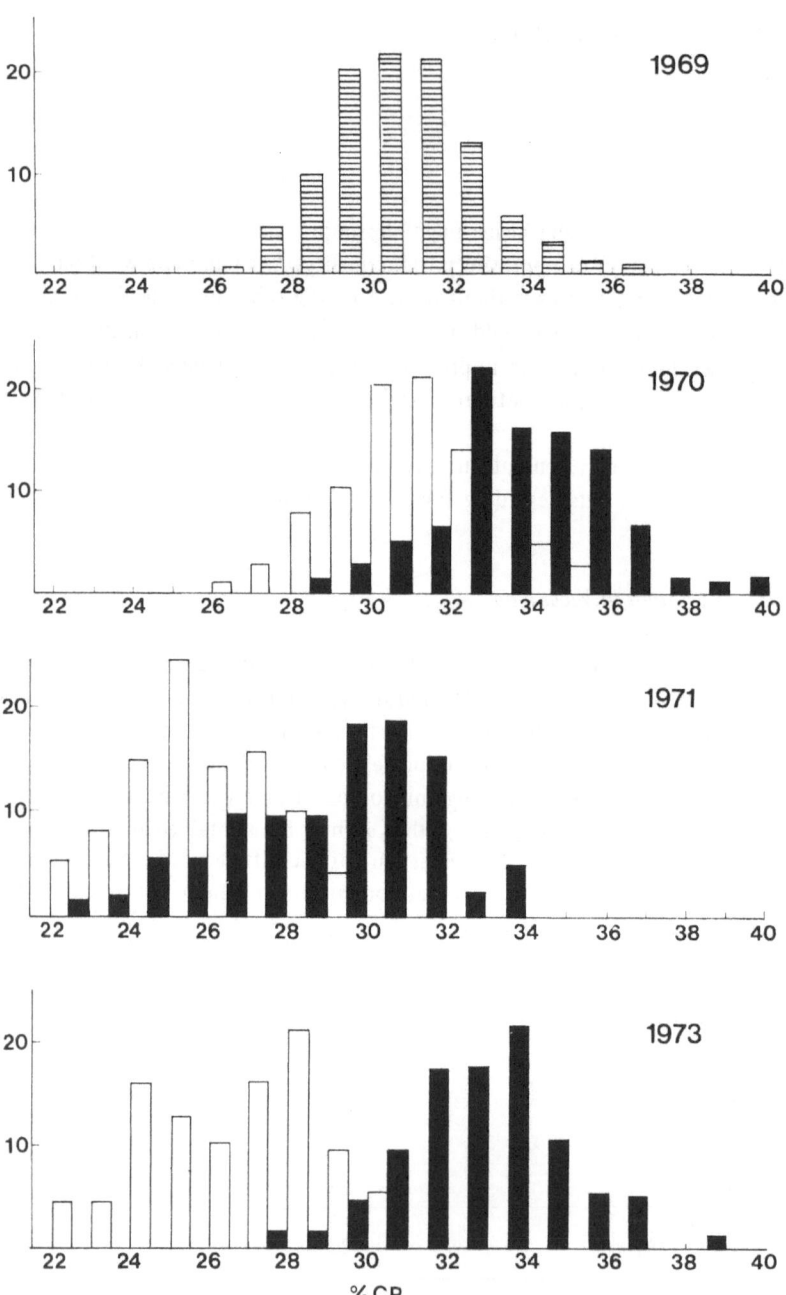

Fig. 1. Variations in populations selected for high and low content of crude protein (CP).

Table 2. Average values for crude protein, number of seeds/plant and thousand grain weight (TGW) in populations selected for high (HPS) and low (LPS) content of protein in field beans.

| Year | | % crude protein | Number of seeds/plant | TGW (g) |
|------|------|------|------|------|
| 1970 | HPS | 33.7 | 45.2 | 321 |
| Selection | LPS | 31.1 | 47.1 | 324 |
| | Diff. | 2.6$^{+++}$ | - 1.9 | - 13 |
| 1971 | HPS | 29.1 | — | 250 |
| Selection | LPS | 25.9 | — | 250 |
| | Diff. | 3.2$^{+++}$ | — | 0 |
| 1972 | HPS | 29.7 | 63.9 | 219 |
| Selection | LPS | 26.3 | 68.0 | 322 |
| | Diff. | 3.4$^{+++}$ | - 4.1 | -103$^{+++}$ |
| 1973 | HPS | 32.4 | 27.3 | 290 |
| Selection | LPS | 26.3 | 33.7 | 320 |
| | Diff. | 6.1$^{+++}$ | -6.4$^{+++}$ | -30$^{+++}$ |
| 1975 | HPS | 34.6 | 18.1 | 308 |
| Multiplication | LPS | 27.8 | 14.4 | 289 |
| | Diff. | 6.8$^{+++}$ | 3.7$^{+++}$ | 19$^{++}$ |
| 1976 | HPS | 32.5 | 22.7 | 304 |
| Multiplication | LPS | 29.1 | 25.9 | 327 |
| | Diff. | 3.4$^{+++}$ | - 3.2$^{+}$ | - 23$^{+++}$ |
| 1978 | HPS | 32.0 | 11.2 | 226 |
| Multiplication | LPS | 26.5 | 15.4 | 216 |
| | Diff. | 5.5$^{+++}$ | -4.2$^{++}$ | 10 |

+P=0.05, ++ P=0.01, +++ P = 0.001

## Selection for High and Low Content of Crude Protein

To test the possibilities of making selection for crude protein content we used at Svalöv, Sweden, a population (20) which showed a variation between 26 and 36% protein content (22). In this population, Sv 0720, positive and negative selections were made; all plants with more than 33% and all with less than 29% respectively were selected. These selections were repeated during four consecutive years (Fig. 1). The variation in the original population corresponds well with figures of protein content of faba beans grown under our climatic conditions in Svalov, and the data from different authors which have been mentioned. The influence of environment on the level of crude protein content from one year to the other is clear e.g. the variation in 1969 and 1970 ranged between 26 and 36% whereas in 1971, it was between 22 and 34%.

Table 2 outlines the effect of the selections. For each selection cycle the difference in protein content between HPS (high protein selection) and LPS (low protein selection) increased, and when the selection pressure ceased, there was a large and significant difference between the two selected populations which was retained in each following multiplication generation. Griffiths and Lawes (9) also concluded that the reliability of protein content seems to be quite good. These experiments make it obvious that it is realistic to select for increasing protein content and to maintain a selected population on a high level.

## Correlation Between Protein Content and Thousand Grain Weight (TGW)

From a breeding point of view, it is also very important to know how other factors, especially yield components, are influenced by changes in the protein content. The seed weight expressed as thousand grain weight (TGW) is highly environmentally influenced within each subspecies. Table 2 makes it apparent that TGW does not follow the changes in protein content resulting from selection. On the contrary, the TGW seems to be quite independent of the selection pressure applied with regard to the crude protein content.

## Correlation Between Protein Content and Seeds Per Plant

The number of seeds per plant seems to be influenced more (Table 2). There is a tendency towards fewer seeds per plant in the HPS in each comparison. Since seeds per plant, besides seed weight, is the most important yield component, this will of course negatively influence the seed yield. On the other hand, it is also clear that the number of seeds per plant in HPS and LPS respectively follow each other between the different years, indicating that it would be possible to obtain an acceptable number of seeds per plant even in a population with a high content of crude protein.

Cereals show mostly negative correlations between grain yield and protein content e.g. (17) (7). Most legumes investigated seem to show no or slight negative, though rarely significant, relationships between these characters. Very few data are reported from faba bean. Griffiths and Lawes (9) as well as de Vries (25) and Bond (5) reported no such relationships in their *Vicia faba* material.

*Correlation Between the Weight of a Single Seed and Crude Protein Content.*

By using a half-seed technique developed at Svalof (21), we have analysed the relationship between the weight of single seeds and crude protein content on materials cultivated under different conditions e.g. in the glasshouse or in the field, open pollinated or isolated in bags (Fig 2).

All through there is a positive correlation between the weight per single seed and protein content. The strength of this correlation seems to be dependent upon the nature of the material. For example, the correlation is often stronger in selfed material compared with open pollinated. The open pollinated, non-inbred, material showed a near vertical regression line. Howeversmall changes e.g. in some extreme values, could change the relationship. For example, a number of small seeds with high protein content could easily have caused the regression line to switchover to indicate a negative correlation.

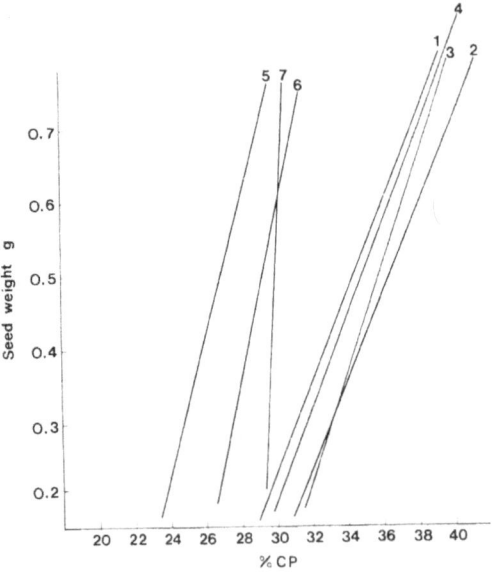

Fig. 2. Correlations between per cent crude protein and seed weight in different lines of faba beans.

*Relationship Between Protein Content and Seed Position Within the Plant*

If selection for protein content is made on single seeds, it is preferable to be aware of the position in the plant of the seeds selected. When checking the protein content and the seed weight of single seeds, Sjödin *et al.* (21) found that the protein content was on the average lower, the higher the node was on the plants. The protein content was lower, the higher the pod position on the peduncle and, within the pod, the distal seed had, on average, the highest protein content. These data indicate that preferably seeds from the same position in the plants have to be chosen in selection work.

324

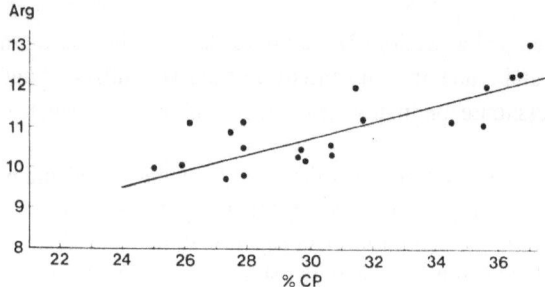

Fig. 3. Correlation between per cent crude protein (CP) and arginine (g/100 protein).

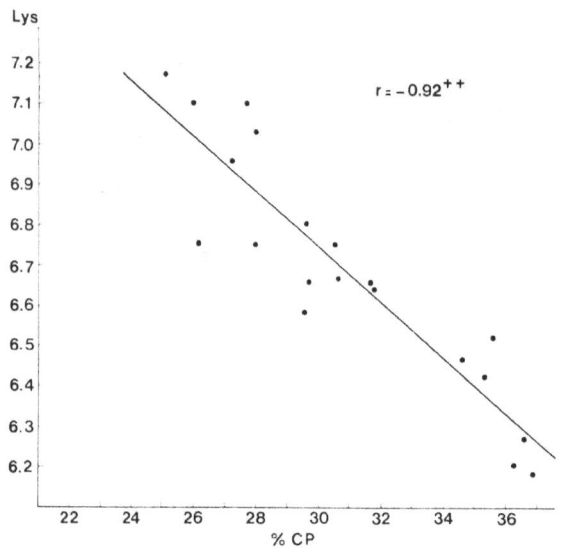

Fig. 4. Correlation between per cent crude protein (CP) and lysine (g/100 protein).

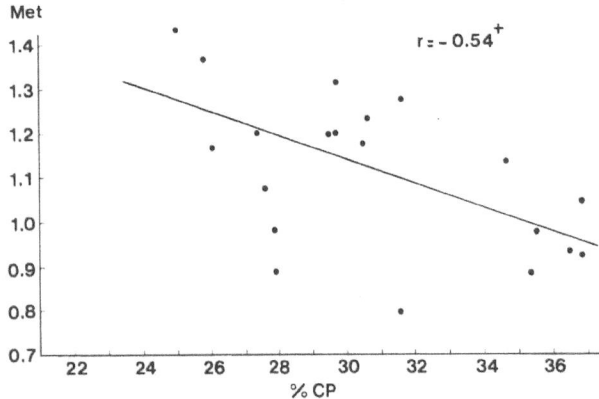

Fig. 5. Correlation between per cent crude protein (CP) and methionine (g/100).

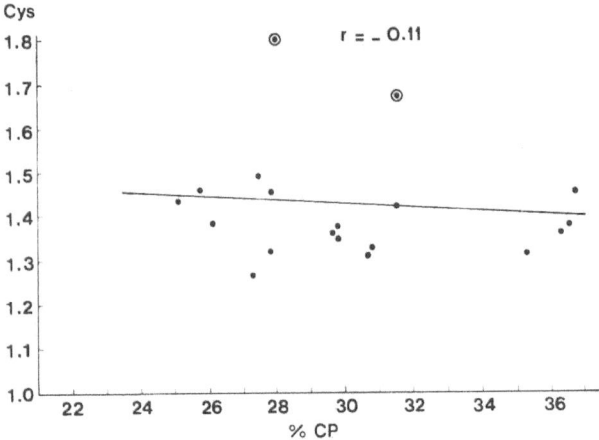

Fig. 6. Correlation between per cent crude protein (CP) and cystine (g/100 protein) The encircled spots are explained in the text.

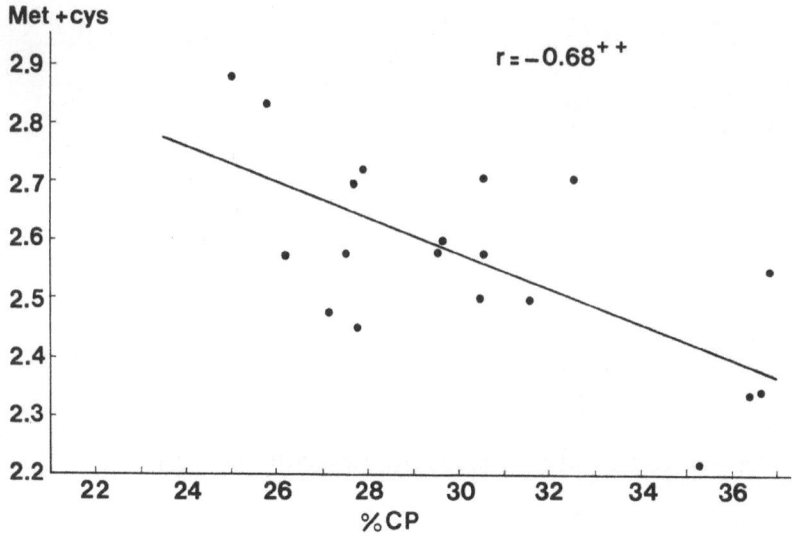

Fig. 7. Correlation between per cent crude protein (CP) and methionine + cystine (g/100 protein).

## Conclusions

It is possible to make progress in selecting for high crude protein content and it is also possible to maintain this high level during the following multiplication generations. Further there seems to be no correlation, or a very weak one, between seed weight per plant and the protein content. It is possible to obtain slight positive relationships between the weight of single seeds and protein content. A high protein content seems to influence the number of seeds per plant somewhat negatively. With regard to these data, it should be possible to select for high protein content without losing anything or, at least only a little of the seed yield. From the protein yield point of view it is questionable whether it is more effective to breed for increased seed yield or increased protein content.

## Correlation Between Protein Content and Amino Acid Composition

The amino acid composition of *V. faba* is mainly characterised by a very low content of the two sulphur-containing amino acids, methionine and cystine. Further, the faba bean has, in contrast to cereals, a high content of lysine and also a high content of arginine; an amino acid that seems, frequently, to play an important role in accounting for differences between populations with regard to protein content.

Analysis of the material used for the selection of protein which have been discussed showed a certain increase of methionine content in the HPS-population and a certain decrease in the content of cystine and lysine. These decreases were counterbalanced by a 13% increase of arginine content in the HPS-population.

The amino acid investigation was extended to more variable material. A strong positive and significant relationship ($r = +0.80^{+++}$) was thereby observed between protein content and arginine (Fig. 3) when the arginine, as well as the other amino acids, is expressed as g/100 g protein. On the contrary, a strong and significantly negative correlation ($r = -0.92^{+++}$) was obtained between protein content and lysine (Fig. 4). Also methionine and protein content were negatively correlated ($r = -0.54^{+}$) even though the relationship was weaker (Fig. 5). The content of cystine (Fig. 6) seems to vary independently of the protein content ($r = -0.11$) whereas the two sulphur amino acids together are more strongly negatively correlated to the protein content ($r = -0.68^{+++}$) (Fig. 7). These relationships have been observed earlier in *V. faba* populations (4) (11) (8) (16) and also in other legumes e.g. (19) (2) (24) (1).

If hypothetic figures are applied to the regression lines and a positive effect of selection for protein content is assumed, it will be found that this selection implies a reduction in the percentage of the sulphur amino acids. However if these amino acids are measured as kg/ha, it will mean an increase.

## Selection for Sulphur Amino Acids

Porter *et al.* (14) and Radford *et al.* (15) have reported strong positive correlations between the content of sulphur amino acids and sulphur in the protein. In the Svalov material, we also obtained a close relation ($r = +0.84^{+++}$) between these two factors. When we wanted to estimate the possibilities of selecting for higher content of methionine and cystine, we made positive and negative selections in four consecutive years with promille sulphur in crude protein as the selection criterion (23). No attention was paid to the level of protein content. During each year, there was a distinct and significant difference between the plants selected for high and low content of sulphur (Table 3). In the populations raised from the selected seeds, there were also significant differences even though they did not increase for each selection cycle. The average values of crude protein content in the selected plants showed each year that those plants with high content of sulphur had the lowest content of crude protein.

Table 3. Selection for high (H) and low (L) content of sulphur in crude protein (CP)

| | | Selected plants 0/00 S in CP | % CP | Pop. from selected plants 0/000 S in CP | % CP |
|---|---|---|---|---|---|
| 1975 | H | 6.12 | 34.8 | 6.75 | 24.3 |
| | L | 4.90 | 36.3 | 6.13 | 25.6 |
| | Diff. | 1.22+++ | -1.5 | 0.62[+++] | -1.3[+++] |
| 1976 | H | 7.68 | 22.4 | 6.02 | 34.1 |
| | L | 5.41 | 28.3 | 5.83 | 32.3 |
| | Diff. | 2.27[+++] | -5.9[+++] | 0.19[++] | 1.8[+++] |
| 1977 | H | 6.79 | 30.0 | 6.21 | 33.3 |
| | L | 5.20 | 35.6 | 5.71 | 31.2 |
| | Diff. | 1.59[+++] | -5.6[+++] | 0.50[+++] | 2.1[+++] |
| 1978 | H | 7.21 | 29.9 | 6.52 | 30.8 |
| | L | 4.56 | 33.4 | 5.98 | 30.3 |
| | Diff. | 2.65[+++] | -3.5 | 0.54[+++] | 0.5 |

++ p = 0.01,  +++ P = 0.001

The fact that no obvious selection effect was obtained, is understandable with the present knowledge of the negative correlation between protein content and sulphur amino acids (Fig. 7). The effects obtained were only a shift along the regression line in the sense that the true effect of the selection was a change in protein content.

The failure of these experiments shows that a better way to make selection must be to choose a constant level of protein content and then select for high and low content of sulphur for plants at this protein level. The importance of working with known protein content levels has also been pointed out by Mosse and Baudet (11).

Another condition for successful selection is that the variation of the character in question is of sufficient magnitude. In the present material, the variation in sulphur was as great as the variation of crude protein and might therefore be considered as equally sufficient. On the other hand, the level of the sulphur amino acids, especially that of the methionine, was very low.   Rudolph et al. (16) also reported a narrow variation of methionine content but somewhat greater variation in cystine content in the 450 faba bean populations they investigated. Although they found a few positive variants for each of the amino acids separately, types with a higher content of both sulphur amino acids were rare. Schuster and Posselt (18) were doubtful about the possibility of making a successful selection for sulphur amino acids, even in soybean.

*Correlation Breakers*

In 1976, the correlation was estimated between protein content and content of sulphur in the population with high sulphur (Fig. 8). The correlation coefficient was -0.62[+++]. On that occasion, two plants (81 and 88 marked with circles in the figure) deviated from the general pattern by having a high content of both protein and sulphur. The progenies from these plants showed the same characteristics and when the amino acid composition was analysed, it was shown that the main reason of deviation for these correlation breakers depended upon a distinctly higher content of cystine (these two plants are also indicated by dots with circles in Fig. 6). Since the relationship between protein content and cystine is comparatively weaker than with methionine, it must be logical to expect other correlation breakers in the relationship between cystine and protein content in the future.

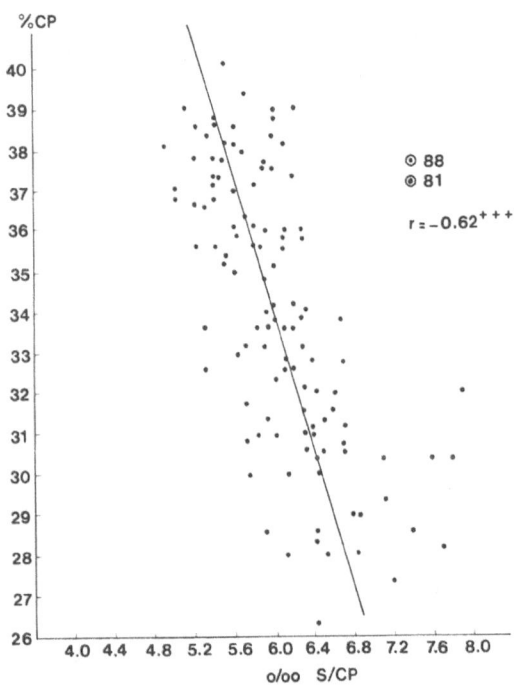

Fig. 8. Correlation between crude protein (CP) and sulphur in crude protein (S/CP). The encircled spots are explained in the text.

*Selections in Legumin and Vicilin Fractions*

Out of the globulins, the legumin fraction is of special interest from the breeding point of view since it contains about twice the content of sulphur amino acids compared to the vicilin fraction (6). An increase of the legumin:vicilin ratio would therefore probably result in a higher content of methionine and cystine in the protein. Mårtensson (10) determined the variation of this ratio in six cultivars harvested in two different years. He then found that the range of ratios varied between the two years but the rank order between the cultivars was nearly the same. He also demonstrated that the increase of the ratio legumin:vicilin was paralled by an increase in the protein content of the seeds.

Mårtensson (10) also determined the variation between single plants with regard to the same ratio. He selected some extreme plants and checked the amino acid composition. He found that the high legumin content unfortunately was due to a high protein content but a lower content of the two sulphur amino acids as well as of lysine. The content of arginine had increased. This implied that the same thing had happened as described above for the relationship between sulphur amino acids and protein content i.e. the real effect of the selection was again a shift along the regression line for the correlation between protein content and each of the amino acids mentioned. The only factor that really changed was the protein content.

*Final Conclusions*

With regard to available data from *Vicia faba* and other legumes it seems possible to select and maintain a higher protein level after some generations of selection. The weak correlation between protein content and seed yield, especially in *Vicia faba* , speaks in favour of the possibility of increasing the protein level and still keeping an acceptable seed yield level. However a fair warning, of the positive relationship between lateness and protein content (13) may be appropriate. Increasing the protein level means a certain decrease of the level of sulphur amino acids as well as of lysine.

The procentual variation of sulphur amino acids is sufficient, but the level of the content in the material known today is so low that even with a very successful selection program, it is doubtful if the new levels reached are of practical interest. New material and new variation such as in presently unknown wild forms or mutants would be opportune. Technically, it would probably be possible to make positive selection using total sulphur as the selection criterion if the protein content in the selected material was kept constant. Under the same conditions it would probably also be possible to manipulate the globulin fractions.

## REFERENCES

1. Adams, M.W. 1973. On the quest for quality in the field bean. In "Nutr. Imp. of Food Legumes by Breeding". PAG-symp. Rome 3-5 July 1972, 143-149.
2. Bliss, F.A. 1973. Cow peas in Nigeria. In "Nutr. Imp. of Food Legumes by Breeding". PAG-symp. Rome. 3-5 July 1972, 151-158.
3. Bond, D.A., Toynbee-Clarke, C. 1968. Protein content of spring and winter varieties of field beans (*Vicia faba* L.) sown and harvested on the same dates. J. Agric. Sci. 70 : 403-404.
4. Bond, D.A. 1970. The development of field beans as a crop in Britain. Proc. Nutr. Soc. 29: 74-79.
5. ---------------- 1977. Breeding for zero-tannin and protein yield in field beans (*Vicia faba* L.). In "Protein quality from leguminous crops". A seminar held at Dijon, France, November 3-5 1976. EUR 5686 EN 348-360.
6. Derbyshire, E., Wright, D.J., Boulter, D. 1976. Legumin and vicilin, storage proteins of legume seeds. Review Phytochemistry 15, 3-24.
7. Focke, R., Heinrichs, F., Lou, D. 1977. Ergebnisse und Probleme bei der Zuchtung von Sommerfuttergerster. Tag.-Ber., Akad. Lantwirtsch. -Wiss. DDR, Berlin. 158, 45-55.
8. Frölich, W.G., Pallmer, W.G., Christ, W. 1974. Variation of the contents of protein and of methionine and cystine in *Vicia faba* L. Z. Pflanzenzüchtg. 72, 160-165.

9. Griffiths, D.W., Lawes, D.A. 1978. Variation in the crude protein content of field beans *(Vicia faba)* in relation to the possible improvement of the protein content of the crop. Euphytica 27, 487-495.

10. Mårtensson, P. 1980. Variation in legumin:vicilin ratio between and within cultivars of *Vicia faba* L. var. *minor*. Proc. E.E.C. seminar of seed legumes, Cambridge 27-29 June 1979.

11. Mosse, J., Baudet, J. 1977. Relationship between amino acid composition and nitrogen contents of broad bean seed. In "Protein quality from leguminous crops". A seminar held at Dijon, France, November 3-5 1976. EUR 5686 EN 48-57.

12. Picard, J. 1977. Some results dealing with breeding protein content in *Vicia faba* L. In "Protein quality from leguminous crops". A seminar held at Dijon, France, November 3-5 1976. EUR 5686 EN 339-347.

13. ---------- 1979. Some reflections on problems and prospects in *Vicia faba* breeding. In "some current research on *Vicia faba* in Western Europe". A seminar held at Bari, Italy, 27-29 April 1978.

14. Porter, W.M., Maner, J.H., Axtell, J.D., Keim, W.F. 1974. Evaluation of the nutritive quality of grain legumes by an analysis for total sulphur. Crop Sci. 4, 652-654.

15. Radford, Jr. R.L., Chavengsaksongkram, C., Hymowitz, T. 1977. Utilization of nitrogen to sulphur ratio for evaluating sulphur-containing amino acid concentrations in seed of *Glycine max* and *G. soja*. Crop Sci. 17, 273-277.

16. Rudolph, A., Hanelt, P., Lehmann, CH., Muntz, K., Scholz, F. 1975. Protein-screening am Werzen- und Ackerbohnen-Sortiment in Gatersleben. Die Nahrung 19, 793-807.

17. Scheibe, A., Gunter, I., Zoschke, M. 1969. Die Anwendung der Ruckkreuzungsmethode zur Verbesserung von Eiweissgehalt und Eiweissqualitat bei winterannuellen Futtergersten. Z. Pflanzenzuchtg. 61, 301-338.

18. Schuster, W., Posselt, U. 1977. Protein content and protein quality of some soybean varieties on different locations. In "Protein quality from leguminous crops". A seminar held at Dijon, France, November 3-5 1976. EUR. 5686 EN. 324-338.

19. Silbernagel, M.J. 1971. Bean protein improvement work by USDA - bean and pea investigations. Rep. of the Tenth Dry Bean Res. Confr. Davis, Calif. Aug. 12-14 1970. Agric. Res. Serv. U.S. Dept. of Agric. 70-83.

20. Sjödin, J. 1971. Induced morphological variation in *Vicia faba* L. Hereditas 67, 155-180.

21. Sjödin, J. Mårtensson, P., Magyarosi, T. 1977. The variation in crude protein content and seed weight within field bean plants *(Vicia faba* L.) analysed with the half-seed technique. J. Swedish Seed. Ass. 87, 33-42.

22. ------------------------------------------------- 1981a. Selection for increased protein quantity in field beans (*Vicia faba* L.) Z. Pflanzenzuchtg. 86, 210-220

23. ------------------------------------------------- 1981b. Selection for improved protein quality in field beans (*Vicia faba* L.). Z. Pflanzenzuchtg. 86, 221-230

24. Tandon, O.B., Bresani, R., Scrimshaw, N.S., Le Beau, F. 1957. Nutritive value of beans. Nutrients in Central American beans. J. Agric. Food Chem. 5, vol. 2, 137-142.

25. Vries, A.Ph. de, 1979. In search of characters to be used for indirect selection on grain and protein yield in *Vicia faba* L. In "Some current research on *Vicia faba* in Western Europe". A seminar held at Bari, Italy, 27-29 April 1978. EUR 6244 EN 324-341.

# 35. ANTINUTRITIONAL FACTORS IN FABA BEANS

LAILA A. HUSSEIN

*National Research Center, Giza, Dokki, Egypt*

Raw faba bean seeds contain a number of biologically active factors that produce physiological damage when eaten in large quantities over a long period (29) (22) (23). These "antinutritional factors" include protease inhibitors, tannins, lectins and flatus producing factors.

## Protease Inhibitors

Substances that have the ability to inhibit the proteolytic activity of certain enzymes are found in raw faba beans. The trypsin inhibitor is located in the cytoplasm in the 2S fraction of water-extractable protein, and is a compound of the minor proteins which are found outside the protein bodies (44). In the plant, the protease inhibitors might function to protect the cytoplasm from rupture of the protease-containing bodies or as part of the defense mechanism against invading microbes and insects (34). A protease inhibitor-inducing factor (PIIF) is believed to induce the accumulation of the protease inhibitor in the plant (24).

Information on the deleterious effects of protease inhibitors ($P_I$) in the seeds of faba beans has been reported in several papers (39) (43) (16) (4) (42) (26). However, Borchers *et al.* (8) and Hussein *et al.* (17) failed to detect any inhibitors.

Warsy *et al.* (42) were able to fractionate four protease inhibitors from the seeds of faba beans by using a combination of extraction with 2.5% trichloracetic acid, ammonium sulphate fractionation, column exchange chromatography on CM 52-cellulose and CM-50 sephadex. The authors described the properties of two purified preparations, $BBP_I$-1 and $BBP_I$-2. Their respective isoelectric points were 8.5 and 7.5 with similar molecular weight of 11,000, and both were regarded as isoinhibitors.

They inhibit a large range of proteases from widely different sources including trypsin, chymotrypsin, pronase, thrombin and papain. Pepsin, carboxypeptidase A and B and subtilisin were not affected by either inhibitor. The purified inhibitors conform to the general pattern of legume inhibitors in being capable of the simultaneous and independent inhibition of both trypsin and chymotrypsin in the fashion of double headed inhibitors (7). These double headed inhibitors were evolved from an ancestral gene by internal gene duplication and subsequent mutation, causing differences in the reactive sites.

Major contributions to our understanding of the mechanism of the action of protease inhibitors has come from work done on the Kunitz and Bowman-Birk soybean trypsin

G. Hawtin & C. Webb (Eds.): Faba Bean Improvement
©1982 ICARDA. ISBN -13:978-94-009-7501-9

inhibitors. Trypsin inhibitors function by combining stoichiometrically with the active enzyme to form tightly bound enzyme substrate-like complexes. These inhibitors differ from normal substrates in that the complex formed is stable because of a very complementary fit, and the accumulation of weak, non-covalent bonds at the contact zone. According to Green (13), the affinity of trypsin for the inhibitor is much greater than that for natural substrates.

The complex formation between trypsin and soybean trypsin inhibitor has an association constant of $10^9$ compared with figures of $10^2$ and $10^5$ for the association constants for natural (casein) and synthetic (benzoyl arginine p-nitroanilide) substrates respectively. A common feature for all protease inhibitors is the presence of a disulphide linkage containing a lysine-X or arginine-X bond in a conformation such that this bond is readily exposed to tryptic attack. Consequently the reduction of the disulphide linkage of the faba bean inhibitor with 2-mercaptoethanol results in the immediate dissociation of the enzyme-inhibitor complex. The rate of trypsin liberation on reduction of faba bean protease inhibitors is much faster than the time kinetic observed with other protease inhibitors reflecting a weak association constant between the faba bean protease inhibitor and the trypsin (16).

With regard to the level of protease inhibitor in faba bean seeds, considerable differences have been reported within and between cultivars. A 16-fold difference in trypsin inhibitor activity among different faba bean cultivars had been reported by Bhatty (6) whereas, Sjodin (cited in 26) was able to select from 100 faba bean plants, two lines, one containing a high level and the other a low level of trypsin inhibitors. Marquardt *et al.* (26) were able to demonstrate variations in the trypsin inhibitor activity among eight faba bean cultivars from six locations in Saskatchewan, Canada, Table 1.

Table 1. Mean trypsin inhibitor level of the whole seed, cotyledon and hull of different cultivars. After Marquardt *et al.* (26).

| | Whole seed | Cotyledon | Hull |
|---|---|---|---|
| | Trypsin inhibitor units/g* | | |
| Ackerperle | 3.2 | 2.9 | 6.7 |
| Bell | 3.2 | 2.7 | 5.9 |
| Blue rock | 3.2 | 2.7 | 6.4 |
| Diane | 4.3 | 4.0 | 5.9 |
| Erfordia | 2.7 | 2.4 | 4.0 |
| Poecnvyje | 2.7 | 2.4 | 4.0 |
| Herra | 3.2 | 2.9 | 5.3 |
| Kleinkornige | 3.2 | 2.9 | 5.9 |
| Average | 3.2 | 2.9 | 5.6 |

*The amount of inhibitor that would inhibit one trypsin unit by using the synthetic substrate BAPA, according to the method of Sambeth *et al* (35).

A comparison of trypsin inhibitor level of whole faba bean (3.2 units/g) with that of whole soybeans (27 units/g) shows that whole soybeans contain about a nine-fold higher concentration of trypsin inhibitor.

Many workers have tried heat treatments of raw faba bean seeds to inactivate the protease inhibitors. The extent to which this activity is destroyed by heat is a function of pH value, temperature and duration of heating. Complete destruction of residual trypsin inhibitor has only been achieved by adjusting the pH of the inhibitor preparations to 12.0 before exposure to heat treatment at $70^{\circ}C$ for 1 h. In acid media between pH 2.5 and 4.0, the protease inhibitor has maximum stability (42, 26), so that heating at $100^{\circ}C$ for 1 h had no effect on the antitryptic activity.

*Nutritional Significance of Protease Inhibitors*

Pancreatic enlargement takes place in the rat and chicks in response to dietary trypsin inhobitors. Pancreatic hypertrophy is the consequence of release of the hormones pancreozymin (PZ) or cholecystokinin-pancreozymin (CCK-PZ) from the intestinal wall in response to the dietary trypsin inhibitors (38). The release of the humoral agents stimulates the trypsin enzyme secretion of the rat pancreas. Trypsin inhibitor failed to produce pancreatic enlargement in calves, dogs and growing pigs (20, 30). A direct relationship has been established between the size of the pancreas and the sensitivity of response to the trypsin inhibitor. The pancreas of animal species whose weights exceed 0.3% of the body weights became hypertrophic, whereas; those whose weights were below this value did not respond to the trypsin inhibitor. Abbey *et al.* (Table 2) reported pancreatic and intestinal enzymatic changes.

In addition, Barnes and Kwong (5) studied the effect of trypsin inhibitor on the metabolism of several amino acids by using radioactive isotopes. The workers concluded that the inhibitor interferes with the catabolism of threonine, valine and with the incorporation of cystine into proteins by blocking the enzyme cystathionine synthetase.

Recently, Rackis *et al.* (33) finished long-term feeding studies in rats fed commercially toasted defatted soy flour diets from weaning to adulthood. Trypsin inhibitor content of the soy diets was 178mg/100 g, corresponding to $338.2 \times 10^3$ trypsin inhibitor units/100 g (approaching the concentration found in the raw faba beans, Ref. 26).

The dietary protein level was 20% with soy protein representing 75% of the total protein. Rats fed soy diets initially grew at an equal rate to those fed a comparable casein-corn control diet. With continued feeding for about 300 days, body weights of the soy flour-fed rats ceased to increase, unless the diet was supplemented with vitamin $B_{12}$. The workers stressed that more research was needed to determine whether the nutrient requirements of rats changed substantially during continuous consumption of trypsin inhibitors.

*Tannins*

The seed coats (hulls, testae) of numerous varieties of faba beans contain condensed tannins of the proanthocyanidin types, which depress organic matter and nitrogen digestibilities of diets in rats, chicks and ducklings (25) (14) (27) and inhibit tyrpsin.

Tannins may be classified as any polyphenolic substance that has a molecular weight higher than 500 (37). They are enzymatically hydrolyzed to a sugar residue and a phenolcarboxylic acid. Certain faba bean cultivars contain high levels of condensed tannins (polymeric flavoids) in their hulls. Martin-Tangue *et al.* (27) elucidated the chemical structure of condensed tannin in eight different cultivars of faba beans from different localitites. Condensed tannins were isolated by the ether precipitation method of Quesmel (32) whereby four different fractions, varying in their catechin, gallocatechin and flaven-3,4-diols (leucocyanidin, leucodelphinidin) contents were prepared (Table 3).

Table 2. Nutritional Significance of Protein Inhibitors

| DIET | Intake/7 days T.I.U. | Intake/7 days food | N-Digest. | PER | Pancreas weight g/kg | Pancreatic Trypsin | Pancreatic Chymotrypsin* | Intestinal Trypsin | Intestinal Chymotrypsin** |
|---|---|---|---|---|---|---|---|---|---|
| Raw faba bean | 32.5 | 89 | 84 | 0.7 | 7.9 | 81 | 49 | 306 | 285 |
| Autoclaved faba bean | | | | | 5.7 | 140 | 77 | 143 | 130 |
| Casein + FB-$P_I$*** | 65.6 | 128 | 91 | 1.9 | 6.5 | 193 | 136 | 314 | 260 |

*Enzyme activity/whole pancreas in g/kg body weight
**Enzyme activity/intestine in g/kg body weight
***Protease inhibitor isolated from faba bean was included in the casein diet at a level of 1.25 g $P_I$/kg casein diet. (1) (2).

Table 3. Total tannin content, it four fractions (A-D) and the nutritional value of different faba bean cultivars.

| Cultivars | Origin | Tannin content | | | | | N-digestibility | B.V.* | NPU** |
|-----------|--------|------|------|------|------|-------|-----------------|-------|-------|
| | | A | B | C | D | Total | | | |
| | | mg/g | | | | | % | % | % |
| S 45 | | | | | | | | | |
| Raw | England | 4.6 | 38 | 12 | 0 | 54.6 | 71 | 34 | 47.4 |
| Steam cooked | | | | | | | 77 | 40 | 51.1 |
| G 77 | France | 3.9 | 25 | 7 | 0 | 36.0 | 79.5 | 61 | 76.2 |
| Bianka | Holland | 0 | 0 | 0 | 0 | 0 | 80.5 | 62.5 | 77.3 |
| 972 D | France | 10.0 | 40 | 9 | 0.9 | 60 | | | |
| 48 B | France | 8.1 | 29 | 12 | 2.1 | 51.2 | | | |
| Ad 23 | England | 4.0 | 20 | 10 | 0 | 34 | | | |
| G 107 | Germany | 0.9 | 18 | 9.5 | 0 | 28.4 | | | |
| Po Ad 74 | Poland | 6.1 | 15 | 7.7 | 1.1 | 30 | | | |

*Data taken from Pereira and Pion (31)

*B.V. = $\dfrac{\text{N-retained}}{\text{N-ingested}} \times 100$

**NPU = $\dfrac{\text{N-retained}}{\text{N-absorbed}} \times 100$

The authors claimed that the tannin quality in fractions C and D was very potent in decreasing the % laying rate of ducklings and the egg weights.

Elias et al. (11) found that tannin concentration was high in colored seed coats, and low in white coated beans. Moreover the same workers found a correlation between tannin concentration in the seed coat and trypsin inhibitor activity. This trypsin inhibitor activity or "the heat resistant factor" in the seed coat was found in the cooked beans and their broth. The tannins (m-digallic acid) and their glucosides (hydrolyzable tannins) inhibit a large variety of enzymes as shown by Tamir and Alumot (40) and Haslam (15). They exert their action either by direct inhibition of the proteolytic enzymes, or by forming indigestible complexes with the food proteins.

The faba bean hulls are much higher in their trypsin inhibitor levels (average 5.6 units/ g) compared to the cotyledon (2.9 units/g). Probably most of the trypsin inhibitor activity of the hulls is attributable to tannins. However the trypsin inhibitor activity of faba beans was destroyed by autoclaving at $120^{\circ}C$ for 20 min, extrusion cooking at $152^{\circ}$ C, or by microwave radiation at $107^{\circ}C$ for 30 min.

*Lectins (hemagglutinins)*

The presence of glycoproteins in legumes which have the ability to agglutinate the blood cells has been recognized for a long time. The only common characteristic of the plant lectins is that all are proteins. The *Vicia faba* lectin molecule consists of two apparently identical sub-units (3). Each sub-unit contains two metal-binding sites and another site for a sugar residue. The physiological role of lectins (from the Latin legere, "to choose") in the plant is by acting as "carbohydrate catchers" for the transport and immobilization of specific sugar types (9). By using immunological methods, Mialomier *et al.* (28) localized the bean lectin in the cytoplasm of the cotyledon and the embryo, and found that it disappeared during germination.

*Mechanism of action :* As already mentioned, the lectins are proteins that have a specific affinity for certain sugar molecules. Since carbohydrate moieties exist in most animal cell membranes, they attach to the receptor groups if the specific structure of the latter fits the former. The observation that *Vicia faba* lectins react differently towards the erythrocytes of different animal species, aroused the interest of immunologists with regard to their potential use as blood typing reagents. The interaction of lectin components with glycoproteins on the cell surface, is manifested in vitro by an agglutination of the cells. Studies with rat, mouse and human erythrocytes have demonstrated that faba bean lectins specifically interact with D-mannose and D-glucoseamine residues on the surface of erythrocytes, whereas, soybean lectins react with D-galactose amine residues (26). Further, the relative hemagglutinating activities of faba beans differ from that of raw soybeans when tested with red blood cells from different animals. (Table 4). Mouse, turkey and rat erythrocytes were precipitated in the presence of faba bean but not soybean extracts. Both soybean and faba bean extracts precipitated erthrocytes from pigs and rabbits.

Table 4. Agglutination of avian and mammalian erythrocytes by extracts from faba beans and soybeans (26).

| Species | Faba bean cv. Diana 1000 HU*/g seed | Soybean cv. Portage |
|---|---|---|
| Mouse | 1.1 | 0 |
| Pig | 0.5 | 1.1 |
| Turkey | 1.1 | 0 |
| Rabbit | 4.4 | 8.2 |
| Rat | 1.1 | 0 |

*Hemagglutinating unit

In *Phaseolus* beans, the lectin type A, the most active in agglutination of blood cells and the most toxic, is inherited as a single dominant factor when crossed with D-type beans (36). Marquardt *et al.* (26) screened eight faba bean cultivars for the presence and level of hemagglutinating activity. (Table 5).

In Egypt, a lower level of only 640 hemagglutinin units/g raw faba bean was found (17). All hemagglutinin activity is in the cotyledons.

Table 5. Hemagglutinating activity of 8 faba bean cultivars. After Marquardt *et al.* (26).

| Cultivars | 1000 HU*/g whole seed |
|-----------|----------------------|
| Ackerperle | 4.4 |
| Bell | 3.9 |
| Blue rock | 3.6 |
| Diane | 4.4 |
| Erfordia | 5.6 |
| Foecnvy | 3.8 |
| Herra | 3.4 |
| Kleinkornige | 3.4 |

*Hemagglutinating unit.

*Nutritional significance of lectins* : It can be concluded that the action of lectin is to combine with the cells lining the intestinal wall, thus causing nonspecific interference with the absorption of nutrients. Support for this hypothesis comes from Etzler and Branstrator (12) who found that a number of different lectins react with crypts and villi of the intestine at different regions, dependent on the specificity of the lectins.

Jaffe (19) reported that the presence of 0.5% hemagglutinin in the rat diet is enough to introduce impairment in the intestinal absorption resulting in death.

The simultaneous existence of protease inhibitors and their stimulation of the excretion of pancreatic juice may also lead to an excessive loss of endogenous protein; obscuring the effect of lectins on intestinal absorption, since both will result in increased fecal nitrogen excretion.

### Flatus Producing Factors

One of the main reasons why people limit their consumption of legumes is the ability of the seeds to produce gas in the gastrointestinal tract. This is referred to as "flatulence". Calloway *et al.* (10) and van Stratum and Rudrum (41), showed that rats and people produce large quantities of intestinal gas composed largely of hydrogen, carbon dioxide and methane, after eating various kinds of beans. The oligosaccharides raffinose and stachyose have been implicated as causing flatus. These oligosaccharides are related by having one or two alpha-D-galactopyranosyl groups attached to sucrose via alpha-1,6-galactosidic linkages.

Owing to the absence of enzymes in the human intestinal mucosa that are capable of hydrolyzing this linkage, the intact oligosaccharides accumulate in the lower intestine where they are fermented by anaerobic bacteria to produce gases which are responsible for the characteristic features of flatulence, namely, cramps, diarrhoea, abdominal rumbling and the associated social discomforts.

As some soybean cultivars have a very low content of raffinose and stachyose (18) it may be possible to reduce the flatulence properties of legumes by selective breeding.

These sugars are partially removed from the beans either by alcohol-water extraction (41) or by limiting proteolysis with microbial proteolytic enzymes (10).

*Other factors*

The presence of phytate (myoinositol hexadihydrogen phosphate) in faba beans has gained great attention in recent years because of its implication in mineral dificiencies. Since the phytate content of the seeds is liable to change according to the storage conditions, temperature and humidity and fermentation period, its level can successfully be controlled by modern means of food technology.

The favism-inducing factors; tentatively identified as vicine, convicine and their aglycones are the subject of a seperate chapter (Chapter 36).

In conclusion, the marked variation in the concentration of antinutritional factors among different faba bean cultivars deserves special attention from the plant breeder to monitor levels in new strains. Further, the findings of Rackis *et al.* (33) that long-term ingestion of small doses of trypsin inhibitors increased the vitamin $B_{12}$ of the animal should be reassessed in human biological experiments.

The marked resistance of lectin to inactivation by dry heat should be brought to the attention of authorities calling for adding raw bean flour to wheat flour in bread formulations. Further, mixtures of partially cooked ground beans and cereals have been recommended in child feeding programs in less developed countries. As cooking such mixtures requires a relatively short heating time, the lectin may not be completely destroyed (21). Furthermore, primative cooking is often done in earthen pots on a wood fire, so that with a tough viscous mass such as cooked beans, heat transfer may be imperfect and the temperature in parts of the preparation may well be too low to destroy the lectins.

## REFERENCES

1. Abbey, B., Neale, R.J., Norton, G. 1979. Nutritional effects of faba bean (*Vicia faba* L.) proteinase inhibitors fed to rats. Br. J. Nutr. 41, 31-38.
2. Abbey, B., Norton, G., Neale, R.J. 1979a. Effects of dietary proteinase inhibitors from field bean on pancreatic function in the rat. Br. J. Nutr. 41, 39-45.
3. Allen, A.K., Desai, N.N., Neuberger, A. 1976. Biochem. J. 155, 127.
4. Anantharaman, K., Carpenter, K.J. 1969. J. Sci. Fd. Agric. 20, 703.
5. Barnes, R.H., Kwong, E. 1965. J. Nutr. 86, 245.
6. Bhatty, R.S. 1974. Can. J. Plant Sci. 54, 413.
7. Birk, Y., Gertler, A., Khalef, S. 1967. Biochem. Biophys. Acta. 147, 402.
8. Borchers, R., Ackerson, C.W. 1947. Archs. Biochem. 13, 219.
9. Boyd, W., Everhart, D., McMaster, M. 1958. J. Immunol. 81, 414.
10. Calloway, D.H., Hickey, C.A., Murphy, E.L. 1971. J. Food Sc. 36, 251.
11. Elias, L.G., Deffernandez, D., Bressani, R. 1979. J. Food Sci. 44, 524.
12. Etzler, M., Branstrator, M.L. 1974. J. Cell Biol. 62, 329.
13. Green, N.M. 1953. J. Biol. Chem. 205, 535.
14. Guillame, J., Belec, R. 1977. Br. Poult. Sci. 18, 573.
15. Haslam, E. 1974. Biochem. J. 139, 285.
16. Hochstrasser, K., Schwartz, S., Illchmann, K., Werle, E. 1968. Hoppe-Seyler's Z. Physiol. Chem. 349, 1449.
17. Hussein, L., Gabrial, G., Morcos, S. 1974. J. Sci. Fd. Agric. 25, 1443.
18. Hymowitz, R., Collins, F. 1974. Agron. J. 66, 239.
19. Jaffe, W.G. 1960. Arzneim. Forsch. 10, 1012.
20. Kakade, M.L., Thompson, R.D., Englesrad, W.D., Behrens, G.C., Yoder, R.O., Crane, F.M. 1976. J. Dairy Sc. 59, 1484.
21. Korte, R. 1972. Écol. Food Nurt. 1, 303.

22. Liener, I. 1973. In: "Nutritional Improvements of Food Legumes by Breeding." Proceed. Symp. Protein Advisory Group. July 1972 FAO, Rome.

23. Liener, I. 1980. "Toxic Constituents of Plant Foodstuffs." 2nd ed., A.P., N.Y.

24. McFarland, D., Ryan, C.A. 1974. Plant Physiol. 54, 706.

25. Marquardt, R., Campbell, L. Stothers, S., McKirdy, J. 1974. Can. J. Anim. Sci. 54,177.

26. Marquardt, R., McKirdy, J.A., Ward, T., Campbell. L.D. 1975. Can. J. Anim. Sci. 55, 421.

27. Martin-Tangue, J., Guillaume, J., Kossa, A. 1977. J. Sci. Fd. Agric. 28, 757.

28. Mialonier, G., Privat, J.P., Monsigny, M., Kahlem, G., Durand, R. 1973. Physiol. Veg. 11, 519.

29. National Academy of Sciences, National Research Council 1973. "Toxicants Occurring Naturally in Foods." 2nd ed., N.A.S.-N.R.C., Washington, D.C.

30. Patten, J.R., Richards, E.A., Pole, H. 1971. Proc. Soc. Exp. Biol. Med. 137, 59.

31. Pereira, E., Pion, R. 1980. In: "Protein quality from leguminous seeds". Ed. Commission Europ. Communities.

32. Quesmel, V.C. 1968. Phytochemistry 7, 1583.

33. Rackis, J.J. McGee, J.E. 1979. J. Am. Oil Chemists' Soc. 56, 162.

34. Ryan, C.A., Santarius, K. 1976. Plant Physiol. 58, 683.

35. Sambeth, W., Nesheim, M., Sefrari, J. 1967. J. Nutr. 92, 479.

36. Schertz, K.F., Jargelky, W., Boyd, W.C. 1960. Proc. Natl. Acad. Sci. 46, 529.

37. Singleton, V.L. Kratzer, F.H. 1973. In: "Toxicants Occurring Naturally in Foods." Natl. Acad. Sci.-Natl. Res. Counc., Washington, D.C. pp. 309-345.

38. Snook, J.T. 1969. J. Nutr. 97, 286.

39. Sohomie, K. Ambe, K.S. 1955. Nature (London) 175, 508.

40. Tamir, M., Alumot, E. 1969. J. Sci. Fd. Agric. 20, 199.

41. Van Stratum, P.G., Rudrum, M. 1979. J. Am. Oil Chemists' Soc. 56, 130.

42. Warsy, A.S., Norton, G., Stein, M. 1974. Phytochemistry 13, 2481.

43. Wilson, B.J., Montab, J.M., Bently, H. 1972. J. Sci. Fd. Agric. 23, 679.

44. Wolf, W., 1972. In: Symp. Seed Proteins. Ed. Inglett, pp. 231. The Avi. Publ. Conneticut.

# 36. FAVISM

RONALD R. MARQUARDT


*Department of Animal Science, University of Manitoba, Winnipeg, Manitoba, Canada*

Favism is used to describe a hemolytic syndrone in susceptible individuals after eating seeds of *Vicia faba*. It is characterised most often by weakness or fatique, pallor, jaundice and hemoglobinuria. In hospitalised patients, large amounts of hemoglobin in the urine persists for 1-3 days. Small amounts of oxyhemoglobin and methemoglobin are also found in the urine, which appears dark brown, red or even black. Death may occur from renal failure (1). In severe cases, erythrocyte counts are markedly reduced and blood transfusions are necessary.

The onset of symptoms may be only a few hours and is usually less than 24 h. In Egypt most of the cases occur among infants on their first exposure to the bean.

Comprehensive reviews on favism and related topics have been published. Dacie (10) emphasised the clinical and hematological aspects of favism and his review is the best source of early and obscure references to favism. Mager, Razin and Hershko (19) cover the biochemical properties of the bean and its effect on red cell metabolism. Belsey (1) has reviewed the epidemiology of favism. A general review of hemolytic anemia in disorders of red cell metabolism has been published by Beutler (2).

## Incidence and Geographic Distribution and Mortality

Favism occurs principally in the Mediterranean countries. In Egypt, Greece and Italy, reports of favism go back to antiquity. The highest incidence of favism has been reported in recent times in Sardinia, with five cases per year per 1,000 population.

Outside the Mediterranean basin, favism has been reported from other Middle Eastern countries such as Iraq and Iran, and from Bulgaria. In 1965, the incidence of favism in three provinces in Iran varied from 2 to 9 per 10,000 people. It has also been reported from Szechwan province in western China and Kwantung province in southern China. In contrast, the disease is observed only sporadically in other European countries, the United States and Canada.

In Sardinia, mortality has been reported to be as high as 2 per 10,000; in Bulgaria and Kwangtung the case mortality was 2.1% and 2.3% respectively.

## Seasonal, Age and Sex Distribution

Whenever favism is endemic, the disease generally has a characteristic seasonal distribution

*G. Hawtin & C. Webb (Eds.): Faba Bean Improvement*
*©1982 ICARDA. ISBN -13:978-94-009-7501-9*

corresponding to the harvesting of fresh beans, although cases also occur in other seasons. In Egypt, cases in the spring are not frequent until several weeks after the fresh beans are ripe and are related to the consumption of stewed dried faba beans (medamass). A second peak occurs in November and corresponds to the time of marketing of faba beans from Manuf that have been stored in a way peculiar to Egypt.

In the Caspian region, fresh beans account for nearly all cases of favism (26-70%) and cooked beans for a much lower incidence (5-27%). In Egypt nearly all cases are associated with dried stewed beans as this is the form in which they are mainly eaten. It has also been reported that inhalation of the pollen of *Vicia faba* flowers will cause favism. Studies by Kathamis *et al.* (15) on a large number of patients fail to confirm the occurrence of pollen-produced cases of favism. In Mazanderan, Iran, no reported cases have occurred in January or February, a period during which the flowers are in bloom.

Favism is generally a pediatric disease. In Greece, 65% of the cases occurred in the 2-5 year age groups, 7.2% in th 6-15 year groups, and 5.5% in infants. In Egypt it has been reported that 50% of the patients were under one and 95% under five years of age.

Favism is also a disease to which male children are predisposed; the ratio of male to female cases ranged from 21:1 in Cyprus to 2.7:1 in Mazanderan, Iran. With increasing age, the proportion of males that suffered from the disease decreased to a much greater extent than it did in females (1).

*Relationship between favism and glucose 6-phosphate dehydrogenase deficiency*

In 1956, Crosby, in reviewing the situation in Sardinia, suggested that hemolysis caused by faba beans resembled what was caused by primaquine in certain individuals with an intrinsic abnormality in erthrocytes. The inherent defect, leading to drug sensitivity, was found to consist of a low concentration and relative instability of reduced glutathione (GSH) in erythrocytes, and of a deficiency of the enzyme glucose-6-phosphate dehydrogenase (G6PD). It has since been found that all persons with a history of favism have G6PD deficiency and that they show the same characteristics of a low level of GSH in erythrocytes and an instability of GSH towards acetyl phenylhydrazine similar to that found in drug-sensitive G6PD-deficient individuals (18). G6PD deficiency which affects more than 100 million people is one of the most prevalent inherited enzymatic defects of clinical significance. Variants of G6PD have been divided into five class (2).

In Mediterranean countries such as Greece, Turkey, Israel, Egypt and Italy, G6PD-Mediterranean (Class 1) appears to be the most prevalent, although careful analysis has shown that a number of different severely deficient variants are also present. Class 1 variant includes those which are regularly associated with nonspherocytic hemolytic anemia and is the only class that is susceptible to favism. They are characterised by absence of spherocytosis, normal osmotic fragility, and normal hemoglobin. It is of special interest that the residual G6PD activity of class 1 variants is often higher than that of class 2 variants in which the severe enzyme deficiency is not associated with the chronic hemolytic diseases, except in unusual instances. The other classes of G6PD also are not associated with hemolytic anemia.

The functional impairment of G6PD activity in class 1 variants is, in many cases, much greater than is suggested by the results of assays of mixed red blood cells under standard conditions. The reason for this discrepancy is not always apparent. In a number of variants, the enzyme shows marked sensitivity to the inhibitory effects of NADPH. In some of these variants, the enzyme is quite active when assayed *in vitro* at low NADPH concentrations, but may be nearly devoid of activity under *in vitro* conditions. Other factors which may play a role are susceptibility to inhibition by ATP and the very high Michaelis constants for $NADP^+$ and/or glucose 6-phosphate. Instability is a characteristic that is particularly common in class 1 variants. In some cases, enough enzyme is present in very young erythrocytes to account for a moderate amount of residual activity in hemolysates, but disappears totally after the red cells reach 5 or 10 days old. The absence of enzyme and corresponding decrease in concentration of GSH content might account for the vulnerability of red cells to destruction in the circulation (2).

Gaetani *et al.* (13) support these observations. They reported that both NADPH/ $NADP^+$ ratio and the level of GSH were reduced after the accidental intake of faba beans by sensitive G6PD-deficient subjects. They concluded that the metabolic effects induced by faba beans may be attributed to oxidative stress probably associated with an inhibitor effect of some unknown metabolite on the pentose phosphate pathway. Decreased excretion of D-glucaric acid also has been reported to occur in persons who are susceptible to the hemolytic effects of the faba bean. This observation and that of impaired formation of salicylamide glucuronide were interpreted as indicating a defect in glucuronide formation in such people (5) (6).

*Genetic aspects of glucose 6-phosphate deficiency :* G6PD deficiency is transmitted by a gene located in the X-chromosome. Accordingly, the enzyme deficiency is fully expressed in the hemizygous (X Y) male, because the mutant gene $(X^1)$ is not counteracted by the normal allele (X). On the other hand, full expression is rather uncommon in females, since it depends on the statistically rare occurrence of a homozygous mutant genotype $(X^1X^1)$ It may be hypothesized that the females with the genotype $(XX^1)$ would have enzyme activities that were of an intermediate nature. However it has been demonstrated, that women with the $(XX^1)$ genotype may have G6PD activity ranging from zero to normal. This seeming anomaly is attributable to the inactivation of one of the two X chromosomes in the female conceptus during early embryonic development. The result is a mosaic of X chromosome activity in the human female; some cells transcribe genes on the maternally-derived and others on the paternally-derived X chromosomes. In the erythroid series, the final proportion may vary between the two extremes and therefore the degree of susceptibility to fabic hemolysis in heterozygous females will be critically dependent upon the relative proportions of G6PD deficient and normal erythrocytes in their blood.

*Genetically inherited and nutritional modifying factors :* It has been postulated that other factors in addition to G6PD deficiency (class 1) influence the susceptibility of an individual to favism. Stamatoyannopoulos *et al.* (26) have shown that in the area of Karditsa with a frequency of G6PD deficiency of about 27%, favism is rare, while in Corfu Island, with an incidence of G6PD deficiency of about 5%, favism is relatively frequent. In both areas, the consumption of faba beans is very common. The researchers concluded that a second genetic factor in addition to G6PD deficiency may also influence the susceptibility of an individual to favism. Bottini *et al.* (3) reported that the susceptibility of G6PD-deficient erythrocytes to hemolysis was highest when the acid phosphatase was of the A phenotype, and lowest with the B phenotypes. Carcassi (4) reported that the incidence of favism was greatly reduced in individuals that also had the thalassemia trait. They concluded that thalassemia may therefore constitute a protective factor against the hemolytic crises. Palmarino *et al.* (25) have demonstrated recently that the thalassemia trait exerts a protective action against favism only in subjects who carry the A allele for acid phosphatase. This data strongly suggests an interaction between thalassemia and acid phosphatase mediated by the habit of eating *Vicia faba;* this interaction conditions the susceptibility to hemolytic favism in G6PD deficient subjects.

Corash *et al.* (8) have also demonstrated that nutritional factors may influence the degree of hemolysis in red cells of G6PD-deficient subjects. High doses of vitamin E supplementation decreased the degree of chronic hemolysis as evidenced by improved red cell life span, increased hemoglobin concentration and decreased reticulocytosis.

### Search for the Causative Agent of Favism

Bein and coworkers (19) demonstrated that faba bean extracts were capable of oxidizing GSH in G6PD-deficient but not normal erythrocytes, when incubated *in vitro* in the presence of glucose. Some of the purified fractions were sparingly soluble in water, and exhibited in neutral solution a rapid loss of GSH-oxidizing activity, concomitant with a change in their spectral characteristics. These and other properties were similar to those described for some pyrimidine derivatives known to occur in faba beans in the form of aglycones of the *B*-glycosides termed vicine and convicine.

*Vicine, convicine and their aglycones :* Fig. 1 gives the structure of vicine, convicine and their hydrolytic products. It shows that the aglycones, divicine and isouramil can be obtaned from the respective glycosides (vicine and convicine) by mild acid hydrolysis or by enzymatic splitting with *B*-glucosidase.

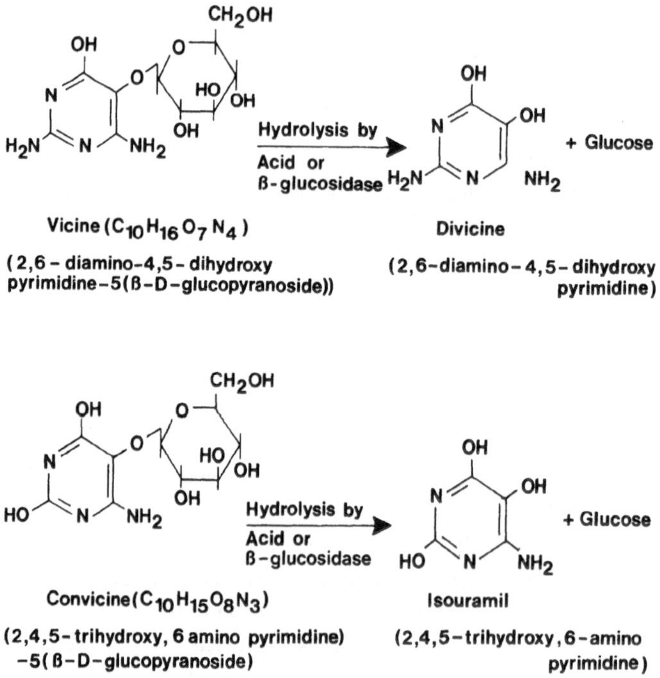

Fig. 1. Structure of Vicine, Convicine and their aglycones.

Both divicine and isouramil are highly active reducing reagents, and are highly unstable in oxygen particularly at an alkaline pH or in the presence of trace quantities of $Cu^{++}$ and other heavy metals. In contrast, the glycosides show none of the reducing properties of their aglycones; are remarkably heat-stable in solution, and their ultraviolet spectra differ significantly from their constituent pyrimidines.

*Possible ethiological role of divicine and isouramil in favism* : The overall pattern of metabolic disturbances resulting from incubation of G6PD-deficient red cells with aglycones of vicine and convicine is essentially identical to that elicited by treatment with acetylphenylhydrazine (2) (18). The powerful capacity for oxidizing GSH exhibited by the pyrimide aglycone *in vitro* is consistent with a possible causative role of these substances in precipitating the favic crises. The free aglycones may arise from the parent faba bean glycosides either in the bean or in the digestive tract through the hydrolytic action of *B*-glucosidase. Variations in conditions for the hydrolysis of the *B*-glucosides, in the lability of the hydrolyzed compounds, and in the nutritional status of the individual might account for the puzzling irregularity that characterises the occurrence of favism in susceptible individuals; irrespective of the degree and frequency of their exposure to the noxious agents.

*Dihydroxyphenylalanine (DOPA)* : Mager *et al.* (19) demonstrated that no appreciable oxidation of GSH took place when a mixture of GSH (2 mM) and DOPA (4 mM) were incubated at $37^{\circ}C$ with continuous shaking in the air. Isouramil (1 mM) caused, under the same conditions, an almost complete disappearance of the intracellular GSH. On the other hand, combined additions of 1 mM DOPA and 0.2 mM isouramil resulted in nearly 80% destruction of the intraerythrocytic GSH, while each compound alone was without perceptible effect. Similar results were obtained with sodium ascorbate and isouramil. The superadditive nature of the combined effects of isouramil and ascorbate on GSH may also play a role in the pathogenesis of favism.

## Animal Models for Studying Favism

Collier *et al.* (7) demonstrated that a diet high in faba beans did not impair growth or feed intake in weanling rats as compared with control animals. Erythrocyte hemolysis and liver GSH levels were also not affected by the diet. Lin and Ling (16) reported that vicine retarded the growth of rats and induced mild hemoglobinuria in juvenile dogs. Other effects did not occur or were not studied. D'Aquino *et al.* (11) reported that a challenge of red blood cells of rats with isouramil resulted in a drastic decrease in GSH in riboflavin-deprived rats. They suggested that the reduced glutathione reductase activity of the riboflavin-deprived red cells can be used as a model mimicking the G6PD-deficient red cells at least for the response of GSH concentration to factors such as isouramil.

Previous studies by several researchers have established that faba beans caused reduced egg size in laying hens. Recent studies in our laboratory (23, 24) have identified and characterized the mode of action of this factor. The initial studies demonstrated that the egg weight-depressing factor was vicine, and possibly convicine. The results also indicated that egg weight depression was mediated via a reduction in the size of the yolk. It may be concluded that the laying hen appears to be much more sensitive to possible favism-producing factors than any other animal species.

Tables 1 and 2 show results of studies by Muduuli (22) and Muduuli *et al.* (23). Table 1 demonstrates that egg weight, egg production rate, egg mass, yolk weight, yolk mass, ratio of yolk height/diameter, percent blood spots on the yolk, plasma lipid levels, plasma vitamin E/lipid ratio, plasma lipid peroxide levels and the weight of liver were affected by dietary vicine. Table 2 shows that vicine depressed egg size, and that this effect was overcome to a small degree by the addition of high levels of vitamin E to the vicine-containing diets. In this study, vicine depressed percent hematocrit and hemoglobin levels, and markedly increased the level of plasma lipids and degree of spontaneous hemolysis of erthrocytes. Vitamin E only slightly affected the results. In addition, dietary vicine markedly depressed fertility and hatchability of eggs which, in contrast to the above results, was reversed to a considerable degree when high levels of vitamin E were added to the vicine-containing diet.

Table 1. Laying hen performance and yolk, plasma and liver characteristics as affected by dietary vicine (from Muduuli *et al.*, 1981)[1].

| Treatment | Control diet | Control diet containing 1% vicine |
|---|---|---|
| Bird performance | | |
| Egg weight, g | $55.5 \pm 0.8^b$ | $48.1 \pm 0.7^a$ |
| Egg production rate (eggs/bird/day)[1] | $100 \pm 3^b$ | $91 \pm 5^a$ |
| Egg mass (weight x production rate)[1] | $100 \pm 3^b$ | $82 \pm 4^a$ |
| Yolk characteristics | | |
| Yolk weight, g[2] | $100 \pm 4^b$ | $88 \pm 1^a$ |
| Yolk mass (weight x production rate)[2] | $100 \pm 4^b$ | $80 \pm 4^a$ |
| Ratio, yolk height/diameter | $0.48 \pm 0.1^b$ | $0.42 \pm 0.01^a$ |
| Per cent blood spots | $0.8 \pm 0.1^a$ | $4.1 \pm 0.02^b$ |
| Plasma characteristics | | |
| Lipid levels (mg/100 ml) | $560 \pm 50^a$ | $1551 \pm 180^b$ |
| Ratio vitamin E/lipids (mg $10^4$/mg) | $6.0 \pm 0.7^b$ | $3.1 \pm 0.4^a$ |
| Hemolysis score | $1.7 \pm 0.3^a$ | $2.5 \pm 0.3^b$ |
| Lipid peroxides, (nmoles MDA/ml)[3] | $1.5 \pm 0.1^a$ | $8.0 \pm 2.5^b$ |
| Protein (mg/100 ml) | $4.5 \pm 0.2$ | $4.6 \pm 0.1$ |
| Liver characteristics | | |
| Wet weight (% of body weight) | $1.51 \pm 0.04^a$ | $1.71 \pm 0.05^b$ |
| Lipid levels, % | $6.4 + 0.9$ | $6.0 \pm 0.6$ |

[1] Mean ± SE with different superscripts are significantly different P <0.05. The number of birds per treatment group were 20.
[2] Values of vicine fed birds as a percent of those fed the control diet.
[3] MDA, malondialdehyde.

These results demonstrate that vicine has a very dramatic and pronounced effect on the metabolism of the laying hens. The susceptibility of erythrocytes to hemolysis, the increased incidence of blood in the yolk, and decreased hematocrit concentration are similar to the hemolytic effects observed following the ingestion of faba beans by G6PD-deficient humans. Hence the current data supports previous conclusions that vicine or its aglycone is

Table 2. Laying hen performance, plasma characteristics and fertility of eggs as affected by dietary vicine and vitamin E (from Muduuli, 1980)[1].

| Treatment | Control diet | Control diet containing 1% vicine | Control diet containing 1.2% vicine and vit. E[2] |
|---|---|---|---|
| Bird performance | | | |
| Egg weight (g)[3] | 100±2[c] | 90±1[a] | 92±1[b] |
| Plasma characteristics | | | |
| Lipids (mg/100 ml) | 762±52[a] | 1334±156[b] | 1289±299[b] |
| Ratio vit. E/lipid (mg x 10^4/mg) | 4.8±0.6[a] | 4.6±0.9[a] | 9.2±0.3[b] |
| Lipid peroxides (nmoles MDA/ml)[4] | 0.88±0.11[a] | 1.63±0.19[b] | 1.58±0.15[b] |
| Hematocrit (%) | 31±1[b] | 22±1[c] | 23±1[a] |
| Spontaneous hemolysis of RBC (%) | 22±5[a] | 64±6[b] | 54±8[b] |
| Hemoglobin (mg/ml hemolysate) | 25.3±7[a] | 22.4±1.0[a] | 22.8±1.0[a] |
| Fertility and hatchability of eggs[5] | | | |
| Percent fertility | 100 | 41 | 82 |
| Hatchability of total eggs (%) | 92 | 23 | 68 |
| Hatchability of fertile eggs (%) | 92 | 56 | 83 |

[1] Means ±SE with different superscripts are significantly different, $P < 0.05$. The number of birds per treatment group were 10.
[2] Vitamin E level was 10-fold higher (55 IU/kg feed) than the amount added to the other diets.
[3] Values of the vicine fed birds as a percent of those fed the control diet.
[4] MDA, malondialdehyde.
[5] A $X^2$ test demonstrated that the differences in fertility and hatchability were significant at $P < 0.05$.

the causative factor of favism in humans. It may also be concluded that the laying hen can be used as a model for studying the mode of action of the hemolytic agent in faba beans (vicine and probably convicine). This would be the first suitable animal model that has been identified. In addition to causing hemolysis of erythrocytes vicine is also responsible for decreased yolk size (mainly composed of lipids) and increased plasma lipid concentrations. These latter results would suggest that vicine, or its hydrolytic products, interfere with lipid transport or fat metabolism.

In addition, vicine markedly depresses fertility and hatchability in eggs. Vitamin E, which has recently been shown to reduce the incidence of chronic hemolysis in G6PD-deficient people, greatly improved fertility and hatchability in eggs. Other biochemical parameters were not as markedly affected by vitamin E supplementation. The increased incidence of lipid peroxides together with the previous observations of Flohe et al. (12) would suggest that divicine is the active compound and that its effects are due to the generation of free radicals. Although the corresponding effects of convicine have not been established, it is conceivable that they would also be similar to that of vicine. Also other nutrients, in addition to vitamin E, may either enhance or reduce the toxic effects of vicine. Additional studies should be done to establish whether dietary status influences vicine toxicity, and to determine its mode of action.

We may conclude that vicine has a dramatic effect on the metabolism of the laying hen and that it affects not only erythrocyte hemolysis but also other biochemical parameters which could lead to serious metabolic disturbances if continued for long periods. A similar situation may also occur in humans when faba beans are eaten, particularly in individuals who are deficient in G6PD.

## Assays for Vicine, Convicine and DOPA

Most assays have involved a nonspecific colorimetric or ultraviolet test, separation of the compounds using paper or thin layer chromatography followed by quantitation using ultraviolet spectrophotometry or, more recently, by gas-liquid chromatography (see (20) for references). All of these methods are either nonspecific or require ardous and time consuming techniques. The gas-liquid chromatographic method for example requires prior derivatization which is not only time consuming but can lead to incomplete derivatization or the formation of optical isomers. Also procedures for the quantitation of vicine and/or convicine in animal tissues and for the detection of the labile aglycone forms of these compounds has not been reported.

Marquardt and Frohlich (20) recently described a simple and specific method involving the use of reverse-phase liquid chromatography for spearating vicine, convicine and their hydrolytic products. The method involves use of a $C_{18}$ bonded column and is superior to other available analytical techniques because of the simplicity of preparing samples and its sensitivity and rapidity. The chromatogram (Fig. 2A) shows the separating of vicine, convicine, DOPA and other compounds. The elution profile (2B) of dehulled faba beans demonstrates the presence of two major peaks, DOPA and tyrosine. In all cases, total analysis time is less than 8 minutes and sample preparation only involves extraction with 5% perchloric acid followed by centrifugation, filtration and injection into the column. The method is capable of detecting as low as 2 ng of vicine or convicine. This new procedure offers a convenient alternative to the conventional methodologies currently employed, and should greatly facilitate future biological studies involving the metabolism of vicine, convicine, DOPA and related compounds in both the plants and animals. This method should be particularly useful for screening faba beans for low levels of vicine and convicine.

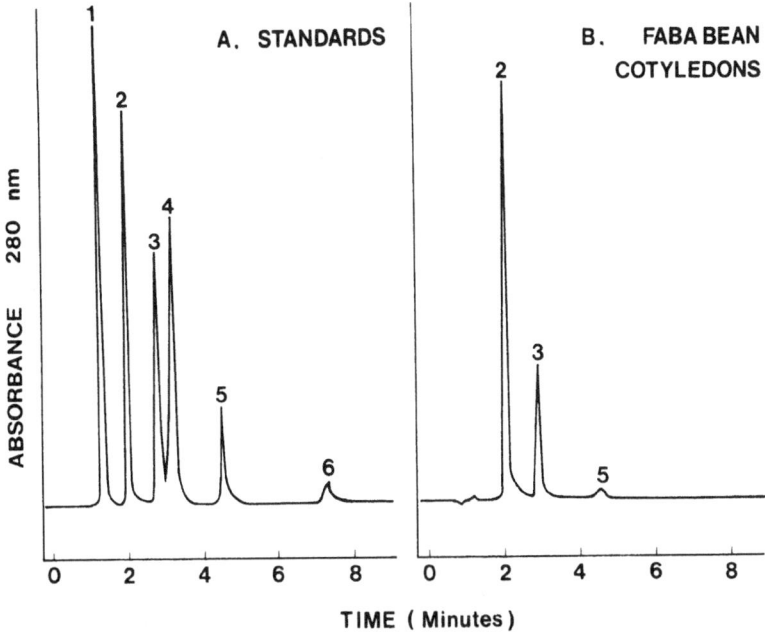

Fig. 2. Chromatogram of a standard solution (A) containing : (1) 0.17 mM cytosine, (2) 0.17 mM (51 μg/ml) vicine, (3) 0.17 mM (50 μg/ml) convicine, (4) 0.17 mM uric acid, (5) 0.33 mM DOPA and (6) 0.33 mM tyr and (B) and extract prepared from the cotyledon portion of faba beans (0.4 g/40 ml).
Compounds eluted in chromatogram B were vicine (2), convicine (3) and DOPA Faba beans and the standard solution, were either extracted or prepared in 5% PCA. Injection volume was 20 μl; flow rate, 2 ml/min., chart speed, 1 cm/min; optical density setting, o.2 and temperature ambient.

*Conclusions*

Favism is a hemolytic sex-linked disease that mainly affects children who have the Mediterranean type of glucose 6-phosphate dehydrogenase deficiency and have eaten faba beans. It appears that certain other genetic factors alter the sensitivity to favism, and that nutritional factors such as vitamin E may either ameliorate or possibly enhance the effects of the hemolytic agents. Although not conclusive, most of the evidence would suggest that the causative agents are vicine; a closely related compound, convicine, or the aglycone forms of these compounds. The aglycones, divicine and isouramil, may initiate the production of a free radical which would not only affect red cell metabolism but also induce perturbations in other tissues.

The identification of a suitable model animal, the laying hen; the development of a rapid and accurate assay procedure (reverse-phase high pressure liquid chromatography) for the assay not only of vicine and convicine but also their algycones, and the ability to prepare large quantities of analytically pure vicine and convicine should facilitate biological studies with regards to mode of action of the hemolytic compounds and means by which their effects can be minimized. Also, the HPLC method will facilitate the selection of faba bean cultivars that are low in these compounds.

## REFERENCES

1. Belsey, M.A. 1973. The epidemiology of favism. Bull. Wld. Hlth. Org. 48: 1-13.
2. Beutler, E. 1978. Hemolytic anemia in desorders of red cell metabolism. Plenum Medical Book Co., N.Y.
3. Bottini, E., Lucarelli, P., Agostino, R., Palmarino, R., Businco, I., Antognoni, G. 1971. Favism : Association with erythrocyte acid phosphatase phenotype. Science 171: 409-411.
4. Carcassi, U.E.F. 1974. The interaction between B-thalassemia, G-6PD deficiency and favism. Ann. N.Y. Academy Sci. 232 : 297-305.
5. Cassimos, C., Malaka-Zafiriu, K. Tsiures, J., Danielides, B. 1974a. Variation in salicylamide glucuronide formation in normal and G-6PD-deficient children. J. Pediatr. 84 : 110-111.
6. Cassimos, C., Malaka-Zafiriu, K., Tsiures, J. 1974b. Urinary D-glucaric acid excretion in normal and G-6-PD-deficient children with favism. J. Pediatr. 84 : 871-872.
7. Collier, H.B., Aherne, F.X., Kennelly, J.J. 1978. Effects of faba bean diet on growth, liver weight, and non-protein thiol levels of erthrocytes and liver in weanling rats. Can. J. Anim. Sci. 58 : 531-532.
8. Corash, L. Spielberg, S., Bartsocas, C., Boxer, L., Steinherz, R., Sheetz, M., Egan, M., Schlessleman, J., Schulman, J.D. 1980. Reduced chronic hemolysis during high-dose vitamin E administration in Mediterranean-type glucose-6-phosphate dehydrogenase deficiency. New England J. Med. 303 : 416-420.
9. Crosby, W.H. 1956. Favism in Sardinia. Blood 11 : 91-92.
10. Dacie, J.V. 1967. The haemolytic anaemia. Second ed., part IV, Churchhill, London, pp. 1061-1077.
11. D'Aquino, M., Gaetani, S., Spadoni, M.A. 1979. A search for an animal model to assay the factor of favism. Nutrition Reports Inter. 20 : 1-9.
12. Flohe, L., Neibch, G., Reiber, H. 1971. The effect of divicine on human erythrocytes. Z. Kiln. Biochem. 9 : 431-437.
13. Gaetani, G.F., Mareni, C., Salvidio, E., Galiano, S., Meloni, T., Arise, P. 1979. Favism : Erthrocyte metabolism during haemolytic and reticulocytosis. British J. Haem. 43 : 39-48.
14. Haidas, S., Zannos-Mariolea, L., Matsaniotis, N. 1975. Red cell 2, 3-diphosphoglycerate levels in children with hereditary haemolytic anaemias. British J. Haem. 31 : 521-530.
15. Kattamis, C.A., Kyriazakou, M., Chaidas, S. 1969. Favism. Clinical and biochemical data. J. Med. Gent. 6 : 34-41.
16. Lin, J.Y. Ling, K.H. 1962a. Studies on favism 2. Studies on the physiological activities of vicine in vitro. J. Formosan Med. Assoc. 61 : 490-494.
17. Lin, J.Y., Ling, K.H. 1962b. Studies on favism. 3. Studies on the physiological activities of vicine in vitro. J. Formosan Med. Assoc. 61 : 579-583.
18. Mager, J., Glaser, G., Razin, A., Izak, G., Bien, S., Noam, M. 1965. Metabolic effects of pyrimidines derived from faba bean glycosides on human erythrocytes deficient in glycose 6-phosphate dehydrogenase. Biochem. Biophys. Res. Comm. 20 : 235-240.
19. Mager, J., Razin, A., Hershko, A. 1969. In : E. Liener (ed.), Toxic constiutents of plant foodstuffs. Academic Press, N.Y. pp 293-318.
20. Marquardt, R.R., Frohlich, A. 1981. A rapid reverse-phase high-performance liquid chromatographic method for the quantitation of vicine, convicine and related pro-

ducts. J. Chromat. 208: 373-379

21. Meloni, T., Pilo, G., Gallisai, D., Dare, A. 1979. Serum glutamic oxaloacetic transaminase, glutamic pyruvic transaminase, gamma-glutamyl transpeptidase and glutamic dehydrogenase levels in favism. Acta. Haemat. 62 : 71-73.

22. Muduuli, D.S. 1980. The quantitative isolation of vicine from faba bean protein concentrate and the involvement of free radicals in the biophysiochemical effects of vicine in the chicken. Ph.D. Thesis, Dept. of Anim. Sci., Univ. of Manitoba, Winnipeg, Canada R3T 2N2.

23. Muduuli, D.S., Marquardt, R.R., Guenter, W. 1981. Effect of dietary vicine on the production performance of laying chickens. Can. J. Anim. Sci. 61 : 757-764.

24. Olaboro, G., Marquardt, R.R., Campbell, L.D. 1981. Isolation of the egg weight depressing factor in faba beans *(Vicia faba* L. var. *minor)*. J. Sci. Food Agric. 32: 1074-80.

25. Palmarino, R., Agostino, R., Gloria, F., Lucarelli, P., Businco, L. Antognoni, G., Maggioni, G., Workman, P.L., Bottini, E. 1975. Red cell acid phosphatase : Another polymorphism correlated with malaria. Am. J. Phys. Anthrop. 43 : 177-186.

26. Stamatoyannopoulos, G., Fraser, G.R., Motulsky, A.G., Fessa, Ph., Akrivakis, A., Papayannopoulou, Th. 1966. On the familial predisposition to favism. Am. J. Hum. Gen. 18 : 253-263.

# 37. COOKING QUALITY OF FABA BEANS

AHMED M. EL-TABEY SHEHATA

*Department of Agricultural Industries, Faculty of Agriculture, University of Alexandria, Alexandria, Egypt*

The cooking quality of faba bean products such as medammis, falafel, nabet and besara, can be evaluated by the following parameters :

*(i) Seed size and weight :* One thousand seeds are weighed and their volume is determined by measuring the change in the level of water in a graduated cylinder. The relative density of beans can be calculated by dividing the weight of 1000 beans by their volume.

As small seeds are believed to be hard to cook, people prefer medium seeds. The weight of seeds is related to their size and chemical composition.

*(ii) Percentage of hulls :* The percentage of hulls affects the nutritional value of products from the whole seeds. The net yields of the dehulled seeds (cotyledons), protein and starch per unit weight of faba beans, are inversely related to the hull content. The percentage of hulls can be determined either on the dry seeds directly or following a few hours of soaking before removing the hull, followed by drying.

*(iii) Hydration coefficient :* The hydration coefficient of raw beans after soaking in distilled water for a defined period, is calculated as the percentage increase in weight of beans :

$$\text{Hydration coefficient} = \frac{\text{weight of soaked beans}}{\text{initial weight of beans}} \times 100 \ (8).$$

The hydration coefficient of cooked beans e.g. stewed whole beans, is calculated by weighing the beans before cooking and after cooking under specified conditions. In case of stewed beans, the raw beans are mixed with tap water (1 : 4) and autoclaved at 120°C for two hours.

*(iv) Swelling coefficient :* The volume of raw beans, before and after soaking, is determined by the absolute displacement method, using water in a graduated cylinder :

$$\text{The swelling coefficient} = \frac{\text{Volume of soaked beans (for a defined period)}}{\text{volume of beans before soaking}} \times 100$$

Both consumers and processors prefer beans that have high hydration and swelling coefficients as these produce greater quantity.

*G. Hawtin & C. Webb (Eds.): Faba Bean Improvement*
©1982 ICARDA. ISBN -13:978-94-009-7501-9

*(v) Colour :* The colour of raw and cooked beans can be measured objectively by colour measuring instruments. Quenzer *et al.,* (19) used the Gardner Color Difference Meter, while Molina *et al.,* (15) used the Lovibond Tintometer type D.

Youssef (26) used a spectrophotometer to measure reflectance of faba beans hulls and cotyledon powders. The C.I.E. characteristics were calculated as described by Mackinney and Little (11).

Colour can be evaluated by panelists using a 5-point scale with highest score indicating the lightest colour which consumers prefer (22).

*(vi) Percentage of germination :* Three hundred seeds are soaked in water for 4 hours before placing them in a container with wet sand or covered with wet cloth and stored at room temperature (about $20^{\circ}C$). Beans are examined every 2 days for 2 weeks, to count and remove the germinated seeds when the radicle is about 5 mm long.

The degree of germination is important for the production of germinated faba beans (nabet).

*(vii) Texture :* Texture is one of the most important cooking characteristics of faba beans. Farmers, processors and consumers value seeds that can be cooked quickly and cooked products which become homogeneously soft without any granulation.

Methods of assessment are based on measuring or evaluating the texture of beans, subjectively or objectively, after cooking under standard specified procedure either for a definite time or for various successive intervals to determine the cooking time.

## Methods for determining cooking time

Muneta (17) cooked faba beans in boiling water for 5 cooking times with 10 minute intervals. The cooked beans were drained, and seed coats were removed before tasting by judges who determined whether the sample was cooked, undercooked or overcooked. The cooking time (C $T_{50}$) was calculated as the time when half of the judges considered that the beans were cooked, or overcooked.

Shivashankar *et al.* (23) cooked 2.5 g of green gram *(Phaseolus aureus* R.) seeds in test tubes with 10 ml. of water. The tubes were placed in a water bath ($96^{\circ}C$) and the seeds were squeezed between the finges at various intervals. Time of cooking was taken when the seed became soft.

In another method, one hundred beans are dropped in 150 ml of boiling distilled water placed in a 250 ml flask and boiled until half of them are split. The time is recorded as cooking time (10).

Cooking time can also be determined by placing 500 g of beans in 1500 ml of water. Fifty beans are removed after 90 minutes of boiling and at 20 minute intervals until 90% or more of the beans are considered to be cooked by squeezing them between the thumb and forefinger (10).

The texture of the cooked beans is evaluated subjectively in all of these methods for determining the cooking time. Burr *et al.,* (3) used an experimental bean cooker to measure the tenderness of cooked beans. He found that a vertical plunger weighing 90 g with a steel needle in its end would penetrate the beans when they become cooked. The cooking time is when the plunger penetrates 50% of the beans.

Methods for determining softness

Youssef (26) used 10 experienced panelists to judge the texture (softness and granulation) according to a 10-point scale, with the highest score for the best texture. Shehata et al., (20) used a 10-point scale for softness. Each judge had to taste 10 cooked faba beans.

Objective measurements with mechanical instruments have been used to determine the softness or hardness of legume seeds.

A tenderometer has been used to measure the softness of cooked faba beans (17) (26). by measuring the force (lbs/in$^2$ or kg/cm$^2$) needed to move the spindles through the sample.

The Universal Penetrometer (7) was adopted to measure the softness of cooked faba beans by using a 100 g weight on top of the penetrating rod (27).

Molina et al., (15) have used the puncture test on 20 beans, and the puncture force is expressed in grams. The same test can be made on raw legume seeds (8).

A cookability index for peas was developed by Chernick and Chernick (4) based on the yield of puree after cooking. Twenty grams of peas are immersed in concentrated sulphuric acid for 30 min., washed and soaked overnight, drained and boiled for 90 min. in Na HCO$_3$ + Na Cl solution (0.05% of each). The mixture is strained through 10 -, 20 -, and 200-mesh sieves. The hard cooking fraction is collected from the first two sieves and the puree from the 200-mesh sieve. Youssef (26) has adapted this method to cooked faba beans.

In beans *(Phaseolus vulgaris)* the degree of cooking has been measured by Kramer shear press (18), while Wang et al. (24) have used the compression test for soybeans.

The texture of purees made from cooked beans can be evaluated by measuring the viscosity with a suitable viscosimeter.

*(viii) Properties of the stewed liquor* : Shehata et al., (21) used different parameters to study the properties of the stewed liquor. They made out the following determinations : weight, volume and relative density, and the content of total solids and soluble solids. Total solids content or viscosity is used to evaluate the thickness of the cooking broth (8).

*(ix) Flavour* : Subjective methods, using experienced panelists, are commonly used for evaluating the flavour of cooked products and their acceptability by consumers.

Relationship between cooking characteristics.

Tenderometer measurements were very high for faba beans (156-200) as compared with other beans (48-69). Tenderometer readings showed considerable variations between replicates, and did not agree with cooking time measurements by taste panelists (17).

Angel and Kramer (1) reported that panel scores were correlated with shear press measurements for canned peas. Penetrometer readings were significantly correlated with panelist scores for 11 samples of faba beans (26).

Shehata et al. (22), after studying 52 faba bean samples from the 1979 crop and 41 samples from the 1980 crop, found that panelist scores for softness, penetrometer readings and hydration coefficients of stewed faba beans were positively correlated with each other The weight of the stewed liquor was negatively correlated with these three characteristics of stewed faba beans. Although total solids and soluble soilds contents of stewed liquor were positively correlated, no significant correlation was found between them and any of the three cooking characteristics of stewed faba beans, (panel score, penetrometer reading, and hydration coefficient). (Fig. 1).

Factors affecting cooking characteristics

a. *Faba bean cultivars and lines* : Although it has been reported that cooking quality of peas is partly genetically controlled, environmental variability and genotype x environmental interactions are quite evident, (4) (6) (5).

Youssef (26) found that the cooking quality of Giza 1 and Giza 2 faba beans that were grown in various regions, varied appreciably.

Samples of 71 lines of faba beans which were supplied by the Food Legume Research Section of the Agricultural Research Center at Giza, Egypt, were evaluated for their cooking characteristics (13). Although these lines were grown under the same environmental conditions, penetrometer readings varied from 41 to 81 with an average of 61. Ten lines had penetrometer readings of more than 71, and 13 lines had readings less than 51.

b. *Fertilizers* : Mattson, et al. (12) reported that higher soluble $P_2O_5$ concentration in the soil increased the phytin content of peas which became easier to cook. Chernick and Chernick (4) concluded that ammonium phosphate and KCl fertilizers did not give significantly different results. Halstead and Gfeller (6) reported a relationship between cooking quality of field peas and additions of P and K fertilizers in a greenhouse experiment, but not in field studies.

Metwally (14) reported that addition of calcium superphosphate at various concentrations to Giza 2 had an effect on the seed weight, crude protein content and oil content.

Addition of urea at the rate of 7.5 to 225 kg per feddan and inoculation of seeds with *Rhizobia* , had no significant effect on the cooking characteristics (penetrometer reading and hydration coefficient of stewed beans) of faba beans (27).

Shehata et al. (21) studied 93 samples of faba beans which were collected in 1979 and 1980. Addition of phosphate fertilizer was negatively correlated with the softness score but not with penetrometer readings. Also nitrogen fertilizer had a negative effect on the panel score.

c. *Location and number of irrigations* : Muneta (17) reported that beans from Michigan required 230-240 minutes to cook while the same cultivars of dry beans from Idaho needed only 60-80 minutes after they had been stored for a long time.

Faba beans grown in Upper Egypt produced stewed beans which had higher penetrometer readings and hydration coefficients than those grown in Lower Egypt. Latitude was negatively correlated with penetrometer reading, and the hydration coefficient of stewed faba beans (21).

Weight and volume of faba bean seeds and percentage of hulls were positively correlated with latitude but negatively correlated with the number of irrigations (21).

d. *Soil properties* : Particle size categories (i.e. clay, silt and sand fractions) were found to be correlated with softness score, and correlated with penetrometer readings. Clay and silt fractions were positively correlated, while the sand fraction was negatively correlated with

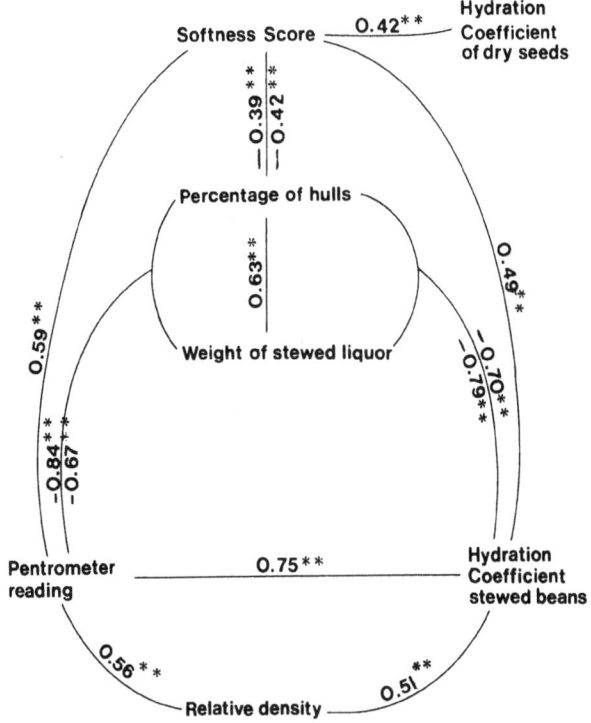

Fig. 1. Correlation coefficients between some of the cooking characteristics of faba bean.

cooking characteristics (21). Hydration coefficient and weight of stewed liquor were similarly affected.

Table 1. Correlation coefficients between certain cooking characteristics of stewed faba beans and soil properties.

| Soil Properties | Softness score | Penetrometer reading | Hydration coefficient | Weight of stewed liquor |
|---|---|---|---|---|
| Salinity | -0.50** | -0.43** | -0.40** | 0.37* |
| Clay fraction | 0.50** | 0.34* | 0.50** | -0.51** |
| Silt fraction | 0.39** | 0.34* | 0.30* | -0.32* |
| Sand fraction | -0.50** | -0.36* | -0.49** | 0.50** |
| Saturation | 0.55** | 0.32* | 0.42** | -0.39** |
| Soluble Ca+ | -0.62** | -0.49** | -0.39** | 0.39** |
| Soluble Mg+ | -0.53** | -0.49** | -0.42** | 0.39** |
| Soluble Na+ | -0.34* | -0.35* | -0.37* | 0.34* |
| Soluble $SO_4^=$ | -0.57** | -0.55** | -0.48** | 0.52** |
| Available K+ | 0.39** | 0.30* | 0.38* | -0.30* |

Available phosphates in soil had no significant correlation with any of the studied characteristics of stewed faba beans (21).

In a pot experiment, Wassimi *et al.* (25) found that cooking quality of lentils was significantly influenced by the supply of major and trace elements. A high level of K with an adequate supply of other nutrients gave the fastest cooking lentils. In a survey of the 1980 faba bean crop (41 samples), Shehata *et al.*, (21) reported a correlation between available K in the soil and the softness score of stewed beans. They also found that available K in soil was correlated with penetrometer reading, hydration coefficient and weight of stewed liquor.

Salinity and soluble Ca and Mg in soil were found to be negatively correlated with the softness score, penetrometer reading and hydration coefficient of stewed faba beans (21). Soluble Na was negatively correlated with the previously mentioned cooking characteristics of stewed faba beans.

Colour score of stewed faba beans was positively correlated with the ratio of monovalent to divalent cations (Na + K/Ca + Mg) but negatively correlated with the additon of animal manure.

Cooking quality of faba beans can be evaluated by measuring several parameters in the dry seeds and in the cooked product. These parameters affect the method of processing, nutritive value and acceptability of the cooked products. Cooking characteristics have shown wide variations among various samples from the 1979 and 1980 crops (Table 2). These characteristics may be taken into consideration during breeding faba beans. However, the role of environmental factors on cooking characteristics of faba beans has been demonstrated. Through research we are continuing to have a better understanding of these relationships. Breeding for certain improved cooking characteristics and knowing the role of environmental factors, would help the farmers to produce faba beans of the highest desirable quality.

Table 2. Range of cooking characteristics in 93 faba bean samples that were grown commercially in Egypt in 1979 and 1980.

| Cooking characteristics | Range |
|---|---|
| **For dry seeds** | |
| Weight of 1000 seeds (g) | 430.2 – 925.8 |
| Volume of 1000 seeds (ml) | 340  – 800 |
| Weight of hulls (%) | 12.5  – 19.6 |
| Hydration coefficient (after 8 h) | 161  – 195 |
| Swelling coefficient (after 8 h) | 172  – 219 |
| Degree of germination (%) | 88  – 100 |
| **For Stewed Beans** | |
| Colour score (out of 5) | 2.1  – 4.0 |
| Softness score (out of 10) | 5  – 10 |
| Granulation score (out of 10) | 0.0  – 9.1 |
| Penetrometer reading | 17.6  – 80.3 |
| Hydration coefficient | 220.0  – 291.8 |

REFERENCES

1. Angel, S., Kramer, A. 1965. Physical methods of measuring quality of canned peas. Food Technol., 19, 96-98.
2. Bourne, M.C. 1967. Size, density, and hardshell in dry beans. Food Technology, 21, 335-338.
3. Burr, H.K., Kon, S., Morris, H.J. 1968. Cooking rates of dry beans as influenced by moisture content and temperature and time of storage. Food Technol., 22, 336-338.
4. Chernick, A., Chernick, B.A. 1964. Studies of factors affecting cooking quality of yellow peas. Can. J. Plant Sci., 43, 174-183.
5. Gfeller, F., Halstead, R.L. 1967. Selection for cooking quality in field peas. Can, J. Plant Sci., 47, 631-634.
6. Halstead, R.L., Gfeller, F. 1964. The cooking quality of field peas. Can. J. Plant Sci., 44, 221-228.
7. Hartman, G.H. 1976. Evaluating cultured product quality with the penetrometer. Cultured Dairy Products J., 11, 20-22, 24 and 28.
8. Hulse, J.H., Rachie, K.O., Billingsley, L.W. 1977. Nutritional standards and methods of evaluation for food legume breeders. IDRC - T S7e, Ottawa, Canada, pp. 100.
9. Hyde, E.O.C. 1954. The function of the hilum in some Papilionaceae in relation to the ripening of the seed and the permability of the testa. Annals of Bot., 18, 241-256.
10. ICAITI, 1978. Determination of cooking time for beans. Central American Standard, Part 8, 2 pp.
11. Mackinney, G., Little, M.S. 1962. Colour of foods. AVI Publish. CO., U.S.A. pp. 308.
12. Mattson, S. Akerberg, E., Erikson, E., Koutler-Anderson, E., Vahtras, K. 1950. Factors determining the composition and cookability of peas. Acta. Agric. Scand. 1, 40-61.

13. Mesallam, A.S., Shehata, A.M. El-Tabey, Youssef, M.M., El-Banna, A.A. 1981. Variability in processing quality of varieties and strains of dry faba beans. (in preparation).

14. Metwally, M.A. 1973. Study of the effect of irrigation and fertilization on yield and technological properties in field bean *(Vicia faba* L.). M.Sc. Thesis, College of Agriculture, Al-Azhar Univ., Cairo.

15. Molina, M.R., Baten, M.A., Gomez-Brenes, R.A., King, K.W., Bressani, R. 1976. Heat treatment : A process to control the development of the hard-to-cook phenomenon in black beans *(Phaseolus vulgaris).* J. Fd. Sci., 41, 661-666.

16. Morris, H.S., Olson, R.L., Bean, R.C. 1950. Processing quality of varieties and strains of dry beans. Food Technol., 4, 247-251.

17. Muneta, P. 1964. The cooking time of dry beans after extended storage. Food Technol., 18, 130-131.

18. Quest, D.C., de Silva, S.D. 1977. Temperature dependence of the cooking rate of dry legumes. J. Fd. Sci., 42, 370-374.

19. Quenzer, N.M., Huffman, V.L., Burns, E.E. 1978. Some factors affecting pinto bean quality. J. Fd. Sci., 43, 1059-1061.

20. Shehata, A.M. El Tabey 1979. The use of legumes as human food in Egypt. "Grain Legume Workshop" 3-4 Dec. 1979, Singapore - sponsored by IDRC.

21. Shehata, A.M. El-Tabey, Youssef, M.M., El-Rouby, M.M., Mesallam, A.S., El-Banna, A., El-Shimy, H. 1981a. Relationship between cooking characteristics of faba beans and agronomical variables. (in preparation).

22. Shehata, A.M. El-Tabey, Youssef, M.M., Mesallam, A.S., El-Rouby, M.M., El-Banna, A. 1981. Cooking characteristics of faba beans *(Vicia faba* L.) cultivated in Egypt. (in preparation).

23. Shivashankar, G., Rajendra, B.R., Vijayakumar, S., Sreekantaradhya, R. 1974. Variability for cooking charactersitics in a collection of green gram *(Phaseolus aureua* Rorb.) J. Food Technol., 11, 232-233.

24. Wang, H.L., Swain, E.W., Hesseltine, C.W., Heath, H.D. 1979. Hydration of whole soybeans affects solids losses and cooking quality. J. Fd. Sci., 44, 1510-1513.

25. Wassimi, N., Abu-Shakra, S., Tannous, R., Hallab, A.H. 1978. Effect of mineral nutrition on cooking qualtiy of lentils. Can. J. Plant Sci., 58, 165-168.

26. Youssef, M.M. 1978. A study of factors affecting the cookability of faba beans *(Vicia faba* L.) Ph.D. Thesis, College of Agriculture, Univ. of Alexandria, Egypt.

27. Youssef, M.M., Shehata, A.M. El-Tabey, Mesallam, A.S., El-Banna, A.A. 1981. Quality attributes of faba beans cultivated with different levels of nitrogen fertilizers. Alex. J. Agric. Res.

# 38. HARD SEEDS IN FABA BEAN

FAROUK AHMED SALIH

Shambat Research Station, P.O. Box 30, Khartoum North, Sudan

Hard seededness is a type of seed dormancy resulting from the impermeability of the seed coat to water or gases, or physical resistance to embryo expansion. Hard faba bean seeds fail to germinate readily when all necessary conditions are provided and adversely affect the cooking quality. However, the impermeable seed character can be beneficial in maintaining high seed viability under conditions that severely reduce the viability of normal permeable faba bean seeds. No research work has been reported on the possible causes of this character in seeds of faba beans from the physical, biochemical and genetical stand points.

*Seed coat structure and anatomy* : McEwen *et al.* (15) examined the cotyledons and seed coats of faba bean variety Ackerperle with a scanning electron microscope. The photomicrographs showed no discontinuity in the thick seed coat. Cross-section of the seed coat showed characteristic palisade, parenchyma, tracheid and hour glass cells which were similar to those of other legumes.

The seed coat which constitutes about 15% of the weight of the faba bean, provides considerable protection for the enclosed cotyledons and embryo. Its outer surface is smooth. The hilum or seed scar, the point of attachment of the funiculus to the seed, is about 1 to 5 mm long and is slightly raised above the surface of the seed coat. The hilum is characterised by the black or light colour of the surface tissue and the central groove running through its length. This groove opens or closes according to the relative amounts of moisture inside and outside the bean; thus playing an active role in the dehydration process. The hilum, along the mid-line of the bean, lies just beyond the end of the axial tip of the embryonic root or radicle. A cross-section of the seed coat in the hilum region shows several distict types of tissues (Fig. 1). Immediately under the flaky surface layer of the hilum is a double layer of fibrous palisade cells. The outer layer appears to be continuous with the single palisade layer found throughout the seed coat. Below the central groove of the hilum are the tracheid cells through which nutrients enter the maturing bean. Around the tracheids are loosely and irregularly structured parenchyma cells. Closer to the inner surface of the seed coat, these parenchyma cells become highly compact. The parenchyma cells are highly irregular and are arranged loosely whereas tracheid cells are in a mesh-like arrangement.

*G. Hawtin & C. Webb (Eds.): Faba Bean Improvement*
©1982 ICARDA. ISBN -13:978-94-009-7501-9

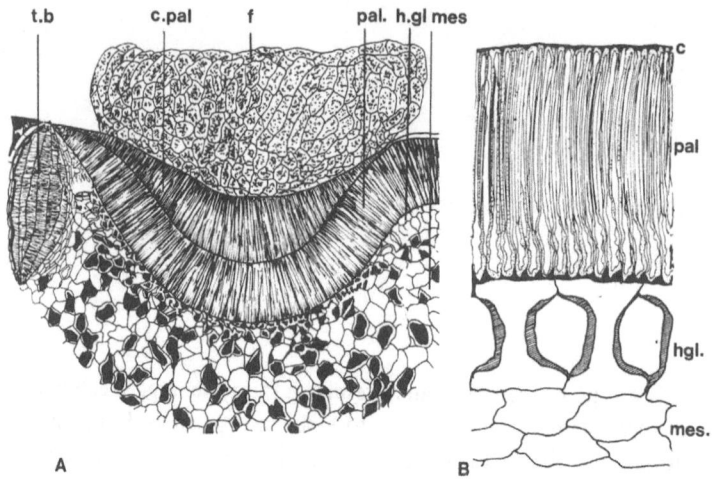

Fig. 1 A. *Phaseolus aureus* : median L.S. of testa through hilum. B. *Vicia faba:* L.S. of testa away from hilum. c.pal., Counter-palisade; f., funicle; h.gl., hour glass cell; mes., mesophyll; pal, palisade cells; t.b., tracheid bar. (x135). after : Chowdhury, K.A. and Buth, G.M. 1970 (8).

Away from the hilum, between the palisade and parenchyma cells, is another type of cell, referred to as hour glass cells which have parallel ridges along the longitudinal axis.

Chowdhury and Buth (8) found that *V. faba* had the largest palisade and hour glass cells of 14 Indian pulses studied. These cells are primarily responsible for the thickness of the seed coat.

*Seed coat physical and chemical properties :* Faba bean seed has a much higher crude fibre level than that of many other legumes (6). This fibre is mainly in the seed coat, and detracts from the use of faba beans as a feed and food. The seed coat accounts for 13 to 17% of the seed dry weight.

However, the seed coat contains 89% of the seed crude fibre and it is imperative that any reduction in seed size be accompanied by at least a corresponding reduction in the seed coat contribution (7). Rowland and Fowler (22) found that cultivars of faba bean differed significantly in seed coat thickness. Also the position of the seed on the plant had no effect on this character. Seed weight and seed coat thickness were not correlated within cultivars, but were highly correlated between cultivars ( r = 0.70). Rowland (22) reported significant differences in seed coat thickness, per cent seed crude fibre, 100 seed weight and seed protein content among 49 cultivars of faba bean. 1000-seed-weight was negatively correlated with seed crude fibre ( r = -0.52), and positively correlated with seed coat thickness ( r = 0.66). Stepwise multiple regression analysis showed that seed coat thickness could provide a partial prediction of seed crude fibre. Rowland and Fowler (21) concluded that variation in thickness within the faba bean seed coat provided an important restriction to selection for thinner seed coats. Also, the fact that seed coat apparently gets thinner as seed weight is genetically reduced is important to breeding programs as it should mean that small seeded cultivars need not produce a meal with higher fibre content than large seeded cultivars.

*Inheritance of hard seed in Vicia species* : Neither the exact inheritance nor cause of hard seededness in faba bean has been determined. Studies on the mode of inheritance have concentrated on the other *Vicia* species which are not used as a human food. Characteristics, including hard seed, of an interspecific cross *V. sativa* x *V. angustifolia* were studied (9). The *V. angustifolia* in the cross was later reclassified *V. cordata* (10). Two *V. sativa* lines used as female parents, had no hard seed (9). One group of $F_1$ plants averaged 39% hard seed while a second group averaged 12% . Pollen sterility of $F_1$ plants was more than 90% . Fertility was restored in a *V. sativa* type $F_4$, and many *V. sativa* type $F_6$ lines were genetically stable for a high percentage of hard seed (11).

Donnelly *et al.* (10) studied the inheritance of hard seed coat in *V. sativa* by using advanced generation lines with a hard seed coat recovered from the interspecific cross, *V. sativa* x *V. cordata* (11) (10). These results showed that hard seededness could be explained if inheritance was to be controlled at two loci. Gene A acts as a simple dominant for hard seededness. Gene B is dominant for soft seededness when the A locus is homozygous recessive (aa). The double recessive genotype (aabb) is hard seeded. However there were apparent exceptions to this theoretical model; two $F_2$ populations of the cross Alabama 1894 x 79 - 6 and one $F_2$ population of the cross Alabama 8 - 43 x 255 - 10 all fitted a 7:9 ratio. Classically, this type of gene interaction is double recessive epistasis rather than the single dominant epistasis outlined above. However, the authors feel these exceptions could have been due to random deviation.

*The procedure used for testing for permeability* : To assay for permeability, 100 g of seeds of each treatment or genotype tested, replicated six times, were placed in water in small Kilner Jars for 48 hours. At the end of this period, seeds that had not germinated or swelled were counted as hard seed. Usually the test for permeability is run in the laboratory under room temperature. This procedure was used in all trials at Hudeiba Research Station, Sudan.

*Genetic factors* : Salih (23) found considerable variation among cultivars. The percentage of hard seeds ranged from 22.8% in the cross RB40 x 188 to a mere 0.1% in triple white. There was a negative association between the hard seed percentage and seed weight ( r = -0.58). This result was contrary to the findings of Rowland and Fowler (21). Both Salih (23) and Rowland *et al.* however, found that final seed yield was not associated with seed coat thickness. This may be explained by the fact that even hard seeds eventually germinate and produce seedlings (16).

The overall means of 12 cultivars in four experiments in the Nile and Northern Provinces showed that the hard seed percentage ranged from 2.52 at Zeidab to 11.56% at Burgeig (Table 1). Cultivars were inconsistent in their hard seed percentage from location to location (27).

Three out of four lines showed a progressive reduction in the percentage of hard seed as the generation advanced (Table 2), but there were again considerable differences between the lines in any generation (24).

In general, the performance of samples collected from farmers in the far North of Sudan in percentage of hard seed after planting at Hudeiba, were inconsistant with their hard seed situation when collected. The hard seed percentages after planting ranged from 1.45% for a seed sample collected from Burgeig, to 13.55% for one from the river bank of Hassa Village in the Nile Province. Highly significant differences in hard seed percentages were indicated. Thirteen samples had a hard seed percentage between 9 and 13%(37).

These studies showed that there was considerable genetic variation between cultivars and collected samples, suggesting that it may be possible to obtain high yielding cultivars with low hard seed percentage.

Table 1. Regional cultivar trial at different locations (38).

| Cultivar | Hard Seed Percentage (%) | | | | |
|---|---|---|---|---|---|
| | Shendi | Zeidab | Hudeiba | Burgeig | Mean |
| NEB 153 | 5.15 | 1.95 | 2.33 | 14.05 | 5.87 |
| NEB 424. S | 4.70 | 1.42 | 2.42 | 7.47 | 4.00 |
| NEB 425. S | 7.95 | 1.30 | 4.47 | 13.50 | 6.80 |
| NEB 428 | 6.05 | 3.97 | 4.45 | 12.50 | 6.74 |
| NEB 423 | 6.75 | 4.58 | 3.98 | 15.97 | 7.82 |
| NEB 152. S | 4.68 | 1.37 | 2.48 | 11.03 | 4.89 |
| NEB 424 | 5.02 | 3.62 | 2.62 | 12.53 | 5.95 |
| NEB 425 | 7.67 | 2.68 | 2.65 | 13.72 | 6.68 |
| 188 x Giza 1 | 9.82 | 2.47 | 2.18 | 10.50 | 6.24 |
| BF 2/2 | 4.65 | 1.45 | 2.75 | 5.15 | 3.50 |
| H.72 | 6.65 | 3.75 | 5.30 | 10.43 | 6.53 |
| BM 9/3 | 7.68 | 1.75 | 4.57 | 11.92 | 6.48 |
| Mean | 6.40 | 2.52 | 3.35 | 11.56 | 5.96 |
| SE | ±0.86 | ±0.58 | ±0.46 | ±0.98 | |

Table 2. Effect of selected generation on the hard seed percentage (%) of four selected lines of faba bean (25).

| LINE | Generation | | | |
|---|---|---|---|---|
| | F 1 | F 3 | F 5 | Mean |
| 188 x Siliam | 14.47 | 7.18 | 7.00 | 9.55 |
| 188 x RB 34 | 10.88 | 4.15 | 3.45 | 6.14 |
| Baladi x G. 1 | 4.77 | 2.85 | 2.77 | 3.46 |
| Baladi x BF 2/2 | 13.48 | 9.38 | 15.13 | 12.66 |
| SE | | ±0.77 | | ±0.45 |
| Mean (± 0.18) | 10.89 | 5.89 | 7.09 | |

*Environmental factors :* The increase in hard seed percentage with maturity in Lupin *(Lupinus varius* L.) might be attributed to the increase in the hardness of seed coat tissue as a result of drying and fluctuations of humidity and temperature during this period and their effect on the permeability of the seed coat (19) (20). Baciu-Miclaus (5) showed that the incidence of hard seeds was correlated particularly with low air humidity and this in turn was the cause of increasing hard seed percentage during development. On the other hand, the detrimental effects of adverse environmental conditions, particularly high temperature and rain, on soybean seed quality during the post-maturation, preharvest period have been reported by Mondragon and Potts (17).

*Crop husbandry factors : a. Sowing date :* In several experiments, sowing faba bean after mid-October generally resulted in a higher percentage of hard seed (26)(27)(35). In one cultivar trial there was a progressive increase in the percentage of hard seed with successive sowing dates (Table 3). All cultivars showed the same general trend.

Table 3. Effect of cultivar and sowing date on the hard seed percentage (%) (27) (28).

| Sowing Dates | CULTIVAR | | | Sowing Date Mean |
|---|---|---|---|---|
| | BF 2/2 | H. 72 | BM 9/3 | |
| Oct. 15 | 4.48 | 2.84 | 3.88 | 3.73 |
| Oct. 25 | 5.38 | 4.86 | 6.64 | 5.63 |
| Nov. 4 | 6.88 | 4.40 | 5.64 | 5.64 |
| Nov. 14 | 4.70 | 7.00 | 5.14 | 5.61 |
| Nov. 24 | 6.22 | 7.06 | 10.62 | 7.97 |
| S.E. | | ±0.64 | | ±0.35 |
| Variety Mean (±0.26) | 5.53 | 5.23 | 6.38 | |

*b. Watering :* Early termination of irrigation and hence enforced earlier maturity, increased the hard seed percentages (31). However, increasing the watering intervals from 8 to 23 days had very little effect on the occurrence of hard seeds in the cultivars BF 2/2 and Siliam. (2) (3). Contrary to this result, Salih (31) reported that the seeds from plots irrigated every 10 days at Zeidab and Hudeiba had 53 and 19% more hard seed respectively than the seed lots from 20 day watering intervals.

*c. Population density :* There was a progressive increase in the percentage of hard seed with increasing plant population in both Zeidab and Hudeiba, but the increases were more pronounced in the Zeidab than in Hudeiba (34).

In another experiment in which two plant populations were studied, 166,600 plants/ha and 333,200 plants/ha, the hard seed percentages were similar; 6.80% and 6.52% respectively (34). The interaction of plant population x variety was highly significant (Table 4).

Table 4. Effect of cultivar and population density on the hard seed percentage (%) (34).

| Cultivar | Population Density (plant/ha) | | Mean |
|---|---|---|---|
| | 166,600 | 333,200 | |
| NEB 152.5 | 4.36 | 6.38 | 5.37 |
| NEB 425 | 4.61 | 6.26 | 5.43 |
| NEB 423 | 8.11 | 7.65 | 7.88 |
| 188 x Giza 1 | 10.11 | 5.80 | 7.96 |
| | ±0.93 | | ±0.66 |
| Mean ± 0.47 | 6.80 | 6.52 | |

*Fertilizer :* The effect of different levels of potassium fertilizer on the germination and hard seed percentage of faba bean seeds during development and storage was studied in Egypt by El Bagoury (14) who found that seeds were able to germinate after 49 days from anthesis, while hard seeds were obtained after 56 days. The germination and hard seed percentages increased markedly with maturity, though the number of germinable seeds was much greater than of hard seed. The germination percentage increased considerably with lengthening storage period, while hard seed percentage decreased. The highest germination percentage was recorded without applying any fertilizer, while the highest hard seed percentage was obtained with the rate of 476 kg/ha of potassium sulphate.

In Sudan, Ayoub (3) found that fertilizer treatments including nitrogen, phosphorus, potassium, gypsum and animal manure had an effect on the percentage of hard seeds; ranging between 15 and 23% . Gypsum - treated plots had the lowest hard seeds, about 15%. The hard seed content of the two cultivars in the trial was about the same.

Salih (36) found that different levels of nitrogen and phosphorus applied alone or in combination with each other at sowing, had no significant effect, but the different methods of applying fertilizer significantly affected the hard seed percentage (Table 5).

Table 5. Effect of different fertilizer treatments and method of application on the hard seed percentage in cv. Hudeiba 72 (36).

| Phosphorus rate kg/ha | Hard Seed % | Nitrogen rate kg/ha | Hard Seed % | Method of Fertilizer Application | Hard Seed % |
|---|---|---|---|---|---|
| 0 | 8.28 | 0 | 8.00 | Broadcast | 9.64 |
| 43 | 8.93 | 43 | 9.06 | Band | 7.58 |
| | | 86 | 8.76 | Placed with seed | 8.61 |
| S.E. | ±0.95 | | ±0.55 | | ±0.55 |

*Inoculation :* Inoculation with Wad Medani and Aleppo cultures of *Rhizobium* had no effect on the percentage of hard seed (30) and these percentages were similar to those observed in the control and no inoculation treatments.

*Time of harvest :* For the cultivar Hudeiba 72, there was a progressive decline in the percentage of hard seed with successive harvest dates (1)(24). The hard seed percentage decreased dramatically with prolonged storage (Table 6). This phenomenon should not be a serious problem with regard to emergence percentage of seeds in the field in view of the fact that sown seeds are generally more than 200 days old, but, of course, the problem arises when the beans are eaten.

Table 6. Effects of harvest date and storage on hard seed percentages in cv. Hudeiba 72(24)

| Plant Age at Harvesting (days) | Hard Seed Percentages | | |
|---|---|---|---|
| | 60 days after harvest | 180 days after harvest | 360 days after harvest |
| 80 | 34.5 | 16.1 | 1.7 |
| 90 | 33.9 | 15.8 | 1.3 |
| 100 | 28.4 | 10.1 | 1.2 |
| 110 | 22.7 | 9.1 | 1.1 |
| 120 | 16.6 | 8.5 | 0.6 |
| SE | ±1.2 | ±0.6 | ±0.2 |

*Seed size :* In two of the three cultivars tested the large seed category contained the highest percentage of hard seed. This result was confirmed by Salih (31)(32), who found that small seed categories contained the lowest hard seed percentage and the large seed within each cultivar contained the highest.

*Possible solutions*

i.  Storing the seeds of a cultivar known to have a high percentage of hard seed for a period of time may help to reduce the problem. The results of El Bagoury and Niyazi (13) in Egyptian clover, El Bagoury (14) in faba bean and Quinlivan (19) in *Lupinus* species support this contention.

ii. Because hard seed is considered undesirable in faba bean, breeding efforts should be directed towards the production of high yielding cultivars with improved water-imbibing and cooking qualities. Information on the genetic control of seed coat thickness would appear to be an essential pre-requisite for such improvement.

iii. More advanced research is needed in the area of seed physiology to reveal causes of the hard seed character.

iv. The problem can be reduced by appropriate cultural practices such as sowing date, water requirement, plant population, seed rate and time of harvest.

v.  The seed coat barrier may be broken in several ways, for example by mechanical scarification or use of alcohol or other solvents, by concentrated acid, or by exposing seeds to alternating temperatures.

## REFERENCES

1. Ageeb, O.A.A., 1976. Field bean time of harvest experiment. Annual Report, Hudeiba Research Station, Sudan.
2. Ayoub, A.T. 1971. Ful Masri watering experiment. Annual Report, Hudeiba Research Station.
3. --------------- 1971. Ful Masri (Broad bean) fertilizer experiment. Annual Report, Hudeiba Research Station.
4. -------------- 1972. Ful Masri fertilizer experiment. Annual Report, Hudeiba Research Station, Sudan.
5. Baciu-Miclaus, D. 1970. Contribution to the study of the hard seed and coat structure properties of soybean. Proc. Int. Seed Test. Ass. 35, 366-617.
6. Ball, J.M. 1975. Utilization of protein supplements in animal feeds. Pages 633-656 in J.T. Harapiak ed. Oil seed pulse crops in Western Canada - a symposium. Western Co-operative Fertilizers LTD.
7. Cerning, J.A. Saposnik, Guilbot, A. 1975. Carbohydrate composition of horse beans *(Vicia faba)* of different origins. Cereal Chem. 52 : 125-138.
8. Chowdhury, K.A., Buth, G.M. 1970. Seed coat structure and anatomy of Indian pulses. Bot. J. Linn. Soc. Suppl. 1, 63 : 169-179.
9. Donnely, E.D., Clark, E.M. 1962. Hybridization in the genus *Vicia*. Crop Sci. 2 : 141-146.
10. Donnelly, E.D., Hoveland, C.S. 1966. Interspecific reseeding *Vicia* hybrids for use on summer perennial grass sods in southeastern U.S.A. Proc. of the Tenth Int. Grassl. Cong. Helsinki, Finland pp. 679-683.
11. Donnelly, E.D. 1971. Breeding hard-seeded vetch using interspecific hybridization. Crop Sci. 11 : 721-724.
12. Donnelly, E.D., Watson, J.E., McGuire, J.A. 1972. Inheritance of hard seed in *Vicia*. Journal of Heredity 63 : 361-365.
13. El Bagoury, O.H., Niyazi, M.A. 1973. Effect of different fertilizers on the germination and hard seed percentage of Egyptian clover seeds *(Trifolium alexandrium* L.) Seed Sci. and Technol. 1 : 773-779.
14. El Bagoury, O.H. 1975. Effect of different fertilizers on the germination and hard seed percentage of broad bean seeds *(Vicia faba)*. Seed Sci. and Technol., 5 : 569-574.
15. McEwen, T.J., Dronzek, B.L., Bushuk, E. 1974. A scanning electron microscope study of faba bean seed. Cereal Chem. 51, 750-757.
16. Mohamed, A.S.A. 1972. Study on the hard seed character in two varieties of field bean grown under Shambat conditions. A dissertation in partial fulfilment of the requirements for the degree of B.Sc. (Agric., Honours, Faculty of Agriculture, University of Khartoum).
17. Mondragon, R.L., Potts, H.C. 1974. Field determination of soybeans as affected by environment. Proc. Assn. Off. Seed Anal. 64 : 63-71.
18. Nakamura, S. 1962. Germination of legume seeds. Proc. Int. Seed Test Ass., 27, 694-710.
19. Quinlivan, B.J. 1968. The softening of hard seed of sand plain lupin *Lupinus varius* L.) Aust. J. Agric. Res. 19, 507-515.
20. ----------------- 1970. The interpretation of germination tests on seeds of *Lupinus* species which developed impermeability. Proc. Int. Seed Test. Ass. 35, 349-359.
21. Rowland, G.G., Fowler, D.B. 1977. Factors affecting selection for seed coat thickness in faba beans *(Vicia faba* L.) Crop Sci. 17, 88-90.
22. Rowland, G.G. 1977. Seed coat thickness and seed crude fibre in faba beans *(Vicia faba)*. Can. J. Plant Sci. 57 : 951-953.

23. Salih, F.A. 1976. Broad bean standard variety trial. Annual Report, Hudeiba Research Station, Sudan.

24. -------------- 1977a. Field bean time of harvest. Annual Report, Hudeiba Research Station, Sudan.

25. ----------- 1977b. Effect of selected generation on seed yield of selected number of faba bean bulk selections. Annual Reports, Hudeiba Research Station, Sudan.

26. ---------- 1976. Effect of seed size on yield and some other characteristics of three varieties of broad bean. Annual Report, Hudeiba Research Station, Sudan.

27. ---------- 1977. Seed size/sowing date experiment. Annual Report, Hudeiba Res. Sta.

28. ---------- 1978a. Seed size/sowing date experiment. Annual Report, Hudeiba Res. Sta.

29. ------------1978b. Broad bean variety and sowing date experiment. Annual Report, Hudeiba Res. Sta., Sudan.

30. ----------- 1979a. Faba bean fertility and inoculation trial. ICARDA, International Nursery Program.

31. ---------- 1978c. Broad bean water closure experiment. Annual Report, Hudeiba Res. Sta., Sudan.

32. ---------- 1979b. Broad bean water closure experiment. Annual Report, Hudeiba Res. Sta., Sudan.

33. ---------- 1979c. Effects of sowing date, methods of sowing on the seed yields of four varieties of broad beans. Annual Report, Hudeiba Res. Sta., Sudan.

34. ---------- 1979a. Interactions between plant population density and watering intervals. Annual Rept., Hud. Res. Sta., Sudan.

35. --------- 1979e. Effect of different nitrogen and phosphorus levels and different methods of applications on broad bean yield and its components. Annual Report, Hudeiba Res. Sta., Sudan.

36. ---------- 1979f. Sudanese collection of broad bean. Annual Report, Hud. Res. Sta. Sudan.

37. --------1978c. Broad bean regional variety trial. Annual Report, Hud. Res. Sta. Sudan.

38. -------- 1979g. Effects of sowing date, seed size and seed rate on seed yield of broad bean. Annual Report, Hudeiba Research Station, Sudan.

# 39. THE EGYPTIAN NATIONAL PROGRAM

ALI ABDEL-AZIZ IBRAHIM

*Field Crops Research Institute, Agricultural Research Center, Giza, Egypt*

In Egypt, food legumes are grown on about 4.3% of the annual cropped area; faba bean being by far the most important. It is produced on more than half of the total area under food legumes, including soybeans.

Stewed faba beans, faba bean cakes and curry are very important staple foods for most of the people; balancing the deficiencies of the basically cereal diet and supplying the bulk of the required protein, especially in the diets of the predominantely rural population.

Faba bean is planted in October and November and harvested in late April and early May. It is grown throughout the country after rice, corn or cotton in the Delta, and cotton corn or sorghum in Middle and Upper Egypt. About 30% of the acreage is in the Nile Delta; 43% in Middle Egypt and 25% in Upper Egypt (Table 1).

The area under faba beans has progressively reduced from 148,200 ha to 108,000 ha between 1955-59 and 1975-79. This has been mainly in the Delta and Upper Egypt where the area dropped from 59,100 ha to 34,600 ha and from 48,100 ha to 26,800 ha respectively. This was mainly due to competiton from other winter crops, mainly wheat and Egyptian clover; particularly in Upper Egypt where introduction of the canal irrigation system has changed the crop pattern.

On the contrary, the acreage under the crop in Middle Egypt increased considerably during the same period, when most of the farmers plant faba beans as a cash crop, preceding cotton in the main cotton rotation.

The average yield/ha increased by 41.4% between 1955 and 1979. The total production increased by 2.94% in spite of an average decrease of 37.2% in the bean acreage.

Until the late 1960's, Egypt was almost self-sufficient of faba bean, but during the last few years consumption increased substantially as a result of the rapidly growing population which rose from about 31 million in 1967 to about 40 million in 1979. Consequently, considerable amounts are imported. Between 1974 and 1979, annual imports of faba bean ranged from 16,000 to 112,000 metric tons, with an average of 48,000 tons.

## Production contraints

The major constraints to faba bean production are :

A gap between the yields of the recommended cultivars in experimental plots and national average yield. Problems of soil salinity associated with poor drainage and inefficient water management, and time consuming and costly manual operations in land

*G. Hawtin & C. Webb (Eds.): Faba Bean Improvement*
*©1982 ICARDA. ISBN -13:978-94-009-7501-9*

374

Table 1. Average area, production and yield/ha of faba bean in Egypt, 1955-1979.

| Period | Lower Egypt | | | Middle Egypt | | | Upper Egypt | | | Total | | |
|---|---|---|---|---|---|---|---|---|---|---|---|---|
| | Area 1000/ha | Produc. 1000/Tons | Yield (kg/ha) | Area 1000/ha | Produc. 1000/Tons | Yield (kg/ha) | Area 1000/ha | Produc. 1000/Tons | Yield (kg/ha) | Area 1000/ha | Produc. 1000/Tons | Yield (kg/ha) |
| 1955-59 | 59.1 | 85.6 | 1447 | 41.0 | 69.6 | 1698 | 48.1 | 82.8 | 1720 | 148.2 | 238.0 | 1605 |
| 1960-64 | 67.4 | 111.3 | 1650 | 41.7 | 85.2 | 2045 | 44.3 | 85.2 | 1919 | 153.4 | 281.6 | 1834 |
| 1965-69 | 58.0 | 97.1 | 1654 | 52.6 | 124.6 | 2351 | 36.0 | 77.0 | 2126 | 146.5 | 298.6 | 2037 |
| 1970-74 | 36.7 | 69.6 | 1905 | 53.6 | 134.7 | 2502 | 28.5 | 75.9 | 2661 | 118.7 | 280.3 | 2362 |
| 1975-79 | 34.6 | 66.4 | 1919 | 46.6 | 104.3 | 2240 | 26.8 | 74.3 | 2779 | 108.0 | 245.0 | 2270 |

preparation, weeding, harvesting and threshing result in considerable losses.

. The small average size of the land holdings.

. Several diseases and insect pests :

(i) Chocolate spot *(Botrytis fabae)*, rust *(Uromyces fabae)* and the parasitic Angio-sperm broomrape *(Orobanche spp.)* are considered the main pathological problems. Root rot/wilt diseases cause considerable losses in high water table and badly aerated soils.

(ii) Aphids and bruchids are the main insect pests.

. High percentages of flower and pod shedding.

. Stress weather conditions i.e. frost and high temperatures at critical growth stages.

. Pricing and marketing constraints related to governmental policy.

. Lack of modern seed cleaning equipment to produce high quality seed; and short-comings in the improved cultivars.

. Lack of funding for basic research supplies, field and laboratory equipment and trans-port.

. Segmentation of various research and production efforts into various institutes.

. Lack of communication between research and extension counterparts.

### The National Research Program

Research on faba bean has been done by the Ministry of Agriculture since 1929. Most of the work has been in the field of applied and adaptive research oriented at solving problems of crop production. The national research program comprises :

. Breeding and cultivar improvement.

. Agronomic research covering planting dates, methods and densities; water and nutri-tional requirements; rate, forms and time of chemical fertilizer application and ef-ficiency of *Rhizobium* inoculation.

. Plant pathology research covering surveys of diseases; studies of environmental condi-tions favouring disease development; causes of the diseases; investigating the behaviour of causal organisms, their life cycle and control measures through resistant varieties; chemical control and/or cultural practices.

. Plant protection research including regional surveys of insects and pests attacking the crop in the field and in storage, examining their life histories and studying control measures.

. Weed control research to determine the effective rates and application techniques of available herbicides to control weeds and *Orobanche.*

. Economic research covering the economics of production and consumption and esti-mation of crop yields by sampling methods and agricultural statistics.

### Faba bean breeding and varietal improvement

The faba bean breeding and varietal improvement program began as early as 1929. The dominant breeding procedures have been selection in landraces and in segregating genera-tions following crosses. The success has been limited by the narrow germplasm pool of landraces and lack of broad genetic base.

However this handicap has been largely eliminated as a result of the provision of a large germplasm collection of diverse geographical origin through the ALAD/IDRC/

ICARDA regional cooperative legume improvement program, initiated in 1972. Some entries showing good resistance to chocolate spot and rust, the major diseases of the crop, have been identified. Other entries with high protein contents, lodging resistance and/or desirable yield components have also been found.

In 1975, an expanded breeding program was initiated to combine these attributes with the adaptation to the local environment shown by native cultivars. This scheme involves hybridization followed by compositing promising lines of early generations for cross pollination by honey bees to produce improved populations. However the traditional pedigree and improved bulk methods will be retained to produce populations for use in hybridization and pure line breeding.

The breeding work has resulted in the release of the following faba bean cultivars :

| Cultivar | Year of release | Special characteristics |
|----------|-----------------|-------------------------|
| Rebaya 8 | Early 1930's | Recommended for all the country. |
| Rebaya 34 | Mid 1930's | Replacement for Reb. 8 in the Delta and Middle Egypt. |
| Rebaya 40 | Late 1960's | Replacement for Reb. 8 in Upper Egypt. |
| Giza 1 | Early 1960's | Tolerant to chocolate spot and rust, replacement for Reb. 34 in North Delta. |
| Giza 2 | Early 1960's | Wide adaptability, replacement for Reb. 34 in South Delta and Middle Egypt. |
| Giza 3 | Late 1970's | Replacement for Giza 1. |
| Giza 4 | Late 1970's | Replacement for Reb. 40. |
| Fam. 402 | Under release | Moderately resistant to *Orobanche*. |

Production of certified seed of the recommended cultivars began during the first 5-year development plan (1960-61 to 1964-65). The general seed policy is to produce and distribute every year enough certified seed to cover one-third of the area under the crop. The foundation and registered seed are produced on the state farms under the supervision of the plant breeders.

The Ministry of Agriculture's seed department is responsible for the production, testing and distribution of certified seeds to farmers.

*Agrotechniques, crop protection and weed control research*

Recommendations arising from the research include :

| | |
|---|---|
| Planting dates | Mid-October in Upper Egypt; late October in Middle Egypt, and early November in the Delta. |
| Plant density | 350,000 plts/ha, distributed on ridges 60 cm apart in 2-seeded hills 20 cm apart on both sides of the ridge. |
| Fertilizers | 71.4 kgs. $P_2O_5$/ha broadcasted and incorporated in the soil during ploughing + 35.7 kgs N/ha dressing 30 days after planting. |
| Irrigation | Four or five waterings at 25-30 day intervals. |
| Weed control | Linuron ½— ¾ kg/acre after planting and before irrigation, or 1-2 hand weedings. |
| *Orobanche* control | Application of Lancer (glyphosate); three sprays, at the rate of 0.064 (a.e.) kg/ha on the plant canopy at 21 day intervals starting just after flowering. |

| | |
|---|---|
| Aphids | Complete control through the application of suitable insecticides. Dimethoate is recommended. |
| Store insects | Fumigation and mixing the seeds with beetle-killer dust is recommended. |
| Intercropping | Intercropping of faba beans with sugar cane is recommended. |

## Research priorities

Many of these areas are already receiving attention by the National Program through IDRC, and ICARDA/IFAD Nile Valley projects :

(i)    Breeding for physiologically efficient, high nodulating and disease- and pest-resistant cultivars.

(ii)   Developing early high yielding varieties to suit the current intensive cropping system.

(iii)  Developing physiological studies to identify the reasons for excessive flower and pod shedding, the basis for tolerance to frost and salinity, and screening techniques for the genetic improvement to combat stress factors.

(iv)   Continuing studies on epidemiology of *Botrytis fabae* and *Uromyces fabae*, the creation of artificial epiphytotics, methods for disease rating, and chemical control methods.

(v)    Continuing studies on the control of *Orobanche* through developing resistant cultivars, cultural practices, use of herbicides and synthetic germination stimulants.

(vi)   Development of techniques for screening for resistance to aphids and studies on storage pests *(Callosobruchus chinensis)*, and their control.

(vii)  Precise determination of quality specifications and study factors bearing on market preference. Consumer acceptance to faba bean depends largely on particular seed characteristics e.g. time of cooking, taste, colour and texture of the seed after cooking.

(viii) Protein and animo acid content as indicative of nutritive value of faba beans is important where animal proteins are beyond the reach of the masses. Breeding high quality cultivars is yet to be included in the program.

(ix)   Exploring and developing field machinery to fit current cropping systems and small holdings will lead to better timing, economic and efficient soil management and agonomic practices.

(x)    Implementing seed production and management practices.

## Research structure

The faba bean national development program is conducted by the Ministry of Agriculture, where research on problems relating directly to the farmers needs is done in the following institutions of the Agricultural Research Center (ARC), at Giza :

.    Field Crops Research Institute for breeding, agronomy and seed technology aspects.

.    Soil and Water Research Institute is responsible for research in the areas of water and plant nutritional requirements and nitrogen fixation.

.    Plant Pathology Research Institute covers all pathological studies related to faba bean.

.    Plant Protection Research Institute undertakes research related to insects, weeds and *Orobanche* control.

.    Agricultural Economics Research Institute is responsible for the areas of economics and production.

378

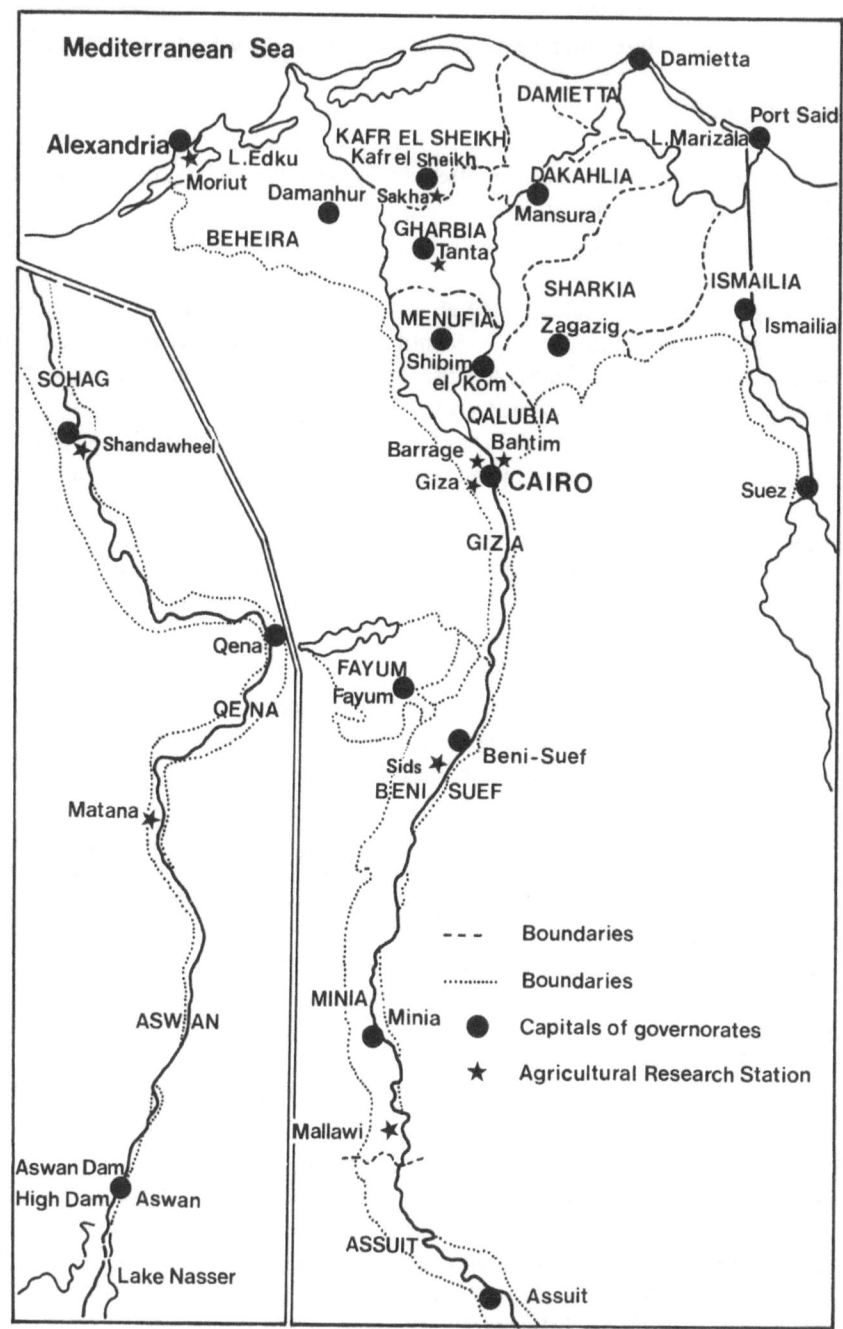

Fig. 1. Map showing agricultural research stations in Egypt.

These Research Institutes are assisted by nine regional sub-research Stations covering the different agroecological regions of the country : Sakha, Nubaria (representing the newly reclaimed lands), Gemmeza and Bahteem in Lower Egypt, Sids and Mallawy in Middle Egypt, and Shandawheel, Koom-ombo and Maatana in Upper Egypt.

Through coordination, among the different Research Institutes, the national research program, is conducted at the different regional sub-research stations and at the farmers' fields.

The Ministry of Agriculture extension service carries the results of research in the form of recommendations for better cultural practices through a fairly well established countryside network of agricultural extension units.

Research related to faba bean is also done at the National Research Center, mostly in the fields of plant phsiology, entomology and food technology, and at the Agricultural Colleges of the Universities, where most of the research is on agronomy and food technology.

*Cooperation with foreign and international centers/agencies*

The following agencies and organisations have been cooperating with the National Research Program :

.   From 1972 to 1975 Ford/ALAD/IDRC assisted through the regional cooperative programs in assembling and providing germplasm collections both from the region and elsewhere, and in training junior scientists.

.   Under the US-PL 480 program, a research project for controlling broomrape *(Orobanche spp.)* in field crops was conducted between 1966 and 1978.

.   Since 1977, the IDRC has provided considerable finicial support for various components of research on major food legumes (faba bean, lentil and chickpeas).

Since 1977, ICARDA has been providing nurseries and breeding material for the National Program. During 1979, IFAD financed the ICARDA/IFAD Nile Valley project on faba bean aiming at increasing the productivity of the crop. The major emphasis of the project is to evaluate the available recommended varieties and cultural practices at the farm level through a series of on-farm trials. These trials provide important feedback for the design of different research trials for experimental work on the research stations. The project also provides funds for training and strengthening the core research and crop improvement program.

The project made it possible to initiate strong cooperation among the different disciplines of research in the institutions under the ARC. The National Program is currently organised and administered through the National Program coordinator and the National Program research leader, in the Field Crops Research Institute, in collaboration with the senior scientists who are responsible for faba bean research in various institutes.

# 40. THE SUDANESE NATIONAL PROGRAM

IBRAHIM A. BABIKER

*Hudeiba Research Station*

The scope for development of Sudan's agricultural land is enormous. The main food legumes in the northern Sudan are faba beans *(Vicia faba)*, dry beans *(Phaseolus vulgaris)* and lentils *(Lens culinaris)*. Recently, there has been a marked increase in the area and production of these food legumes.

About 95% of faba beans in Sudan are grown in the Northern and Nile provinces. The area under faba beans is estimated as 22,000 ha with a likely increase to about 30,000 ha during the next few years. In these provinces, faba beans are grown under irrigation with some areas under flood irrigation in the Nile basin.

Cash cropping in the Northern province has increased during the past decade. The proportion of land under cereals has declined while that under fruits, legumes, spices and other specialized crops has increased. However, it is being recognised that there may be problems associated with increasing acreage under faba bean and other high value crops, particularly with regard to fuel supply, labour availability and marketing. The high prices being paid for faba beans should still make them a profitable crop, with good incentives for production.

The basic rotation is the typical 2-year rotation :

| Summer | Winter | Summer | Winter |
|--------|--------|--------|--------|
| Sorghum | Food Legume | Fallow | Wheat |

However, in reality, a more complex cropping pattern is being practiced according to locality, soil, and marketing prospects.

There has been a substantial shift from wheat to beans in both provinces as returns from faba beans have become significantly higher due to the rapid growth in urban demand, which has outstripped the supply. Both provinces have become specialized in the production of faba beans for consumption locally and in central Sudan. In addition to the relatively high farming standards, this trend has been helped greatly by the long season, and favourable physical and climatic conditions. The production of faba beans is expected to continue, and the high person:land ratio will be an added incentive to maximise the output/unit area.

Yields do not seem to have increased signficantly in the last decade, despite the increasing use of fertilizers and crop protection measures.

*G. Hawtin & C. Webb (Eds.): Faba Bean Improvement*
©1982 ICARDA. ISBN -13:978-94-009-7501-9

382

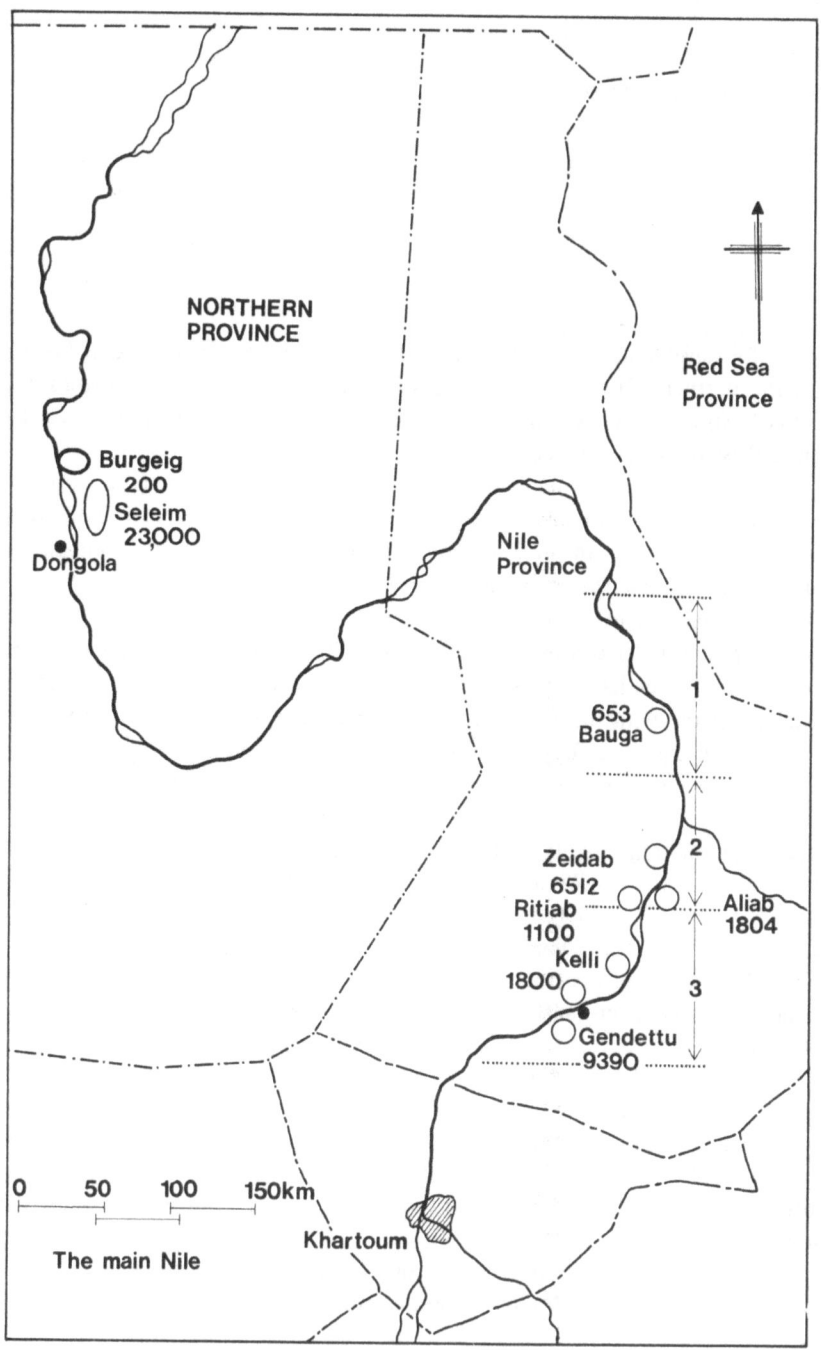

Fig. 1. Faba bean distribution in Northern Region, Sudan
   O   indicates main pump schemes, with area (ha) under faba beans.

Table 1. Relative proportion under different crops, by area in the northern Provinces

|                        | 1967 | 1979 |
|------------------------|------|------|
| Legumes                | 14   | 38   |
| Cereals                | 73   | 37   |
| Fodder                 | 9    | n.a. |
| Vegetables and spices  | 2    | 15   |
| Others                 | 2    | 10   |
| Total                  | 100  | 100  |

*Marketing and prices*

Faba bean is used almost exclusively as dry seeds, sometimes with other foodstuffs or spices added to give it the flavour preferred. Previously eaten mainly in urban areas, it has recently become a national dish for all.

Most marketing is done by traditional merchant families who have powerful control on every aspect from production to retailing. Under these circumstances, the farmer is not getting the full reward for his work. The village merchant may be an agent for merchants in urban areas. While the present structure may be reasonably effective in supplying products in demand for urban centers within the country, government advice and assistance is required to promote the development of export markets and protect the farmer against exploitation. There is a great need for good marketing management and a price information service to exploit these opportunities for the farmers' benefit. This should involve farmer organizations, private merchants and traders, and local processing factories.

A very small amount of faba bean is canned at the Karima Tomato Factory when its equipment is not being used for other commodities.

Absence of a reasonably good infrastructure is a great obstacle in the way of marketing farmers' produce.

*Seed production and availability*

The Plant Propagation Administration of the Ministry of Agriculture is the responsible organisation for the production of certified seeds. Cultivars are released by the Crop Variety Committee on which the Research Corporation is adequately represented. The seeds of certain crops are multiplied in specialized state farms or sometimes in the fields of selected farmers under the supervision of the propagation staff.

The production and distribution of pulses, wheat and vegetable seeds in the Northern region is the responsibility of the propagation station at Hudeiba, but due to lack of facilities, mainly land and funds, the station only distributes about 10% of improved seeds in any of the above crops. Consequently most farmers use the seeds from their own produce. This traditional practice results in the use of inferior seeds and hence reduced yields. Farmers are aware of the importance of improved seed and may even be willing to pay more for it, but unfortunately this is not available.

The present system for producing improved faba bean seeds is inadequate. The situation could be improved by :

a.    The production of improved seeds within the existing government schemes "seed farms", or

b.    Contracting selected farmers to produce improved seeds.

As the importation of certified seeds has disadvantages, the need for improved seed production cannot be overemphasised.

*Constraints to production*

Poor cultural practices by many of the farmers are a major cause of low yields in Sudan. Research has clearly shown the importance of watering as a major factor in determining the crop yield, either alone or in combination with other factors. Early sowing, optimum weeding and growing on ridges have all shown their superiority.

Fungal diseases which attack faba beans in the Sudan, include wilt, root rot and powdery mildew in different degrees of severity. Plants may exhibit some degree of tolerance to wilt which appears to be a major factor limiting further expansion of production southwards as well as being a challenge to breeders to produce more tolerant cultivars. Powdery mildew generally appears around December (pod-filling stage) with the drop in temperatures. As plants may show some degree of tolerance at that stage, the effect may not be reflected in any appreciable loss in yield. The problem of poor stands with which wilt and root rot have always been associated is awaiting concerted effort. Both phyllody and mosaic virus diseases, which are important in reducing yields can be minimized by following the recommended practices, particularly with respect to planting date.

The faba bean aphid, *(Aphis craccivora)* is the main pest on faba beans in both provinces. It appears normally at the end of December, being helped by the onset of cooler weather. It is effectively controlled by several insecticides. Prophylatic sprayings are sometimes necessary.

The lesser army worm *(Spodoptera exigua)* may inflict large crop losses 2-3 weeks after sowing (mid-October). If this insect is detected early enough, insecticide control is possible, but complete losses in farmers' fields still occur. Storage pests are still quite a threat to legume seeds as well as those of other crops. However little research is being carried out on these pests at present.

Other major constraints to faba bean production, include the availability of suitable high yielding cultivars, and the transfer of research results to the farmer through an efficient extension system.

*Research on faba beans*

Research on faba beans in Sudan has a long history, both at the University of Khartoum and the Agricultural Research Corporation (ARC). Hudeiba Research Station is the main ARC Station conducting research on faba beans although other stations and sub-stations are also involved.

Recognising the gap in faba bean yields between those achieved on research stations and farmers fields, the Nile Valley Project, funded by IFAD, was started in 1979 in collaboration with ICARDA and faba bean scientists in Egypt. The major objective of the Project is increasing production at the farmers level through:

a.    testing and demonstrating improved packages of production technology and

b.    undertaking basic research on major constraints to production.

The project also aims at strengthening and coordinating current and future research on faba beans, and includes the provision of funds for equipment, supplies, travel, training, consultancies, meetings etc.

## Conclusions

Involvement of farmers in the Project has great merit, but it has to be supervised very closely and maintained in close relationship with scientists. This emphasises the desperate need for a highly efficient and dynamic extension service which we, unfortunately, lack. For without that, the hoped for breakthrough in faba bean production may take some time to materialize.

The approach to the big problem of low yields of faba beans in the Sudan can be made by first looking into the factors of production under the different localities in the Sudan and evaluating them at a special conference at which all parties are represented. Sometimes the reasons leading to poor yields may be outside the scope of the agricultural sector. Secondly, yields are very low in relation to those obtained in Egypt. Our Egyptian colleagues have been consulted on this matter but still more integration is needed before substantial progress can be made.

The Nile Valley Project has made possible a wider scope for research. New experiments have been initiated this season to cover areas of crop water requirements, sowing methods, quality tests and consumer preferences. Greater depth of research has also become possible in the light of new data being collected.

The interaction of the environment on plant population and the testing of practices under different farm managements has also been possible this season. It is now recognised that plant populations in our experiments may be below the optimum being experienced elsewhere. With this view, a series of treatments was designed to provide relevant data. It is normal to recommend a high plant population for marginal or less fertile soils; a situation which definately needs further experimentation. The shift of our recommended sowing date according to farmers' experience may warrant special attention and an in-depth review on this subject may be held before the beginning of next season.

There is a great need to mechanize the cultural operations of faba beans, especially in planting. Research on water management, soil tests, fertilizers and their residues has to be intensified. This will, invariably, involve the use of sophisticated and expensive equipment which, consequently, will necessitate the establishment of a maintenance unit; possibly jointly between ARC and the University.

For quality tests and consumer preference the facilities available at the Food Processing Center are being tapped. More research involvement and cooption of extra scientists to the center should be considered. Unfortunately we have been unable to secure an agricultural economist, despite the urgent need for socio-economic surveys.

With more data available at the end of this season and reviewing the long history of faba bean production in the traditional area of the Nile and Northern provinces, we would be in a better position to extend southwards into Khartoum and Gezira provinces. The crop performance at Shambat and Gezira is, of course, a good indicator of how far our ambition may be satisfied. The major limiting factors to southward expansion are likely to be the availability of suitable cultivars, length of the season and diseases.

A positive move could be taken after full evaluation of these factors but the extent of any such development will be determined by the extent of our present resources. Similarly, it may be argued that such a move should be delayed in favour of an increased emphasis in the traditional environment of the

APPENDIX (I). Areas under faba beans in Nile and Northern Provinces.

| Season | 1976-77 | | 1977-78 | | 1978-79 | | 1979-80 | | Mean | |
|---|---|---|---|---|---|---|---|---|---|---|
| | Area Feddan | Yield kg/F | Area Feddan | Yield kg/F | Area Feddan | Yield kg/F | Area Feddan | Yield kg/F | Area Feddan | Yield kg/F |
| **Nile Province** | | | | | | | | | | |
| faba beans | 18827 | 600 | 20933 | 100 | 19868 | 700 | 30477 | 800 | 22776 | 775 |
| Haricot beans | 11699 | 800 | 10086 | 1000 | 6609 | 600 | 7258 | 600 | 3913 | 750 |
| **Northern Province** | | | | | | | | | | |
| faba beans | 23773 | 800 | 31881 | 1000 | 23010 | N.D. | 21762 | 880 | 25105 | 893 |
| Haricot beans | 150 | 300 | 50 | 500 | 19 | N.D. | 150 | 960 | 92 | 753 |

faba bean. Whichever way we look at it, we are facing a real challenge for which we have to be fully equipped.

REFERENCES

1. Proceedings of ICARDA IFAD Nile Valley Project on Faba Bean, Annual Coordination Meeting, 1979/80 Cairo, Egypt, 25-27.
2. Reappraisal of the Northern and Nile provinces pump schemes part (1), Main Report 1979.
3. Agricultural Reports Nile province Damer 1975-1980.
4. Agricultural Reports Northern province Dongola 1975-1980.
5. Northern province Agric. production Corporation.
6. Reports 1977-80 (NPAPC) Ed Damer.

# Index

## A

Abscisic acid 179
Acetylene reduction 131
Ackerperle 363
*Acrothecium* 132
*Acyrthosiphon gossypii* 229, 285
   *pisum* 263, 285, 289
   *sesbaniae* 277
Adaptability 24, 81
Additive gene effects 71
*Aegilops* 93
Afghanistan 5, 98
Africa 16, 96
*Agrobacterium* 94, 163
Agronomic characters 12, 73, 91
   practices, 117, 124, 177, 369, 385
Agronomy 109, 117
*Agropyron* 93
Albino mutants 84
Albumin/globulin ratio 313
Aldicarb 266, 267
Aldrex T 227
Alexander technique 58
Alfalfa mosaic virus 233, 236
Algeria 5, 24, 237
Alkaline soils 123
Allogamous species 72, 80
Allogamy 15, 96
Allomaintainers 64
Allopatric species 92, 95
Alloploid 91, 94, 95, 105
*Alternaria* 132, 213
America 1, 251
Amex 192
Amino acid 87, 311, 327, 377
Ammonium nitrate 219
Animal manure 358, 368
Anthesis 57
Anthocorid bug 278
Antiascorbic factor 313
Antigens 260

Antinutritional factors, 314, 319, 333, 340
Aphids 98, 117, 237, 292, 377
*Aphis craccivora* 229, 271, 277, 285, 384
   *fabae* 12, 163, 285
   *gossypii* 229, 277
*Apion vorax* 237
Apomictic development 103
Aquadulce 24, 146
Arginine 87, 317, 327
*Ascochyta* blight 19, 98, 267
   *fabae* 12, 24, 261
Ascorbate 347
Asexual reproduction 94
Aspartate 130
Aspartic acid 87
*Aspergillus* 132
Assimilation rate 114
Australia 1, 251
Autofertile lines, 22, 49
Autofertility 16, 47, 49, 84
Autogamous species 72
Autogamy 20, 96
Autosterility 49, 80
Auxin 179
Auxotrophic mutants 131
*Avena fatua* 193

## B

Bahrain 287
Baladi 16, 311
Banner 46, 47, 48
Barley 20, 58, 186
Bavistan 221
Bean flour 340
Bean leafroll virus 236, 263, 266
Bean weevil 272
Bean yellow mosaix virus 233, 236, 266, 271
Bee hives 50
Bees 16, 49, 73
*Bemisia tabaci* 277